Estimation and Monitoring of Forest Biomass and Fuel Load Components

Estimation and Monitoring of Forest Biomass and Fuel Load Components

Guest Editors

Haikui Li

Weisheng Zeng

MDPI

Basel • Beijing • Wuhan • Barcelona • Belgrade • Novi Sad • Cluj • Manchester

Guest Editors

Haikui Li
Institute of Forest Resource Information Techniques
Chinese Academy of Forestry
Beijing
China

Weisheng Zeng
Academy of Inventory and Planning
National Forestry and Grassland Administration
Beijing
China

Editorial Office
MDPI AG
Grosspeteranlage 5
4052 Basel, Switzerland

This is a reprint of the Special Issue, published open access by the journal *Forests* (ISSN 1999-4907), freely accessible at: https://www.mdpi.com/journal/forests/special_issues/6XZ3R7LRO1.

For citation purposes, cite each article independently as indicated on the article page online and as indicated below:

Lastname, A.A.; Lastname, B.B. Article Title. *Journal Name* **Year**, *Volume Number*, Page Range.

ISBN 978-3-7258-5125-6 (Hbk)
ISBN 978-3-7258-5126-3 (PDF)
https://doi.org/10.3390/books978-3-7258-5126-3

Contents

forests

MDPI

Editorial

Advances in Estimation and Monitoring of Forest Biomass and Fuel Load Components

Haikui Li [1,*] and Weisheng Zeng [2]

[1] Institute of Forest Resource Information Techniques, Chinese Academy of Forestry, Beijing 100091, China
[2] Academy of Inventory and Planning, National Forestry and Grassland Administration, Beijing 100714, China; zengweisheng0928@126.com
* Correspondence: lihk@caf.ac.cn

Forests play a pivotal role in global carbon sequestration, biodiversity conservation, and climate change mitigation. Accurately quantifying forests' biomass and fuel load components is essential for sustainable forest management and carbon accounting. This Special Issue of *Forests*, titled "Estimation and Monitoring of Forest Biomass and Fuel Load Components," includes 13 research papers addressing methodological advancements, regional case studies, and novel applications in biomass modeling, carbon storage, stand structure optimization, and regeneration dynamics. In general, these contributions underscore the importance of integrating multi-scale data, advanced statistical methods, and ecological considerations to estimate and monitor forests' biomass and carbon storage.

Methodological Innovations in Biomass and Carbon Stock Modeling

A recurring theme across multiple studies is the development of robust models to estimate biomass and carbon storage. Zeng et al. established a three-level model system for China's diverse range of forests, integrating weighted regression, dummy variables, and simultaneous equations. Their hierarchical framework—spanning forest categories (Level I), types (Level II), and sub-types (Level III)—achieved remarkable precision, with determination coefficients (R^2) exceeding 0.95 at Level III. This system addresses the limitations of earlier models; for example, three major forest types exhibited considerable uncertainties (total relative error (TRE): −20% to 74%) [1]. Similarly, Yin et al. devised additive biomass models for poplar plantations using seemingly unrelated regression (SUR), demonstrating the significance of tree height as a predictor and revealing carbon allocation shifts from competitive growth to self-stabilization as trees mature [2].

Soil variables and structural diversity, which are often overlooked in traditional models, were focused on by Guo et al. and He et al. [3,4]. Guo's soil-sensitive Weibull distribution model for *Larix principis-rupprechtii* plantations incorporated available potassium and nitrogen, improving the parameter accuracy (the root mean squared error (RMSE) was reduced by 10.4%) [3]. He et al. quantified carbon sink dynamics in mixed forests by integrating stand structural diversity and carbon growth grades, projecting a 51% increase in Jilin Province's carbon stocks by 2060 compared with age-only models. These studies highlight the necessity of multidimensional approaches for precise carbon accounting [4].

Harnessing Technology and Machine Learning

Remote sensing and machine learning are revolutionizing forest monitoring. Xu et al. contrasted the Random Forest (RF) and nonlinear mixed-effects (NLME) models for predicting tree height–diameter relationships in mixed coniferous–broadleaf forests. The

RF model outperformed NLME, achieving an R^2 of 0.970, underscoring its adaptability for complex, high-dimensional data [5]. Yang et al. employed LiDAR-derived variables and Gaussian process regression to predict the height-to-crown-base (HCB) in *Picea crassifolia* forests, identifying spatial competition indices as critical drivers of HCB variability [6]. These methodologies enable rapid, non-destructive assessments that are critical for large-scale inventories.

Climate resilience modeling has also benefited from technological integration. Meng et al. developed a climate-sensitive transition matrix model for natural forests in Chongqing, predicting forest growth under three RCP (representative concentration pathways) scenarios [7]. Their findings showed minimal divergence in carbon stocks across emission pathways but highlighted successional shifts from pioneer to climax species, stressing the need for adaptive management policies.

Optimizing Stand Structure and Management Practices

Several studies explored silvicultural interventions to enhance the productivity and resilience of forests. Zhou et al. demonstrated that restorative thinning in *Quercus variabilis* plantations increased the spatial heterogeneity, reducing DBH and density autocorrelation distances while amplifying tree height gradients [8]. Such interventions mimic natural disturbance regimes, promoting structural complexity. Similarly, Sheng et al. classified mixed broadleaf forests into developmental stages (establishment, competition, and quality selection) and optimized their spatial structure using a harvest intensity of 10%, quadrupling the structural quality index (Q(x)) [9]. These findings demonstrate the need for stage-specific management to maximize ecological and economic outcomes.

Seedling regeneration dynamics, which are crucial for forest sustainability, were analyzed by Hu et al. [10]. Their study of *Quercus* forests in Beijing revealed species-specific regeneration capacities (*Q. aliena* and *Q. variabilis* outperformed others) and identified soil factors (e.g., exchangeable calcium) and stand diversity as key determinants of later-stage regeneration. This underscores the interplay between biotic and abiotic factors in sustaining forest health.

Regional Applications and Model Validation

Region-specific insights emerged as a cornerstone of this Special Issue. Huang et al. incorporated the stand age into allometric equations for eucalypt hybrids, revealing age-driven shifts in biomass allocation (e.g., the stem biomass increased from 40% to 60% with age) [11]. Wang et al. addressed bamboo biomass estimation, constructing models for *Dendrocalamus brandisii* that balance simplicity and accuracy—a critical step for managing economically vital bamboo plantations [12]. Meanwhile, Yao et al. proposed a crown-width-based algorithm to locate tree pith positions, reducing estimation errors to 7.6%, which enhances dendrochronological and biomass studies [13].

Conclusions and Future Directions

The studies in this Special Issue collectively advance the precision and scope of forest biomass and fuel load monitoring. Key innovations include the integration of soil variables, structural diversity, and climate scenarios into predictive models; the adoption of machine learning for complex datasets; and the validation of silvicultural practices across diverse forest types. However, challenges remain, particularly in scaling localized models to larger regions and reconciling field data with remote sensing outputs. Future research should prioritize the following topics:

Cross-scale validation: Testing models' generalizability across biogeographic regions.

Dynamic modeling: Incorporating temporal shifts in climate, species composition, and disturbance regimes.

Policy integration: Translating scientific findings into adaptive management frameworks for carbon markets and wildfire mitigation.

As global forest ecosystems face unprecedented pressures from climate change and land-use shifts, the methodologies and insights presented here provide a robust foundation for evidence-based stewardship. By fostering interdisciplinary collaboration and technological innovation, we can ensure that forests continue to serve as vital carbon sinks and biodiversity havens for generations to come.

Acknowledgments: We extend our gratitude to all the authors, reviewers, and editors who contributed to this Special Issue.

Conflicts of Interest: The authors declare no conflict of interest.

References

1. Zeng, W.; Zou, W.; Chen, X.; Yang, X. A Three-Level Model System of Biomass and Carbon Storage for All Forest Types in China. *Forests* **2024**, *15*, 1305. [CrossRef]
2. Yin, M.; Gao, T.; Zhao, Y.; Ni, R.; Zheng, P.; Zhao, Y.; Zhang, J.; Li, K.; Li, C. Carbon Stock Estimation of Poplar Plantations Based on Additive Biomass Models. *Forests* **2024**, *15*, 1829. [CrossRef]
3. Guo, H.; Liu, X.; Liu, D. Soil-Sensitive Weibull Distribution Models of *Larix principis-rupprechtii* Plantations across Northern China. *Forests* **2024**, *15*, 1562. [CrossRef]
4. He, X.; Guo, H.; Lei, X.; Gao, W.; Li, Y. Quantifying the Effects of Carbon Growth Grade and Structural Diversity on Carbon Sinks of Natural Coniferous–Broadleaved Mixed Forests Across the Jilin Province of China. *Forests* **2025**, *16*, 227. [CrossRef]
5. Xu, Q.; Yang, F.; Hu, S.; He, X.; Hong, Y. Tree Height–Diameter Model of Natural Coniferous and Broad-Leaved Mixed Forests Based on Random Forest Method and Nonlinear Mixed-Effects Method in Jilin Province, China. *Forests* **2024**, *15*, 1922. [CrossRef]
6. Yang, Z.; Yang, H.; Zhou, Z.; Wan, X.; Zhang, H.; Duan, G. LiDAR-Based Modeling of Individual Tree Height to Crown Base in *Picea crassifolia* Kom. in Northern China: Comparing Bayesian, Gaussian Process, and Random Forest Approaches. *Forests* **2024**, *15*, 1940. [CrossRef]
7. Meng, X.; Ma, Z.; Xia, Y.; Meng, J.; Bai, Y.; Gao, Y. A Study on the Growth Model of Natural Forests in Southern China Under Climate Change: Application of Transition Matrix Model. *Forests* **2024**, *15*, 1947. [CrossRef]
8. Zhou, M.; Wang, Y.; Jin, S.; Wang, D.; Yan, D. Spatial Distribution Pattern of Response of Quercus Variabilis Plantation to Forest Restoration Thinning in a Semi-Arid Area in China. *Forests* **2024**, *15*, 1278. [CrossRef]
9. Sheng, Q.; Dong, L.; Liu, Z. Optimizing Stand Spatial Structure at Different Development Stages in Mixed Hard Broadleaf Forests. *Forests* **2024**, *15*, 1653. [CrossRef]
10. Hu, X.; Duan, G.; Jin, Y.; Cheng, Y.; Liang, F.; Lian, Z.; Li, F.; Wang, Y.; Chen, H. Study on the Natural Regeneration Characteristics and Influencing Factors of Typical Quercus Forests in Northern China. *Forests* **2025**, *16*, 250. [CrossRef]
11. Huang, R.; Zhu, W.; Du, A.; Xu, Y.; Wang, Z. Stand Age Affects Biomass Allocation and Allometric Models for Biomass Estimation: A Case Study of Two Eucalypts Hybrids. *Forests* **2025**, *16*, 193. [CrossRef]
12. Wang, Z.; Zeng, W.; Guo, L.; Xu, Z.; Fan, S.; Cai, C.; Hui, C.; Liu, W. Construction and Comparison of Single-Tree Biomass Model for *Dendrocalamus brandisii*. *Forests* **2025**, *16*, 301. [CrossRef]
13. Yao, J.; Shang, X.; Hu, X.; Jin, Y.; Cai, L.; Li, Z.; Li, F.; Liang, F. An Algorithm for Determining Pith Position Based on Crown Width Size. *Forests* **2024**, *15*, 2172. [CrossRef]

forests

MDPI

Article

Spatial Distribution Pattern of Response of Quercus Variabilis Plantation to Forest Restoration Thinning in a Semi-Arid Area in China

Mengli Zhou †, Yuan Wang †, Shanshan Jin, Decai Wang * and Dongfeng Yan

College of Forestry, Henan Agricultural University, Zhengzhou 450002, China; mlzhou@henau.edu.cn (M.Z.); fyk@stu.henau.edu.cn (Y.W.); jsszzz@henau.edu.cn (S.J.); ydflx@henau.edu.cn (D.Y.)

* Correspondence: dcwang@henau.edu.cn; Tel.: +86-0371-6855-5280

† These authors contributed equally to this work.

Abstract: Plantations are increasing in frequency and extent across the landscape, especially in China, and forest thinning can accelerate the development of late-successional attributes, thereby enhancing plantation stand structural heterogeneity. To quantify the effect of forest restoration thinning on the spatial heterogeneity and the structure of *Quercus variabilis* plantations, a restoration thinning experiment in a 40-year-old *Quercus variabilis* plantation by removing trees from the upper canopy level was conducted; two one-hectare sample plots with thinning and a control (i.e., unlogged) were sampled; and geostatistics methods were used to analyze the spatial distribution pattern of the DBH, height, and density of the stand. We found that restoration forest thinning in the *Quercus variabilis* plantation had a significant impact on the average DBH and tree height of the stand. Meanwhile, the coefficient of variation and structure ratio of the DBH, tree height, and stand density in the thinning plot were larger than those in the control plot. The range and spatial autocorrelation distance of the DBH and stand density in the thinning plot were smaller than those in the control plot, but the fractal dimension showed the opposite trend. The range and spatial autocorrelation distance of tree height in the thinning plot were higher than those in the control plot. These findings suggested that, compared with the control plot, the stereoscopic distribution of the DBH and stand density in the thinning plot fluctuated less and changed gentler, and its spatial continuity was not high but its variation was significant; meanwhile, the stereoscopic distribution of the tree height in the thinning plot was highly fluctuating and changed more significantly, with a strong spatial dependence and strip gradient distribution. Hence, forest restoration thinning could improve the distribution of the DBH and stand density and adjust the spatial heterogeneity of the DBH, tree height, and stand density of *Quercus variabilis* plantations.

Keywords: semi-variance analysis; spatial heterogeneity; forest restoration thinning; *Quercus variabilis* plantation

1. Introduction

Plantations need to be conscientiously managed to enhance stand quality and maintain ecosystem function and processes. Sustainable management could promote ecosystem function and accelerate the accumulation of carbon for reserves in a plantation forest, thus requiring basic scientific exploration [1]. Hence, the notion of using silviculture to help sustainably reach multiple management objectives in a sustainable manner has been evoked by generations of foresters. Forest thinning is one of the most commonly and importantly silvicultural methods of enhancing wood production [1], regulating stand structure, decreasing severe wildfire events, and increasing forest resilience for environmental disturbances, particularly during drought [2,3]. In addition, forest thinning has generally been prescribed to capture mortality, bring changes to the stand structure and herbaceous community, supply early financial return to landowners, and reallocate the growing space to fewer trees to increase the future merchantable volume and economic benefit and improve the ecological and recreational values [1,4].

Depending on the purpose of implementation, forest thinning includes restoration thinning and commercial thinning. Among them, commercial thinning mainly promotes the growth of well-growing trees to maximize commercial value by removing branched and stunted trees [5,6]. Meanwhile, restoration thinning is mainly used to reconstruct the stand structure and habitat characteristics, to improve the anti-interference ability, and to re-establish the ecosystem process by the selective removal of trees, especially in degraded stands [7,8]. However, the effect of forest thinning in promoting tree vitality and enhancing the habitat features of ecosystem restoration varies among different and limited forest types [9,10].

There is a general assumption that by redistributing the growing space and reducing competition from neighbors for the surrounding resources, forest thinning improves the growth efficiency of residual trees, compared with the same-sized trees of unthinned stands [11]. However, although forest thinning releases a certain amount of nutrient space, forest thinning can also increase tree mortality by wind damage and water stress [12,13], cause growth stagnation in residual trees, and then decrease the stand volume and increment [14–17]. Recently, correlational studies on the response of forest thinning have mainly concentrated on the growth of individual trees and stands, understory diversity, biomass, carbon sequestration and stock, the physical and chemical properties of soil nutrients, soil microbial community structure, and stand structure. It is generally believed that forest thinning can increase the growth of trees or stands (for example, relative volume growth, cumulative volume growth, and basal area growth) [18–20], enhance biomass accumulation and organic C input into the soil [1], and increase biomass accumulation in the short-term [21]. In terms of the influence of forest thinning on stand structure and diversity and shrub and herb diversity [22], the number of smaller trees is expected to increase rapidly [18].

The spatial pattern is a fundamental attribute of forest ecosystems and can affect many ecological processes and functions [23]; for example, within-stand tree spatial patterns can significantly affect tree recruitment and mortality [24,25]. There is broad agreement that forest management practices should address spatial aspects of forest structure and aim to conserve or restore spatial heterogeneity [26,27]. Spatial heterogeneity is of great importance in the study of populations, communities, ecosystems, and landscapes and is defined either as the variation in the spatial distribution of a point pattern [28,29] or as the variation in qualitative or quantitative values of a surface pattern [30]. It can be caused by habitat factors and their temporal variations [31,32], individual traits [33], and neutral processes [34].

There is broad agreement is that forest management should deal with the spatial aspects of forest structure and aim to conserve or restore the spatial heterogeneity of forests [26,27]. Restoring structural and spatial heterogeneity in forests as a technique to increase the supply of ecosystem services and landscape value has become a major objective of forest management and has garnered a lot of attention in forest research [5,35]. Forest thinning has been promoted as one means of enhancing heterogeneity in secondary forests [5]. Some studies think that traditional thinning methods result in a more uniform tree spacing and a narrower distribution of crop tree sizes [36]. However, more recent approaches such as variable-density thinning create a nonuniform tree spacing, thereby potentially increasing spatial heterogeneity [37]. In addition, the overall influence on the spatial variability of trees and the vertical distribution of canopy layers further depends on the forest thinning method implemented (e.g., thinning from below vs. thinning from above) [38]. In terms of the influence of forest thinning on the spatial heterogeneity, Choe et al. [39] utilized geo-statistics methods to explore the cutting disturbances on the soil nutrient spatial heterogeneity of understory plant diversity in the broadleaved-Korean pine forest in Changbai mountain, and they found that the spatial heterogeneity of soil nutrients was affected and increased by cutting intensity. Kuehne et al. [40], meanwhile, evaluated the effects of thinning on the structural heterogeneity of growth, ingrowth, and mortality in secondary coastal Douglas-fir forests, and they discovered that differences in stand structural heterogeneity between treatments are mostly nonsignificant and the variable-density thinning showed promise in increasing structural heterogeneity in young even-aged stands, despite the short-term nature of this study. In addition, Larson et al. [41] paid attention to the effect of restoration thinning on spatial heterogeneity in a mixed-conifer forest and found that restoration thinning could reduce spatial aggregation and lead to globally random tree patterns consisting of local tree clumps, openings, and widely spaced single trees, similar to reference conditions. Unfortunately, there is a certain knowledge gap on the influence of thinning on the spatial heterogeneity of broad-leaved forests or stands.

The *Quercus variabilis* plantation is a typical deciduous broadleaf forest in China, especially in the semi-arid areas. Meanwhile, *Quercus variabilis* is a valuable economic tree species with developed roots and drought resistance. Further understanding of the effects of thinning on DBH, tree height (TH), and stand density (SD) would help address how the stand structure and the spatial stability of the stand respond to restoration thinning. In this paper, we conducted a comparative experiment to reveal the general responses of the spatial distribution patterns of *Quercus variabilis* plantations to forest thinning. This study aimed to (1) explore the influence of thinning on the spatial heterogeneity and population dynamics; (2) understand in-depth the ecological process and mechanism of community stability of *Quercus variabilis* plantations responding to thinning; and (3) thus propose more effective management measures to improve the growth of *Quercus variabilis* plantations.

2. Materials and Methods

2.1. Study Area and Experimental Design

The research area was located on the Qingliangsi Experimental Forest Farm (112°44′–13°5′ E, 34°26′–34°33′ N), Dengfeng Forestry Bureau, Henan Province, China, which belongs to the Songshan Mountain of Funiu Mountains. The climate type is temperate continental monsoon with drought and little rain all year-round and is a typical semi-arid area. The total area and the maximum elevation are about 12,207.6 ha and 1512 m, respectively. The annual average rainfall, the annual average temperature, and the frost-free period are 614 mm, 14.2 °C, and 238 d, respectively. The area has predominantly brown and cinnamon soil. Meanwhile, the soil condition is rich in potassium, low in nitrogen, highly deficient in phosphorus, and low in organic matter content. In the research area, the main conifer tree species include *Pinus tabuliformis* and *Platycladus orientalis*, while the main broadleaf tree species include *Quercus acutissima*, *Quercus aliena*, and *Robinia pseudoacacia*.

Based on the distribution characteristics and growth status of vegetation, two representative stands of the *Quercus variabilis* plantation whose stand origin is aerial seedlings, with homogeneous stand conditions within each site, were selected as the experimental stands. The age and canopy density of these two stands were 40 years old and 0.9, respectively. Meanwhile, the tree species composition of these two stands was as follows: *Quercus variabilis* was dominant with a small amount of *Quercus aliena* and *Robinia pseudoacacia*. In 2013, a restoration thinning was conducted in one of the two stands, while another stand was taken as the control stand, which was unlogged. Fieldwork was performed in 2019 in the thinning and control stands. During these six years, there was no more forest management treatment for the thinning and control stands.

Field data were collected from 2 one-hectare permanent sample plots (100 m $\times$ 100 m) which were established in these two representative *Quercus variabilis* plantations and named the thinning plot (TP) and the control plot (CP), respectively. The following measurements were made in each plot: tree species, diameter at breast height (DBH), tree height, and relative coordinates of all living trees (DBH $\geq$ 2.0 cm), as well as elevation, aspect, slope, and other variables (Table 1). Hence, there were 3600 observations in total in these two plots. Based on the adjacent grid method, each one-hectare sample plot was divided into 25 continuous square subplots with an area equal to 0.04 ha, and 50 subplots from two one-hectare samples were obtained in total to determine the relative coordinates of each living tree in two 1 ha plots (Figure 1). Hence, the relative coordinates (X, Y, and Z) of each living tree in subplots were determined to the nearest 0.1 m using NTS-332R.

Table 1. General characteristics of the thinning and control plots.

Plot Type	Geographical Coordinate	Altitude (m)	Slope (°)	Aspect	Stand Density (Stems·ha^{-1})
TP	112°56′17″ E, 34°27′13″ N	617	9	Northwest	1487
CP	112°56′21″ E, 34°27′19″ N	630	7	Northwest	2113

Note: TP and CP represent the thinning plot and control plot, respectively, and are the same as those below.

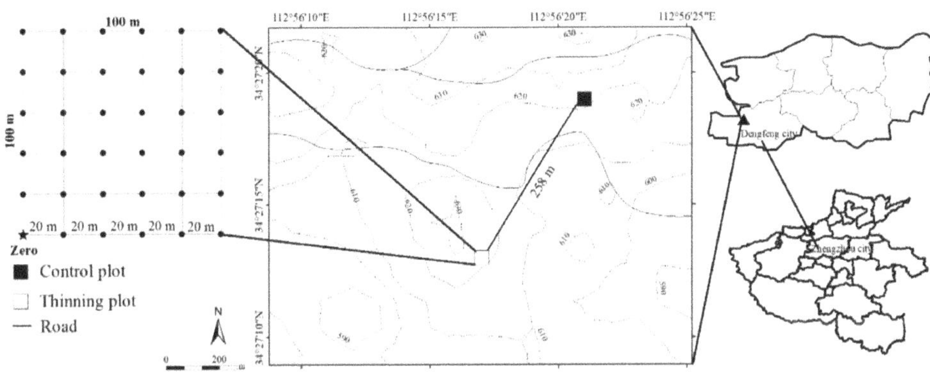

Figure 1. Location of the research area and sampling plots.

2.2. Spatial Autocorrelation and Distribution Analysis

2.2.1. Semi-Variance Function

A semi-variance function can be used for reflecting the overall distribution of the grayscale region and for analyzing the image self-correlation and the relationships between pixels and pixels of the neighborhood [42]. Hence, the formula of the semi-variance function is as follows (Equation (1)):

$$\lambda(h) = \frac{1}{2N(h)} \times \sum_{i=1}^{N(h)} [Z(X_i) - Z(X_i + h)]^2 \quad (1)$$

where $\lambda(h)$ is the value of the semi-variance function; h is the distance between two sample points; $N(h)$ is the number of sample pairs whose distances are equal to h; and $Z(X_i)$ and $Z(X_i + h)$ are the measured values of $Z(X)$ at the spatial positions X_i and $(X_i + h)$, respectively.

The semi-variance function analysis contains three basic parameters: nugget value (C_0), sill value (C_0 + C), and range (a). The nugget value (C_0) is the half-square difference, caused by many factors not equal to 0, when h = 0. The positive intercept value at the intersection of the semi-variance diagram curve and the vertical axis indicates the variation caused by random factors. The sill value (C_0 + C) is the maximum value of $\lambda(h)$ on the semi-variance function curve and it represents the maximum variation in the data. The range (a) is the abscissa value when the half-square tends to be stable and represents the maximum distance of spatial correlation. The structure ratio (SR) (i.e., $C_0/(C_0 + C)$) is the ratio of the nugget value to the sill value, which represents the proportion of heterogeneity caused by random factors in the total spatial heterogeneity. If SR < 25%, the spatial heterogeneity of the regionalized variable is mainly affected by structural factors, and it means that there exists a strong spatial correlation. If SR ranges from 25% to 75%, regionalized variables have a moderate spatial correlation. If SR is greater than 75%, the spatial heterogeneity is mainly influenced by random factors.

The main types of semi-variance function models are the spherical model, linear without still model, linear with still model, exponential model, and Gaussian model. The formulas of these five models are shown in Table 2. According to the sample fitting variance diagram and determination coefficients (R^2) of the constructed model, the optimal semi-variance function models for the DBH, tree height, and stand density of the *Quercus variabilis* plantation were selected.

Table 2. The formal list of the semi-variance function models.

Model Form	Equations
Spherical model	$\lambda(h) = \begin{cases} 0 & , h = 0 \\ C_0 + C \times \left[\frac{3h}{2a} - \frac{1}{2} \times \left(\frac{h}{a}\right)^3\right] & , 0 < h \le a \\ C_0 + C & , h > a \end{cases}$
Linear without still model	$\lambda(h) = \begin{cases} C_0 & , h = 0 \\ C_0 + A \times h & , h > 0 \end{cases}$
Linear with still model	$\lambda(h) = \begin{cases} C_0 & , h = 0 \\ C_0 + A \times h & , 0 < h \le a \\ C_0 + C & , h > a \end{cases}$
Exponential model	$\lambda(h) = \begin{cases} 0 & , h = 0 \\ C_0 + C \times \left[1 - e^{\left(-\frac{h}{a}\right)}\right] & , 0 < h \le 3a \\ C_0 + C & , h > 3a \end{cases}$
Gaussian model	$\lambda(h) = \begin{cases} 0 & , h = 0 \\ C_0 + C \times \left[1 - e^{\left(-\frac{h}{a}\right)^2}\right] & , 0 < h \le \sqrt{3}a \\ C_0 + C & , h > \sqrt{3}a \end{cases}$

Note: $\lambda(h)$ is the value of semi-variance; h is the separation distance; C is the structural variability; C_0 is the nugget variance; (C_0 + C) is a constant; a is the range of variance; A is a constant.

2.2.2. Moran's Index

Moran's index (I) is a method to quantify the spatial autocorrelation of stands. By comparing with the neighbors' values, it is widely used to identify spatial clustering and outlier values of variables. The formula of the global Moran's index is shown in Equation (2):

$$I = \frac{n}{S_0} \times \frac{\sum\limits_{i=1}^{n}\sum\limits_{j=1}^{n} W_{ij}(x_i - \overline{x})(x_j - \overline{x})}{\sum\limits_{i=1}^{n}(x_i - \overline{x})^2} \tag{2}$$

where W_{ij} is the weight between the ith observation and the jth observation; x_i and x_j are the measured values of the variable x at the ith and jth observation points, respectively; and S_0 is the sum of values of all W_{ij}. The value of I is between −1 and 1. When $I > 0$, the spatial correlation is positive and the intensity of the spatial correlation increases with increasing I. When $I < 0$, the spatial correlation is negative and the degree of spatial difference increases with decreasing I. When $I = 0$, the spatial correlation is random.

2.2.3. Fractal Dimension

The formula of the fractal dimension is as follows (Equation (3)):

$$D = \frac{4 - m}{2} \tag{3}$$

where m is the slope of the double logarithm semi-variance curve and D is the fractal dimension. Among them, the closer the value of D is to 2, the smaller the spatial heterogeneity and correlation are; the smaller the value of D is, the greater the spatial heterogeneity and correlation are.

2.2.4. Kriging Interpolation Method

Kriging is a probabilistic interpolation method based on the covariance of a dataset [43–45]. Kriging provides the interpolation estimates and the variances of interpolation errors that contribute to the uncertainty analysis [46]. Spatial Kriging interpolation uses the original data with regionalized variables and the structural characteristics of the semi-variance function to make the linear unbiased optimal estimation of regionalized variable values at the non-sampling points. Hence, giving measured n data values, i.e., $Z(s_1)$, $Z(s_2)$,..., $Z(s_n)$, at $s_1, s_2, \ldots, s_n$ locations, $\hat{Z}(s_0)$ at the unknown location of s_0 is estimated by the following equation (Equation (4)) [47]:

$$\hat{Z}(s_0) = \sum_{i=1}^{n} \lambda_i Z(s_i) \tag{4}$$

where $Z(s_i)$ is the measured value at the s_i location, and λ_i is the unknown weight for the measured value at the s_i location.

2.3. Data Processing

To calculate the stand attributes (i.e., average DBH, average tree height, stand density) and conduct spatial autocorrelation and distribution analysis, firstly, based on the adjacent grid method and relative coordinates, the thinning and control sample plots were divided into 100 subplots with an area equal to 0.01 ha (10 m × 10 m). The average DBH, tree height, stand density, and the center coordinate for each subplot in the thinning and control plots were calculated. Then, the center coordinates of these subplots were taken as the corresponding coordinates of the average DBH, average tree height, and stand density in these subplots. Next, semi-variance function analysis required the data to obey a normal distribution. If the data did not conform to the normal distribution, it would produce a proportional effect and then affect the accuracy of semi-variance and the fractal dimension. Hence, the normality of the DBH, tree height, and stand density of the TPs and CPs was tested by the Shapiro–Wilk test in SPSS 24.0 software. If $P_{S\text{-}W} > 0.05$, the data obeyed a normal distribution; otherwise, the data should be transformed by the Box–Cox transformation to conform to the normal distribution. The calculation and analysis of the semi-variance function were measured by the normal distribution data and the GS^+ 9.0 software, and then a 3D distribution map was obtained. In addition, the 2D distribution map was generated by Kriging interpolation in ArcGIS 10.6.0 software.

3. Results

3.1. Effect of Forest Thinning on the DBH, Tree Height, and Stand Density of Quercus Variabilis Plantation

The average DBH, average tree height, and stand density of the CP were 10.8 cm, 9.0 m, and 0.21 stems/m^2, while those of the TP were 12.8 cm, 11.3 m, and 0.14 stems/m^2 (Table 3). After thinning, the average values of the DBH and tree height were evidently improved, while that of stand density was decreased.

Table 3. Descriptive statistics of community structure attribute in the thinning and control plots.

Plot Types	Index	Average ± SD	Minimum	Maximum	Coefficient of Variation (%)	Skewness	Kurtosis	$P_{s\text{-}w}$	$P^*_{s\text{-}w}$
CP	Average DBH (cm)	10.80 ± 1.31 a	17.10	8.80	12.16	1.55	5.10	0.000	0.455
	Average TH (m)	9.00 ± 0.89 a	12.90	7.30	9.89	1.22	3.20	0.000	0.877
	SD (stem/m^2)	0.21 ± 0.07 a	0.37	0.02	35.31	−0.42	0.18	0.115	
TP	Average DBH (cm)	12.80 ± 1.36 b	15.80	8.40	10.59	−0.38	2.15	0.032	0.054
	Average TH (m)	11.30 ± 1.13 b	13.90	6.70	9.97	−0.47	1.92	0.014	0.138
	SD (stem/m^2)	0.14 ± 0.04 a	0.26	0.05	27.86	−0.07	0.59	0.096	

Note: Different letters mean a significant difference in the community structure attributes between different plots at the 0.05 level. $P_{s\text{-}w}$ is the normal test probability value of the original data; $P^*_{s\text{-}w}$ is the probability value of the normal test after Box–Cox transformation. DBH, TH, and SD represent the diameter at breast height, tree height, and stand density, respectively.

The average DBH and stand density of the TPs and CPs had a moderate variation, and the average tree height of these two sample plots (i.e., TPs and CPs) had weak variations, based on the classification of the coefficient of variation (CV, CV $\leq$ 10%, soft variation; 10% < CV < 100%, moderate variation; CV $\geq$ 100%, substantial variation) (see Table 3). However, the CV values of the average DBH and stand density in the CP were higher than those in the TP, while the CV values of the average tree height in the CP was smaller than those in the TP. In terms of the range, the range values of the average DBH, average tree height, and stand density of the CP were 8.30 cm, 5.60 m, and 0.35 stem/m^2, respectively, while those of the TP were 7.20 cm, 7.40 m, and 0.21 stem/m^2, respectively. Generally, thinning measures made the size distribution of the DBH and stand density more concentrated and the size distribution of the tree height more dispersed. The possible reason for this situation may be that thinning removes the trees which are branched, stunted, or competitive to promote the growth of residual trees and releases a certain nutritional space, and then makes the residual trees in the upper layer receive more light resources to promote their height growth.

In addition, the result of the Shapiro–Wilk test (see Table 3) showed that stand density conformed to the normal distribution while DBH and tree height should be transformed by the Box–Cox transformation into a normal distribution to satisfy the conditions of geo-statistics interpolation.

3.2. *Influence of Forest Thinning on the Spatial Distribution Patterns of DBH, Tree Height, and Stand Density of Quercus Variabilis Plantation*

3.2.1. The Semi-Variance Function Analysis of DBH, Tree Height, and Stand Density of Quercus Variabilis Plantation

The selected semi-variance function models (i.e., exponential model, linear without still model, linear with still model, spherical model, and Gaussian model) were introduced to evaluate the variation in DBH, tree height, and stand density of the TPs and CPs in the *Quercus variabilis* plantation, respectively. From Table 4, the exponential model was used to analyze the spatial heterogeneity of the DBH, tree height, and stand density in the TPs and CPs, with the exception of tree height in the CP and DBH in the TP. The determination coefficients (R^2) of these models, with the exception of the models generated by tree height in the CP, were higher and ranged from 0.880 to 0.976. Hence, the selected models could accurately describe and better reflect the spatial correlation and structure characteristics of the DBH, tree height, and stand density of the TPs and CPs.

Table 4. Types and parameters of semi-variogram model for spatial analysis of DHB, tree height and stand density in the thinning and control plots.

Plot Types	Index	Model Form	Nugget	Sill	Structure Ratio (%)	Range (m)	Determination Coefficients (R^2)	Residual Sum of Squares (RSS)
CP	DBH (cm)	Exponential model	<0.001	<0.001	0.16	28.2	0.880	<0.001
	TH (m)	Spherical model	<0.001	<0.001	0.17	12.6	0.001	<0.001
	SD (stem/m^2)	Exponential model	<0.001	<0.001	0.20	39.3	0.967	<0.001
TP	DBH (cm)	Spherical model	2.1	51.83	4.05	21.9	0.880	18.4
	TH (m)	Exponential model	24.7	264.4	9.34	33.0	0.912	407
	SD (stem/m^2)	Exponential model	<0.001	<0.001	6.54	30.3	0.976	<0.001

Note: DBH, TH, and SD are the same as those in Table 3.

As shown in Table 4, the sill values of DBH, tree height, and stand density in the TP were larger than those in the CP, which indicated that the spatial variability in DBH, tree height, and stand density in the TP was higher than that in the CP. The structural ratio values of DBH, tree height, and stand density in the TPs and CPs ranged from 0.16% to 9.34%, which indicated that the proportions of spatial heterogeneity generated by random factors and structural factors ranged from 0.16% to 9.34% and from 90.66% to 98.84%, respectively. In addition, the structural ratio values of DBH, tree height, and stand density of the TP were all larger than those of the CP. In general, the spatial autocorrelation of the DBH, tree height, and stand density of the *Quercus variabilis* plantation were all strong and caused by structural factors (such as forest stand and soil). Hence, the forest thinning measure could decrease the spatial autocorrelation of the DBH, tree height, and stand density of the *Quercus variabilis* plantation. The main reason for this situation is that thinning reduces the number of neighboring trees among the residual trees and changes the resource allocation in the stand.

In addition, the semi-variance function values of DBH, tree height, and stand density in the TPs and CPs all first increased as the separation distance increased and then remained stable after a threshold (Figure 2). As shown in Figure 2, the threshold values of the semi-variance function curves of DBH, tree height, and stand density in the TP were 21.9 m, 33.0 m, and 30.3 m, respectively, and those in the CP were 28.2 m, 12.6 m, and 39.3 m, respectively. This indicated that the DBH, tree height, and stand density in the TPs and CPs had a strong spatial autocorrelation in this range. From Table 4 and Figure 2, the range of tree height in the TP was higher than that in the CP. In contrast, the DBH and stand density all showed the opposite trend, which indicated that the spatial autocorrelation scale of tree height after thinning was large. However, the tree height of the *Quercus variabilis* plantation after thinning showed homogeneity on a large distance scale. In addition, the range of DBH, tree height, and stand density in the TPs and CPs were all larger than the sampling distance (i.e., 10 m), which indicated that the sampling distance met the requirements of semi-variogram analysis.

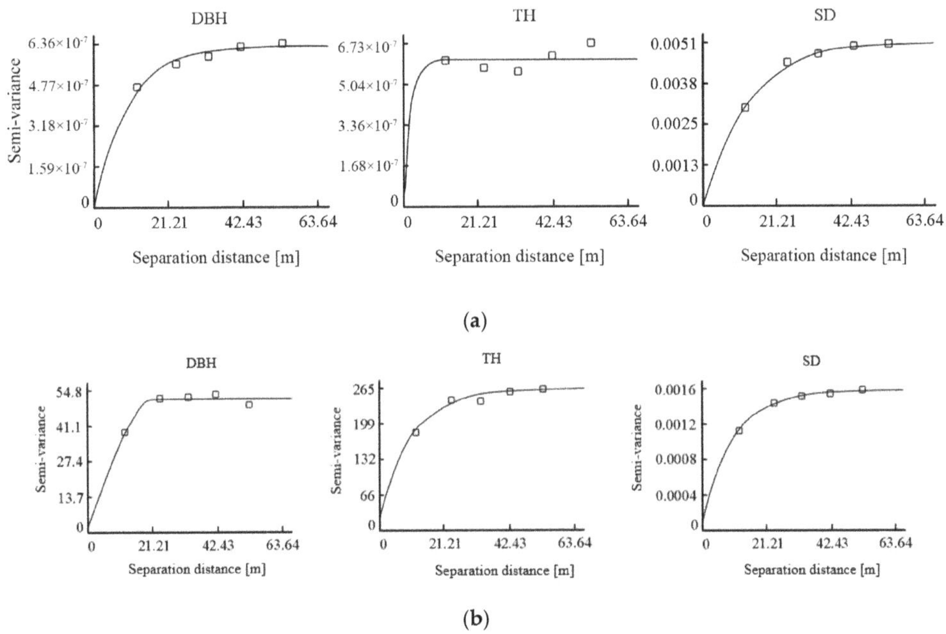

Figure 2. Semi-variograms of DBH, TH, and SD in the thinning and control plots. (**a**) Semi-variances of DBH, TH, and SD of the control plot; (**b**) semi-variances of DBH, TH, and SD of the thinning plot. (DBH, TH, and SD are the same as Table 3).

3.2.2. Fractal Dimension of DBH, Tree Height, and Stand Density of Quercus Variabilis Plantation

As shown in Figure 3, the determination coefficient (R^2) values of these models generated by the separation distance and the fractal dimension of DBH, tree height, and stand density in the TPs and CPs, with the exception of tree height in the CP and the DBH in the TP, ranged from 0.861 to 0.992. The regular pattern of the influence of forest thinning on the fractal dimension was similar to that of the effect of forest thinning on the range for DBH, tree height, and stand density in the *Quercus variabilis* plantation. The fractal dimension values of DBH, tree height, and stand density in the TP were 1.912, 1.877, and 1.886, respectively, while those in the CP were 1.893, 1.969, and 1.822, respectively (see Figure 3). This indicated that forest thinning could improve the spatial dependence and simplify the spatial distribution of tree height while weakening the spatial dependence and complicating the spatial distribution of DBH and stand density in the *Quercus variabilis* plantation. The main reason may be that thinning decreases the number of competitors around residual trees and leads to the release of nutrient space, changing the allocation of light resources to cause a change in tree height growth to some extent.

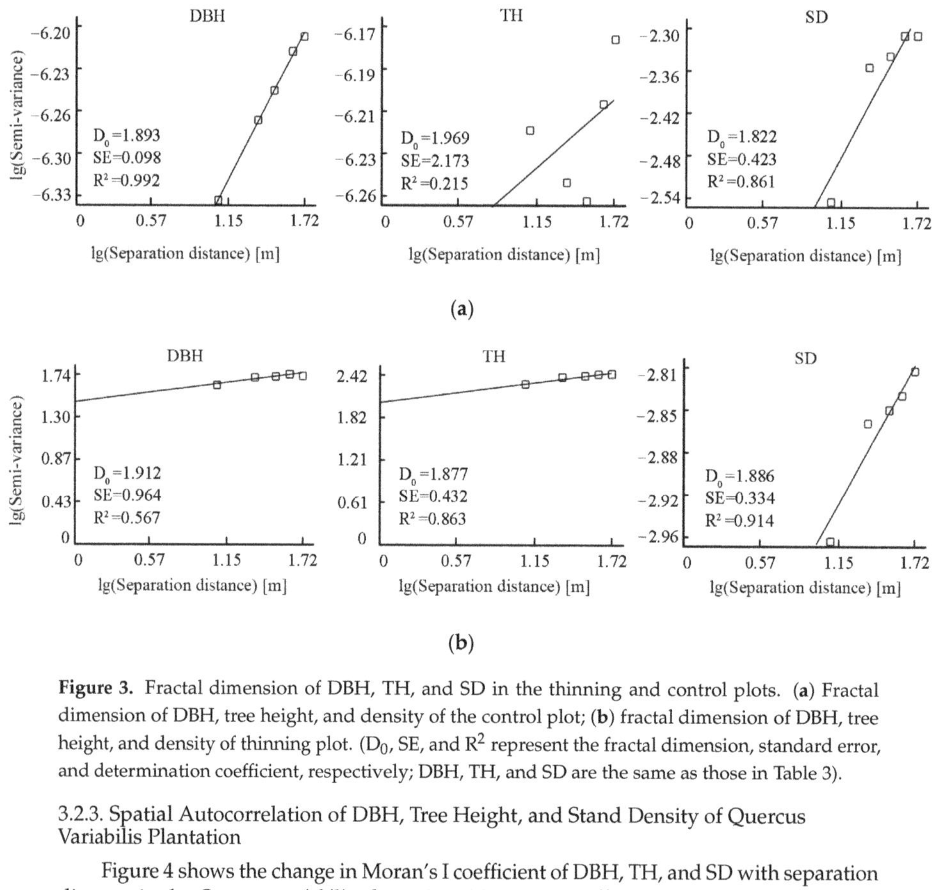

Figure 3. Fractal dimension of DBH, TH, and SD in the thinning and control plots. (**a**) Fractal dimension of DBH, tree height, and density of the control plot; (**b**) fractal dimension of DBH, tree height, and density of thinning plot. (D_0, SE, and R^2 represent the fractal dimension, standard error, and determination coefficient, respectively; DBH, TH, and SD are the same as those in Table 3).

3.2.3. Spatial Autocorrelation of DBH, Tree Height, and Stand Density of Quercus Variabilis Plantation

Figure 4 shows the change in Moran's I coefficient of DBH, TH, and SD with separation distance in the *Quercus variabilis* plantation. Moran's I coefficient values of the DBH and stand density in the CP were basically higher than those in the TP. Meanwhile, tree height showed the opposite trend, which indicated that the spatial correlation of tree height was larger and those of DBH and stand density were all smaller after thinning. In addition, there were two trends in the relationship between Moran's I coefficient and separation distance. One was that Moran's I coefficient of DBH and tree height in the TP and tree height and stand density in the CP all decreased with the increase in separation distance, and their curves had an intersection point with the horizontal axis. The other was that Moran's I coefficient of DBH in the TP and stand density in the CP all decreased rapidly, fell slowly, and finally increased with the increase in separation distance, and their curves had two intersection points with the horizontal axis, which indicated that the DBH in the

TP and the stand density in the CP all had a large plaque. Combined with the range, the plaque of DBH in the TP was about 21.9 units and that of the stand density in the CP was about 39.3 units.

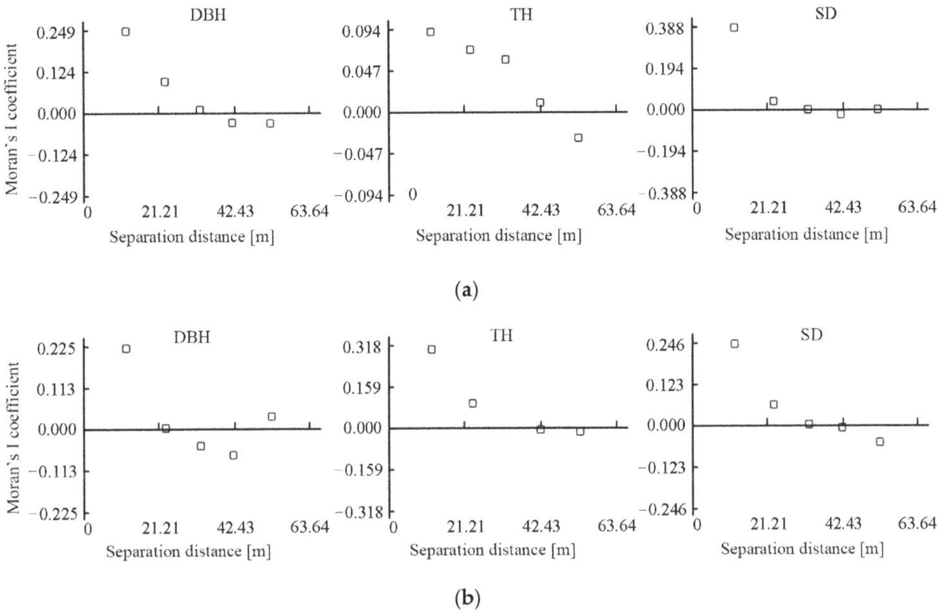

Figure 4. Moran's I coefficient of DBH, TH, and SD in the thinning and control plots. (**a**) Moran's I coefficient of DBH, tree height, and density of the control plot; (**b**) Moran's I coefficient of DBH, tree height, and density of thinning plot. (DBH, TH, and SD are the same as those in Table 3).

3.2.4. Spatial Distribution Pattern of DBH, Tree Height, and Stand Density of Quercus Variabilis Plantation

As shown in Figure 5, the spatial distributions of DBH and stand density in the CP and tree height in the TP had a strip gradient with some patches, while those of the DBH and stand density in the TP and the tree height in the CP had a patch-like gradient. Among them, in the CP, the DBH and tree height increased gradually from northwest to southwest and east, and their high-value area appeared in the east of the study area, while the spatial distribution of the stand density was opposite to those of the DBH and tree height. However, in the TP, DBH increased gradually from southwest to northeast, and its high-value area was in the northeast; tree height decreased gradually from southwest to northwest and southeast, and its high-value area was in the southwest; and stand density increased from northeast to southwest, and its high-value area appeared in the southwest. Hence, the spatial heterogeneity of tree height in the CP and those of DBH and stand density in the TP were relatively large, and their spatial continuity were high on a small scale and had substantial variance.

Further, based on the sharpness of the peaks and the transition degree of each index, the kurtosis of the 3D stereogram of DBH and stand density changed from significant fluctuations to gentle, while that of tree height changed from mild to severe, after thinning (see Figure 6). This result indicated that thinning made the spatial distribution of DBH and stand density uniform with a more negligible overall difference and spatial heterogeneity, while it made the spatial distribution of tree height uneven with a more significant overall difference and spatial heterogeneity.

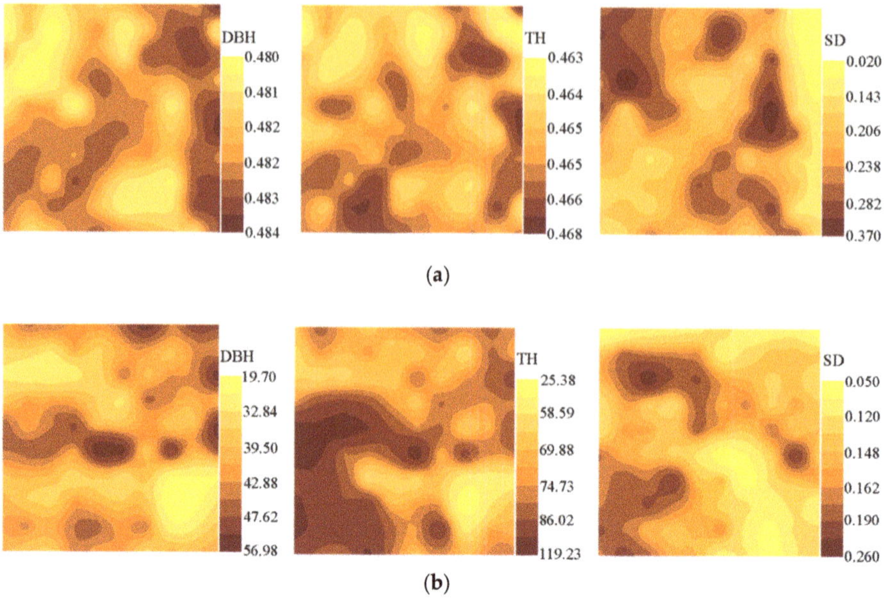

Figure 5. Two-dimensional map of spatial distribution pattern of DBH, tree height, and stand density in the thinning and control plots. (**a**) Spatial distribution pattern of DBH, tree height, and stand density of the control plot; (**b**) spatial distribution pattern of DBH, tree height, and stand density of the thinning plot. (DBH, TH, and SD are the same as those in Table 3).

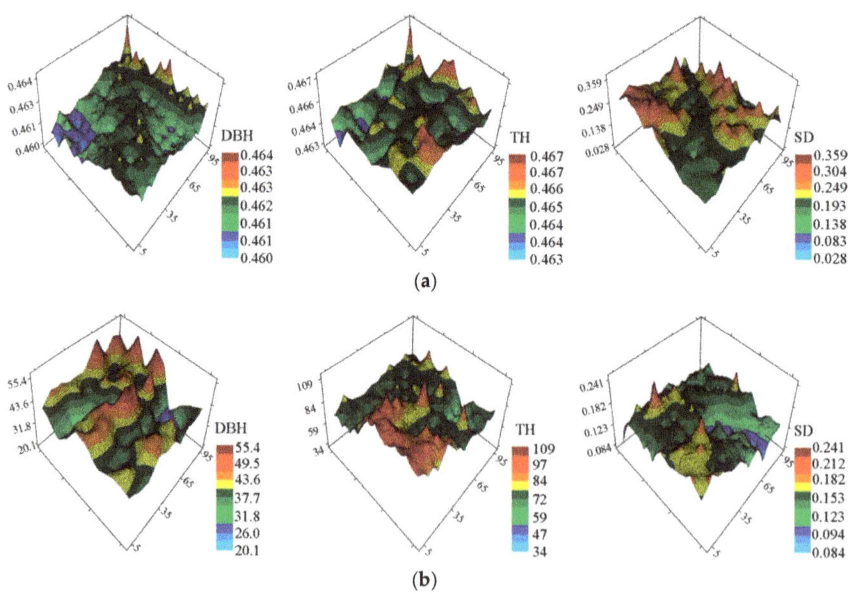

Figure 6. Three-dimensional map of spatial distribution pattern of DBH, tree height, and stand density in the thinning and control plots. (**a**) Spatial distribution pattern of DBH, tree height, and stand density of the control plot; (**b**) spatial distribution pattern of DBH, tree height, and stand density of the thinning plot. (DBH, TH, and SD are the same as those in Table 3).

4. Discussion

Many silvicultural practices such as planting cutover forest lands, vegetation management techniques, and stand thinning have been criticized for creating relatively uniform conditions with forest stands [48]. Pre-commercial thinning, in particular, tends to leave a homogeneous array of trees throughout a treated stand, and this practice contributes to both volume and quality increases in wood fiber on those crop trees selected for superior growth and form during the thinning process [49]. Forest restoration thinning is also a technique that alters the stand structure and ecological succession of understory vegetation. Conventional stand thinning generally leads to positive responses in the biomass of understory vegetation [50–55] and has the potential to increase tree growth [16,56–58] and reduce the competition among trees [59] while generating a gap to increase the diameters of the rest of the trees and the stability of the stand [60]. Due to the forestry characteristics in China, forest restoration thinning could not attain a high timber yield. Hence, it is necessary to analyze the effect of forest restoration thinning on the spatial heterogeneity and structure of the *Quercus variabilis* plantation in China.

4.1. The Effect of Thinning on the Non-Spatial Structure of Quercus Variabilis Plantation

Forest thinning plays an important role in adjusting tree species composition, reducing stand density, improving forest health, optimizing stand structure, and improving the comprehensive benefits of the stand. Our result indicated that the average DBH and average tree height of the TP were significantly higher than those of the CP. In addition, the variation coefficient and range values of DBH and stand density in the TP were smaller than those in the CP, while the variation coefficient and range values of tree height in the TP were more extensive than those in the CP. This situation showed that the distributions of DBH and stand density in the TP were relatively concentrated and better than those in the CP, while the distribution of tree height in the TP was opposite. In general, the influence of thinning on the tree growth and non-spatial structure of the *Quercus variabilis* plantation was positive. Our results are consistent with those of previous research [61,62]. Among them, Gradel et al. [59] researched the response of the non-spatial structure of birch and larch stands in North Mongolia to thinning and found that thinning could positively affect the total yield, although the distribution and coefficient of variation in diameter were not greatly changed by thinning. Nguyen et al. [62] studied the influence of thinning intensity on tree growth in a *Pinus koraiensis* plantation and found that thinning increased tree growth (diameter at breast height and crown size) with the exception of height growth, while the moderately thinning stand had the highest basal area and stock volume at the stand level. Meanwhile, Cabon et al. [61] researched the influence of thinning on tree growth in a Mediterranean evergreen oak coppice, and their result revealed that thinning increased long-term growth at the stem level but decreased the wood biomass at the stand level, while moderate thinning had a sustained beneficial influence on the growth of all stem size classes; by contrast, stronger thinning intensities increased the size asymmetry of competition and their overall effect dropped faster. However, the reason for this situation might be that thinning removed the branched, stunted, or competitive trees around residual trees; decreased the number of neighboring trees around the residual trees; and released a certain amount of nutrition space before adjusting the structure and density of the stand, improving the utilization efficiency of water and light resources, and reducing the proportion of trees with a small DBH and low tree height, resulting in the better growth and development of the remaining trees [63,64]. Due to an excellent correlation between forest growth and stand density, suitable thinning could adjust the stand density, improve the illumination, increase the nutrition of the reserved trees, and effectively promote the growth of trees. Hence, although this research analyzes the impact of thinning from multiple stand features (i.e., DBH, tree height, and stand density), it was necessary to conduct in-depth research on the canopy structure and diversity of the understory plant in the future.

4.2. The Effect of Thinning on the Spatial Heterogeneity of Quercus Variabilis Plantation

Spatial heterogeneity is the variability and complexity of the system or system attributes in space and is the main reason for the formation of spatial patterns. At present, the research on spatial patterns mainly used the crown width, crown height, tree height, DBH, and stand coverage as regionalized variables to analyze the spatial pattern of stands [65]. This study changed the previous spatial distribution pattern analysis methods, such as the distance method, angular scale, and coefficient of variation of the Voronoi method, and used the semi-variance function analysis and Kriging interpolation method to analyze the spatial heterogeneity. Our results showed that the variation range of the DBH, tree height, and stand density of the thinning and control plots were all greater than 10 m, and the sampling distance met the research requirements. The tree height, DBH, and stand density in the thinning and control plots had a strong autocorrelation. The variation ranges and the spatial autocorrelation distances of DBH and stand density in the thinning plot were all lower than those in the control plot. Meanwhile, the variation range and the spatial autocorrelation distance of tree height in the thinning plot were larger than those in the control plot. Compared with the control plot, the spatial distributions of DBH and stand density in the thinning plot fluctuated less and changed gentler; their spatial continuity was not high but their spatial variation was more significant. Meanwhile, the spatial distribution of tree height in the thinning plot was highly fluctuating, with a strong spatial dependence, simple spatial distribution, and strip gradient distribution. The reason might be that forest thinning decreases the number of competitors around the remaining trees; changes the allocation of habitat resources, especially the allocation of light resources in vertical space; and then adjusts the horizontal and vertical distribution of trees in the stand to make stand structure more stable, letting the forest environment become more suitable for plant growth and development, improving the canopy structure and light environment to make the shrubs and grass grow better, accelerating litter decomposition, increasing the soil nutrient and forest growth space, and then obtaining profit for plant growth and development. However, the slope position, topography, intraspecific competition, and other factors would lead to an uneven distribution of tree height after thinning. This reason is supported by some previous studies [65–67]. Among them, Liu et al. [65] studied the spatial heterogeneity of urban street trees in Harbin, China, and found that interspecific and intraspecific competition would affect the spatial heterogeneity of tree height and DBH. A large number of studies showed that the DBH and height of trees were influenced by soil factors. Li et al. [67] studied the spatial heterogeneity of the DBH growth of Korean pine and revealed that terrain factors could explain the spatial distribution of its DBH growth, to a certain extent. Meanwhile, Samra et al. [66] thought that the growth of tree height had some relationship with the soil environment by studying the spatial dependence between soil alkalinity and tree growth. Exploring the spatial heterogeneity of the DBH, tree height, and stand density of plantations in the thinning and control plots could further elucidate the influence of thinning on stand structure by geostatistical methods and provide a basis for reasonable thinning. However, slope position, soil, and human disturbance all had a certain effect on DBH, tree height, and stand density under different thinning intensities. Hence, it is necessary to further study the spatial heterogeneity of DBH, tree height, and stand density under different thinning intensities.

4.3. The Suggestion of Forest Restoration Thinning in the Semi-Arid Area

Forestry surveys have revealed that the allometric relationships among tree height, canopy area, and breast height diameter are mainly driven by light limitations [68]. However, in open forests such as semi-arid forests, the light resource is generally not a limiting factor for tree growth [69]. In semi-arid areas, there are rich radiation and heat resources but insufficient water resources, thus seriously affecting the survival, growth, and management of afforestation. Some researchers have considered that the semiarid climate conditions limit the height growth of trees [70] and generate an overall deficiency in moisture, which considerably influences the elevational forest distribution and may even control the upper forest limit [71–73]. Human-driven forest thinning is of primary importance for sustainable water resource management in semi-arid basins [74,75]. In China, the semi-arid area accounts for 22% of the total land area. Hence, based on the effect of thinning on the spatial heterogeneity of DBH, tree height, and stand density in *Quercus variabilis* plantations, forest restoration thinning in semi-arid areas is the suggested target tree management method taking into account soil and water conservation and confirming the disturbance trees around the remaining trees. Then, the competitive trees or the trees that are not conducive to the reconstruction of the stand structure around the remaining trees are selectively removed, but not the deep-rooted tree species, maintaining the water resources environment in the region. Meanwhile, the intensity of this restoration thinning should be moderate to adjust the stand structure and maintain the ecological function of the forest in the semi-arid area to meet the forest-adaptive management requirements.

5. Conclusions

The influence of forest thinning on the stand structure and spatial heterogeneity of a *Quercus variabilis* plantation was explored by geostatistical methods. Forest thinning could improve the distribution of the DBH and stand density and increase the effect of structural factors on the spatial heterogeneity of the DBH, tree height, and stand density. In addition, forest thinning could reduce the stereoscopic distribution fluctuation in DBH and stand density, reduce the fluctuation change, reduce their spatial continuity, and enhance the significance of their spatial difference. However, forest thinning makes the stereoscopic distribution fluctuation of tree height change severely, improves its spatial dependence, and simplifies its spatial distribution, which shows a banded gradient distribution. Forest thinning improves the allocation of light resources in the stand, promotes tree growth, and enriches the spatial heterogeneity of stand features. For the suggested forest management in the semi-arid area, moderate restoration thinning which utilizes the target tree management method considering soil and water conservation is recommended. Hence, this research will provide a reference for exploring the spatial variation and heterogeneity of stand attributes and formulating forest-adaptive management measures in semi-arid areas.

Author Contributions: Conceptualization, D.W. and D.Y.; methodology, D.W. and D.Y.; software, M.Z.; validation, M.Z., D.W. and D.Y.; formal analysis, D.W. and D.Y.; investigation, M.Z., D.W. and D.Y.; resources, M.Z., D.W. and D.Y.; data curation, M.Z., Y.W. and S.J.; writing—original draft preparation, M.Z.; writing—review and editing, M.Z., Y.W., D.W. and D.Y.; visualization, M.Z.; supervision, D.W. and D.Y.; project administration, D.W. and D.Y.; funding acquisition, D.W. All authors have read and agreed to the published version of the manuscript.

Funding: This research was funded by the Research Project of Henan Science and Technology, grant numbers 201300111400 and 192102110060.

Data Availability Statement: Data are available upon request.

Conflicts of Interest: All authors declare that they have no conflicts of interest.

References

1. Yohannes, Y. Effect of thinning on biophysical soil properties of Cupressus lusitanica stand in Munessa forest, Ethiopia. *Ethiop. J. Sci. Technol.* **2019**, *12*, 233–248. [CrossRef]
2. Makinen, H.; Isomaki, A. Thinning intensity and growth of Scots pine stands in Finland. *For. Ecol. Manag.* **2004**, *201*, 311–325. [CrossRef]
3. Sohn, J.A.; Saha, S.; Bauhus, J. Potential of forest thinning to mitigate drought stress: A meta-analysis. *For. Ecol. Manag.* **2016**, *380*, 261–273. [CrossRef]
4. Curtis, R.O.; Marshall, D.D.; Bell, J.F. LOGS: A pioneering example of silvicultural research in coast Douglas-fir. *J. For.* **1997**, *95*, 19–25. [CrossRef]
5. Bauhus, J.; Puettmann, K.; Messier, C. Silviculture for old-growth attributes. *For. Ecol. Manag.* **2009**, *258*, 525–537. [CrossRef]
6. Di, S.J. River red gum (*Eucalyptus camaldulensis*): A review of ecosystem processes, seedling regeneration and silvicultural practice. *Aust. For.* **2002**, *65*, 14–22.
7. Dagley, C.M.; Berrill, J.P.; Leonard, L.P.; Kim, Y.G. Restoration thinning enhances growth and diversity in mixed redwood/Douglas-fir stands in northern California, USA. *Restor. Ecol.* **2018**, *26*, 1170–1179. [CrossRef]

8. Plummer, J.F.; Keyes, C.R.; Varner, J.M. Early-Stage thinning for the restoration of young redwood—Douglas-Fir forests in northern coastal California, USA. *Restor. Ecol.* **2012**, *2012*, 725827. [CrossRef]
9. Hood, S.M.; Cluck, D.R.; Jones, B.E.; Pinnell, S. Radial and stand-level thinning treatments: 15-year growth response of legacy ponderosa and Jeffrey pine trees. *Restor. Ecol.* **2018**, *26*, 813–819. [CrossRef]
10. Puettmann, K.; Ares, A.; Burton, J.; Dodson, E. Forest restoration using variable density thinning: Lessons from Douglas-fir stands in Western Oregon. *Forests* **2016**, *7*, 310. [CrossRef]
11. O'Hara, K.L. Stand growth efficiency in a Douglas-fir thinning trial. *Forestry* **1989**, *62*, 409–418. [CrossRef]
12. Lagergren, F.; Lankreijer, H.; Kucera, J.; Cienciala, E.; Molder, M.; Lindroth, A. Thinning effects on pine-spruce forest transpiration in central Sweden. *For. Ecol. Manag.* **2008**, *255*, 2312–2323. [CrossRef]
13. Kuehne, C.; Weiskittel, A.R.; Wagner, R.G.; Roth, B.E. Development and evaluation of individual tree-and stand-level approaches for predicting spruce-fir response to commercial thinning in Maine, USA. *For. Ecol. Manag.* **2016**, *376*, 84–95. [CrossRef]
14. Harrington, C.A.; Reukema, D.L. Initial shock and long-term stand development following thinning in a Douglas-fir plantation. *For. Sci.* **1983**, *29*, 33–46.
15. Sharma, M.; Smith, M.; Burkhart, H.E.; Amateis, R.L. Modeling the impact of thinning on height development of dominant and codominant loblolly pine trees. *Ann. For. Sci.* **2006**, *63*, 349–354. [CrossRef]
16. Makinen, H.; Isomaki, A. Thinning intensity and growth of Norway spruce stands in Finland. *Forestry* **2004**, *77*, 349–364. [CrossRef]
17. Moulinier, J.; Brais, S.; Harvey, B.D.; Koubaa, A. Response of boreal jack pine (*Pinus banksiana* Lamb.) stands to a gradient of commercial thinning intensities, with and without N fertilization. *Forests* **2015**, *6*, 2678–2702. [CrossRef]
18. Connell, M.J.; Raison, R.J.; Jenkins, P. Effects of thinning and coppice control on stand productivity and structure in a silvertop ash (*Eucalyptus sieberi* L. johnson) forest. *Aust. For.* **2003**, *67*, 30–38. [CrossRef]
19. Primicia, I.; Artázcoz, R.; Imbert, J.B.; Puertas, F.; Traver, M.C.; Castillo, F.J. Influence of thinning intensity and canopy type on scots pine stand and growth dynamics in a mixed managed forest. *For. Syst.* **2016**, *25*, e057. [CrossRef]
20. Bose, A.K.; Weiskittel, A.; Kuehne, C.; Wagner, R.G.; Turnblom, E.; Burkhart, H.E. Does commercial thinning improve stand-level growth of the three most commercially important softwood forest types in North America? *For. Ecol. Manag.* **2018**, *409*, 683–693. [CrossRef]
21. Zasada, M.; Bronisz, K.; Bijak, S.; Dudek, A.; Bruchwald, A.; Wojtan, R.; Tomusiak, R.; Bronisz, A.; Wróblewski, L.; Michalak, K. Effect of the cutting age and thinning intensity on biomass and carbon sequestration—The Gubin forest district case study. *Folia For. Pol.* **2009**, *51*, 18–24.
22. Li, X.; Li, Y.; Zhang, J.; Peng, S.; Chen, Y.; Cao, Y. The effects of forest thinning on understory diversity in china: A meta-analysis. *Land Degrad. Dev.* **2020**, *31*, 1225–1240. [CrossRef]
23. Franklin, J.F.; Spies, T.A.; Pelt, R.V.; Carey, A.B.; Thornburgh, D.A.; Berg, D.R.; Lindenmayer, D.B.; Harmon, M.E.; Keeton, W.S.; Shaw, D.V. Disturbance and structural development of natural forest ecosystems with silvicultural implications, using Douglas-fir as an example. *For. Ecol. Manag.* **2002**, *155*, 399–423. [CrossRef]
24. Sánchez, M.A.J.; Moore, M.M.; Bakker, J.D.; Parysow, P.F. 108 years of change in spatial pattern following selective harvest of a ponderosa pine stand in northern Arizona, USA. *J. Veg. Sci.* **2009**, *20*, 79–90. [CrossRef]
25. Das, A.; Battles, J.; Stephenson, N.L.; Mantgem, P.J.V. The contribution of competition to tree mortality in old-growth coniferous forests. *For. Ecol. Manag.* **2011**, *261*, 1203–1213. [CrossRef]
26. North, M.; Stine, P.; O'Hara, K.; Zielinski, W.; Stephens, S. *An Ecosystem Management Strategy for Sierran Mixed-Conifer Forests*, 2nd ed.; General Technical Report, PSW-GTR-220; Forest Service: Missoula, MT, USA, 2009.
27. Larson, A.J.; Churchill, D. Tree spatial patterns in fire-frequent forests of western North America, including mechanisms of pattern formation and implications for designing fuel reduction and restoration treatments. *For. Ecol. Manag.* **2012**, *267*, 74–92. [CrossRef]
28. Shaver, G.R. Spatial heterogeneity: Past, present, and future. In *Ecosystem Function in Heterogeneous Landscapes*; Springer: New York, NY, USA, 2005; pp. 443–449.
29. Zhu, X.; He, Z.B.; Du, J.; Chen, L.F.; Lin, P.F.; Tian, Q.Y. Spatial heterogeneity of throughfall and its contributions to the variability in near-surface soil water-content in semiarid mountains of China. *For. Ecol. Manag.* **2021**, *488*, 119008. [CrossRef]
30. Dutilleul, P.; Legendrem, P. Spatial heterogeneity against heteroscedasticity: An ecological paradigm versus a statistical concept. *Oikos* **1993**, *66*, 152–171. [CrossRef]
31. Tscharntke, T.; Steffan-Dewenter, I.; Kruess, A.; Thies, C. Characteristics of insect populations on habitat fragments: A mini review. *Ecol. Res.* **2002**, *17*, 229–239. [CrossRef]
32. Leyequien, E.; Verrelst, J.; Slot, M.; Schaepman-Strub, G.; Heitkönig, I.M.A.; Skidmore, A. Capturing the fugitive: Applying remote sensing to terrestrial animal distribution and diversity. *Int. J. Appl. Earth Obs. Geoinf.* **2007**, *9*, 1–20. [CrossRef]
33. Tilman, D.; Kareiva, P. *Spatial Ecology: The Role of Space in Population Dynamics and Interspecific Interactions*; Princeton University Press: Princeton, NJ, USA, 1997.
34. Rosindell, J.; Wong, Y.; Etienne, R.S. A coalescence approach to spatial neutral ecology. *Ecol. Inform.* **2008**, *3*, 259–271. [CrossRef]
35. Franklin, J.F.; Mitchell, R.J.; Palik, B.J. *Natural Disturbance and Stand Development Principles for Ecological Forestry*; General Technical. Report. NRS-19; USDA Forest Service, Northern Research Station: Delaware, PA, USA, 2007.
36. Nyland, R.D. *Silviculture: Concepts and Applications*; McGraw-Hill: New York, NY, USA, 2002.

37. Churchill, D.J.; Larson, A.J.; Dahlgreen, M.C.; Franklin, J.F.; Hessburg, P.F.; Lutz, J.A. Restoring forest resilience: From reference spatial patterns to silvicultural prescriptions and monitoring. *For. Ecol. Manag.* **2013**, *291*, 442–457. [CrossRef]
38. Peck, J.L.E.; Zenner, E.K.; Brang, P.; Zingg, A. Tree size distribution and abundance explain structural complexity differentially within stands of even-aged and uneven-aged structure types. *Eur. J. For. Res.* **2014**, *133*, 335–346. [CrossRef]
39. Choe, C.H.; Kim, J.H.; Xu, C.Y.; Choe, J.S.; Man, H.S.; Jo, S.M. Effect of soil nutrient spatial heterogeneity by cutting disturbance on understory plant diversity in broadleaved-Korean pine forest in Changbai mountain, China. *Eur. J. For. Res.* **2021**, *140*, 603–613. [CrossRef]
40. Kuehne, C.; Weiskittel, A.R.; Fraver, S.; Puettmann, K.J. Effects of thinning-induced changes in structural heterogeneity on growth, ingrowth, and mortality in secondary coastal Douglas-fir forests. *Can. J. For. Res.* **2015**, *45*, 1448–1461. [CrossRef]
41. Larson, A.J.; Stover, K.C.; Keyes, C.R. Effects of restoration thinning on spatial heterogeneity in mixed-conifer forest. *Can. J. For. Res.* **2014**, *42*, 1505–1517. [CrossRef]
42. Tian, F.; Yan, Y.B. A sift feature matching algorithm based on semi-variance function. *Adv. Mater. Res.* **2013**, *647*, 896–900. [CrossRef]
43. Goovaerts, P. *Geostatistics for Natural Resources Evaluation (Applied Geostatistics)*; Oxford University Press: Oxford, UK, 1997.
44. Montero, J.M.; Fernández-Avilés, G.; Mateu, J. Spatial and Spatio-Temporal Geostatistical Modeling and Kriging. John Wiley and Sons: New York, NY, USA, 2015.
45. Gräler, B.; Pebesma, E.; Heuvelink, G. Spatio-temporal interpolation using gstat. *R J.* **2016**, *8*, 204–218. [CrossRef]
46. Pham, Q.H.; Gagnon, M.; Antoni, J.; Tahan, A.; Monette, C. Prediction of hydroelectric turbine runner strain signal via cyclostationary decomposition and kriging interpolation. *Renew. Energy* **2022**, *182*, 998–1011. [CrossRef]
47. Tziachris, P.; Metaxa, E.; Papadopoulos, F.; Papadopoulou, M. Spatial modelling and prediction assessment of soil iron using kriging interpolation with ph as auxiliary information. *Int. J. Geo-Inf.* **2017**, *6*, 283. [CrossRef]
48. Kimmins, J.P. *Balancing Act: Environmental Issues in Forestry*; University of British Columbia Press: Vancouver, BC, USA, 1992.
49. Johnstone, W.D. Thinning lodgepole pine. In *Lodgepole Pine: The Species and Its Management*; Baumgartner, D.M., Krebill, R.G., Arnott, J.T., Weetman, G.F., Eds.; Washington State University: Spokane, WA, USA, 1985.
50. Crouch, G.L. *Effects of Thinning Pole-Sized Lodgepole Pine on Understory Vegetation and Large Herbivore Activity in Central Colorado*; USDA Forest Service, Rocky Mountain Research, Station Res.: Fort Collins, CO, USA, 1986; pp. RM–268.
51. Hungerford, R.D. *Predicting Response of Understory Vegetation to Stand Treatment: Consequences for Multi-Resource Management*; General Technical Report, USDA Forest Service, INT-237; Intermountain Research Station: Ogden, UT, USA, 1987.
52. Klinka, K.; Chen, H.Y.H.; Wang, Q.L.; Montigny, L. Forest canopies and their influence on understory vegetation in early- seral stands on West Vancouver Island. *Northwest Sci.* **1996**, *70*, 193–200.
53. Stone, W.E.; Wolfe, M.L. Response of understory vegetation to variable tree mortality following a mountain pine beetle epidemic in lodgepole pine stands in northern Utah. *Vegetatio* **1996**, *122*, 1–12. [CrossRef]
54. Thomas, S.C.; Halpern, C.B.; Falk, D.A.; Liguori, D.A.; Austin, K.A. Plant diversity in managed forests: Understory responses to thinning and fertilization. *Ecol. Appl.* **1999**, *9*, 864–879. [CrossRef]
55. Sullivan, T.P.; Sullivan, D.S.; Lindgren, P.M.F. Stand structure and small mammals in young lodgepole pine forest: 10-year results after thinning. *Ecol. Appl.* **2001**, *11*, 1151–1173. [CrossRef]
56. Assmann, E. Grundflächenhaltung und zuwachsleistung bayerischer fichten-durchforstungsreihen. *Forstwiss. Cent.* **1954**, *73*, 257–271. [CrossRef]
57. Gebhardt, T.; Haberle, K.H.; Matyssek, R.; Schulz, C.; Ammer, C. The more, the better? Water relations of norway spruce stands after progressive thinning. *Agric. For. Meteorol.* **2014**, *197*, 235–243. [CrossRef]
58. Olivar, J.; Bogino, S.; Rathgeber, C.; Bonnesoeur, V.; Bravo, F. Thinning has a positive effect on growth dynamics and growth-climate relationships in Aleppo pine (*Pinus halepensis*) trees of different crown classes. *Ann. For. Sci.* **2014**, *71*, 395–404. [CrossRef]
59. Gradel, K.O.; Jensen, U.S.; Schønheyder, H.C.; Østergaard, C.; Knudsen, J.D.; Wehberg, S.; Søgaard, M. Impact of appropriate empirical antibiotic treatment on recurrence and mortality in patients with bacteraemia: A population-based cohort study. *BMC Infect. Dis.* **2017**, *122*, 1–9. [CrossRef] [PubMed]
60. Varmola, M.; Kolström, T.; Mehtätalo, E. The effect of release cutting on the growth and external quality of the dominant trees in a Pinus sylvestris stand established by spot sowing. *Scand. J. For. Res.* **1998**, *13*, 151–159. [CrossRef]
61. Cabon, A.; Mouillot, F.; Lempereur, M.; Ourcival, J.M.; Simioni, G.; Limousin, J.M. Thinning increases tree growth by delaying drought-induced growth cessation in a mediterranean evergreen oak coppice. *For. Ecol. Manag.* **2018**, *409*, 333–342. [CrossRef]
62. Nguyen, T.T.; Tai, D.T.; Zhang, P.; Razaq, M. Effect of thinning intensity on tree growth and temporal variation of seed and cone production in a *Pinus koraiensis* plantation. *J. For. Res.* **2019**, *30*, 95–105. [CrossRef]
63. Park, J.; Kim, T.; Moon, M.; Cho, S.; Ryu, D.; Kim, H.S. Effects of thinning intensities on tree water use, growth, and resultant water use efficiency of 50-year-old *Pinus koraiensis* forest over four years. *For. Ecol. Manag.* **2018**, *408*, 121–128. [CrossRef]
64. Tsai, H.C.; Chiang, J.M.; Mcewan, R.W.; Lin, T.C. Decadal effects of thinning on understory light environments and plant community structure in a subtropical forest. *Ecosphere* **2018**, *9*, e02464. [CrossRef]
65. Liu, X.; Zhou, W.; Wang, W.J. Spatial Heterogeneity of Urban Street Trees in Harbin, China. *Bull. Bot. Res.* **2019**, *39*, 590–597.
66. Samra, J.S.; Gill, H.S.; Richter, J.; Anlauf, R. Spatial dependence of soil sodicity and tree growth in a natric hapustalf. *Soil Sci. Soc. Am. J.* **1990**, *54*, 1228–1233. [CrossRef]

67. Li, C.; Luo, P.; Li, Z.Q. Spatial heterogeneity of diameter at breast height growth for Korean pine natural forest and its relationships with terrain factors. *J. Nanjing For. Univ. (Nat. Sci.)* **2017**, *41*, 129–135.
68. Arzai, A.H.; Aliyu, B.S. The relationship between canopy width, height and trunk size in some tree species growing in the savana zone of nigeria. *Bayero J. Pure Appl. Sci.* **2010**, *3*, 260–263. [CrossRef]
69. Dai, J.Y.; Liu, H.Y.; Wang, Y.C.; Guo, Q.H.; Hu, T.Y.; Quine, T.; Green, S.; Hartmann, H.; Xu, C.Y.; Liu, X.; et al. Drought-modulated allometric patterns of trees in semi-arid forests. *Commun. Biol.* **2020**, *405*, 1–7. [CrossRef]
70. Hartmann, H. Will a 385 million year-struggle for light become a struggle for water and for carbon?—How trees may cope with more frequent climate change-type drought events. *Glob. Chang. Biol.* **2011**, *17*, 642–655. [CrossRef]
71. Miehe, G.; Miehe, S.; Will, M.; Opgenoorth, L.; Duo, L.; Dorgeh, T.; Liu, J. An inventory of forest relicts in the pastures of Southern Tibet. *Plant Ecol.* **2008**, *194*, 157–177. [CrossRef]
72. Liang, E.Y.; Lu, X.M.; Ren, P.; Li, X.X.; Zhu, L.P.; Eckstein, D. Annual increments of juniper dwarf shrubs above the tree line on the central Tibetan Plateau: A useful climatic proxy. *Ann. Bot.* **2012**, *109*, 721–728. [CrossRef]
73. Liu, H.; Williams, A.P.; Allen, C.D.; Guo, D.; Wu, X.; Anenkhonov, O.A.; Liang, E.; Sandanov, D.V.; Yin, Y.; Qi, Z.; et al. Rapid warming accelerates tree growth decline in semi-arid forests of Inner Asia. *Glob. Change Biol.* **2013**, *19*, 2500–2510. [CrossRef]
74. Biederman, J.; Harpold, A.A.; Gochis, D.J.; Ewers, B.; Reed, D.E.; Papuga, S.; Brooks, P.D. Increased evaporation following widespread tree mortality limits streamflow response. *Water Resour. Res.* **2014**, *50*, 5395–5409. [CrossRef]
75. Moreno, H.A.; Gupta, H.V.; White, D.D.; Sampson, D.A. Modeling the distributed effects of forest thinning on the long-term water balance and streamflow extremes for a semi-arid basin in the southwestern US. *Hydrol. Earth Syst. Sci.* **2016**, *20*, 1241–1267. [CrossRef]

forests

MDPI

Article

A Three-Level Model System of Biomass and Carbon Storage for All Forest Types in China

Weisheng Zeng [1,*], Wentao Zou [2], Xinyun Chen [1] and Xueyun Yang [1]

[1] Academy of Inventory and Planning, National Forestry and Grassland Administration, Beijing 100714, China; chenxinyun77@126.com (X.C.); yangxy0892@126.com (X.Y.)

[2] Research Institute of Forestry Policy and Information, Chinese Academy of Forestry, Beijing 100091, China; zouwt@caf.ac.cn

* Correspondence: zengweisheng0928@126.com

Abstract: Forest biomass and carbon storage models are crucial for inventorying, monitoring, and assessing forest resources. This study develops models specific to China's diverse forests, offering a methodological foundation for national carbon storage estimation and a quantitative basis for national, regional, and global carbon sequestration projections. Utilizing data from 52,700 permanent plots obtained during China's 9th national forest inventory, we calculated biomass and carbon storage per hectare for 35 tree species groups using respective individual tree biomass models and carbon factors. We then constructed a three-level volume-based model system for forest biomass and carbon storage, applying weighted regression, dummy variable modeling, and simultaneous equations with error-in-variables. This system encompasses one population of forests, three forest categories (level I), 20 forest types (level II), and 74 forest sub-types (level III). Finally, the assessment of these models was carried out with six evaluation indices, and comparative analyses with previously established biomass models of three major forest types were conducted. Determination coefficients (R^2) for the population average model, and three dummy models on levels I, II, and III, exceed 0.78, 0.85, 0.92, and 0.95, respectively, with corresponding mean prediction errors (MPEs) of 0.42%, 0.34%, 0.24%, and 0.19%, and mean percent standard errors (MPSEs) of approximately 22%, 21%, 15%, and 12%. Models for 20 forest types and 74 sub-types yield R^2 values above 0.87 and 0.85, with MPE values below 3% and 5%, respectively. Notably, the estimates of previous biomass models of three major forest types demonstrated considerable uncertainty, with TRE ranging from −20% to 74%. However, accuracy has improved with larger sample sizes. In total biomass and carbon storage estimations, the R^2 values of dummy models for levels I, II, and III progressively increase and MPSE and MPE values decrease, whereas TRE approximates zero. The tiered model system of simultaneous equations developed herein offers a quantitative framework for precise evaluations of biomass and carbon storage on different scales. For enhanced accuracy in such estimations, applying level III models is recommended whenever feasible, especially for national estimation.

Keywords: biomass conversion factor; simultaneous equations; error-in-variable; dummy variable; carbon storage

1. Introduction

Forest biomass and carbon storage, akin to forest volume, are vital for monitoring forest resources and are integral indicators of forest ecosystem function and productivity [1–3]. As global climate concerns rise, the examination of forest carbon storage and its sequestration potential is increasingly prioritized [4,5]. Forest biomass can be determined through individual tree biomass models [6,7] or by formulating stand-level biomass models and conversion factors [2,6]. Forest carbon storage estimates are derived by multiplying biomass by the forest's average carbon factor [2].

Reviewing work by Luo et al. [8], it is evident that from 1978 to 2013, Chinese researchers developed 5924 tree biomass models covering approximately 200 species. Since 2014, the State Forestry Administration of China has developed tree biomass models and related carbon accounting parameters for the primary tree species or groups, resulting in a series of official standards [9–21]. Yet, comparatively, stand-level biomass models [7,22–41] are less prevalent than individual tree models [7–22,26,27,42–46]. Given that stand-level models are more commonly utilized on large scales, such as national, regional, and global levels, the development of these models for all forest types in China is of substantial practical significance.

Among the stand-level biomass models for China's forest types, Fang et al. [29,30] stand out with volume-derived biomass models for 21 forest types from 418 sample plots, which have seen broad application [35,38]. Wang et al. [32] created hyperbolic relationships between biomass and volume for 16 forest types using 1266 sample plots from forest inventory, and Zhang et al. [41] developed power function models for 10 types with 1828 sample plots from forest inventory, along with 21 biomass models for various regions and forest categories.

Evaluating the performance of these models reveals three main deficiencies: First, the modeling sample sizes are often insufficient, with Fang et al. [30] basing 18 models on fewer than 30 sample plots, among the models for 21 forest types, and only three models for larch (*Larix* spp.), Chinese fir (*Cunninghamia lanceolata*), and Chinese pine (*Pinus tabulaeformis*) were based on more than 30 plots. Similarly, Wang et al. [32] used fewer than 50 plots for 10 models, and Zhang et al. [41] used fewer than 50 plots for two models. Second, the modeling methods tend to be oversimplified, commonly using ordinary least square (OLS) regression, which does not account for the heteroscedasticity of the biomass and volume data. Third, the evaluation indices are typically limited, with only R^2 [30,41] or R [32] provided, lacking other error-related indices, making it difficult to gauge their uncertainty in the application.

Moreover, as the discourse on climate change intensifies and strategies for 'carbon peak and carbon neutrality' are implemented, the precise estimation of forest carbon storage has garnered escalating interest. Researchers often rely on existing biomass models to gauge carbon storage trends, typically using a universal carbon factor of 0.5 or 0.47 for all forest types [37–40]. Nevertheless, carbon factors vary among different forest types, a fact underscored by Zeng et al. [31] who formulated simultaneous models for volume, biomass, and carbon storage across 10 major forest types in northeast China using 2000 plots.

The primary aims of this study are twofold: (1) To develop a tiered system of models for forest biomass and carbon storage based on the measured data of 52,700 sample plots from the ninth national forest inventory (NFI) of China, which includes one population, three forest categories (level I), 20 forest types (level II), and 74 forest sub-types (level III). This system employs weighted regression, dummy variable modeling, and simultaneous equations with error-in-variables [31,47–50]. (2) To assess the evaluation indices of the three-level models, providing a basis for understanding the uncertainty of the models when applied on different scales.

2. Materials and Methods

2.1. Data Description

The dataset for this research is derived from the permanent sample plots in the ninth National Forest Inventory (NFI) of China. Nationwide, 52,700 effective sample plots were analyzed, each with a non-zero stand volume. In-depth analysis of these plots included calculating forest volume and biomass (accounting for above- and below-ground biomass, but not understory vegetation) as well as carbon storage per hectare. These calculations were based on official standards on tree biomass models and carbon factors for major tree species, which were developed using the measurement data of 7534 sample trees from destructive sampling [9–21,43]. On the tree level, 35 species/groups were classified according to the national standard [43], and the estimates of volume, biomass, and carbon storage at the plot level were obtained by summing the estimates of individual trees, which were based on one-variable models. Even though tree height is the second most important variable to estimate biomass, the contribution to above- or below-ground biomass estimation is not highly great [43]. These plots are organized by dominant tree species into three categories and 20 forest types, categorized based on area and stand volume, encompassing a diverse range of species such as fir (*Abies*), spruce (*Picea*), and others, including broadleaved and coniferous types. To ensure robust model development and subsequent validation, the sample plots were divided, with two-thirds utilized for model construction and one-third reserved for validation. Table 1 provides a detailed breakdown of the data used for both modeling and validation across the different forest categories and types.

Table 1. The basic data pertaining to the modeling samples and validation samples for 20 forest types.

Forest Type		Number of Plots	Modeling Samples			Validation Samples		
			Number of Plots	Max Volume (m^3/ha)	Max Biomass (t/ha)	Number of Plots	Max Volume (m^3/ha)	Max Biomass (t/ha)
Coniferous	*Abies* spp.	534	355	1439	764	179	1364	792
	Picea spp.	1353	900	941	564	453	800	487
	Larix spp.	2495	1665	485	381	830	522	379
	Cunninghamia lanceolata	3152	2100	456	257	1052	454	262
	Cupressus spp.	1328	885	511	768	443	388	622
	Pinus massoniana	2607	1740	319	325	867	333	290
	P. tabulaeformis	1186	790	306	377	396	275	343
	P. yunnanensis	766	510	340	250	256	355	242
	Other coniferous	1681	1125	471	323	556	414	327
Mixed	Conifer mixed	1898	1265	881	600	633	789	487
	Conifer–broadleaf mixed	4364	2910	650	606	1454	502	436
	Broadleaf mixed	13,073	8715	732	870	4358	706	723
Broadleaves	*Quercus* spp.	4474	2980	706	875	1494	402	538
	Betula spp.	2201	1465	472	359	736	420	360
	Populus spp.	3924	2615	396	367	1309	380	298
	Robinia pseudoacacia	830	550	188	248	280	193	245
	Eucalyptus spp.	1036	690	261	296	346	195	248
	Hevea brasiliensis	701	465	298	241	236	370	296
	Other hardwood	3368	2245	488	549	1123	443	509
	Other softwood	1729	1150	381	366	579	422	321

2.2. Modeling Methods

2.2.1. Forest Classification

To enhance the practicability of the modeling results, based on the data available, 15 forest types among the 20 were subdivided according to dominant tree species (group) [51] or geographical region [52], excluding Chinese pine, Yunnan pine, Robinia, eucalyptus, and rubber-woods. One type (other coniferous) was divided into 8 sub-types by dominant tree species (group), and two types (other hardwood broadleaved and other softwood broadleaved) were divided into 12 and 6 sub-types, respectively, by both dominant tree species (group) and geographical region, and the remaining 12 types were divided by geographical region from 2 to 6 sub-types. The whole country is categorized into six geographical regions, that is, North China (NC), Northeast (NE), East China (EC), Central South (CS), Southwest (SW), and Northwest (NW). These regions can also be combined as North (N = NE + NC + NW), South (S = EC + CS + SW), West (W = NW + SW), and other regions as necessary. Consequently, 20 forest types were reclassified into 74 forest sub-types (see Table 2).

Table 2. A three-level system of forest classification.

Forest Category	Forest Type	Forest Sub-Type/Level III	
Level I	Level II	Number	Name
Coniferous	*Abies* spp.	2	*Abies* spp. I (N); *Abies* spp. II (SW)
	Picea spp.	2	*Picea* spp. I (N); *Picea* spp. II (SW)
	Larix spp.	3	*Larix* spp. I (NE); *Larix* spp. II (NC); *Larix* spp. III (W)
	Cunninghamia lanceolata	3	*C. lanceolata* I (EC); *C. lanceolata* II (CS); *C. lanceolata* III (SW)
	Cupressus spp.	3	*Cupressus* spp. I (NE + NC); *Cupressus* spp. II (NW); *Cupressus* spp. III (S)
	Pinus massoniana	3	*P. massoniana* I (EC); *P. massoniana* II (CS); *P. massoniana* III (SW)
	P. tabulaeformis	1	*P. tabulaeformis*
	P. yunnanensis	1	*P. yunnanensis*
	Other coniferous	8	*P. sylvestris*; *P. armandii*; *P. densata*; *P.* K.T.D.; Foreign pine; Other pine; *Cryptomeria* spp.; Other coniferous
Mixed	Conifer mixed	4	Conifer mixed I (N); Conifer mixed II (EC); Conifer mixed III (CS); Conifer mixed IV (SW)
	Conifer–broadleaf mixed	5	Conifer–broadleaf I (NE); Conifer–broadleaf II (NC + NW); Conifer–broadleaf III (EC); Conifer–broadleaf IV (CS); Conifer–broadleaf V (SW)
	Broadleaf mixed	6	Broadleaf mixed I (NE); Broadleaf mixed II (NC); Broadleaf mixed III (NW); Broadleaf mixed IV (EC); Broadleaf mixed V (CS); Broadleaf mixed VI (SW)
Broadleaves	*Quercus* spp.	5	*Quercus* spp. I (NE); *Quercus* spp. II (NC); *Quercus* spp. III (NW); *Quercus* spp. IV (SE); *Quercus* spp. V (SW)
	Betula spp.	3	*Betula* spp. I (NE); *Betula* spp. II (NC); *Betula* spp. III (W)
	Populus spp.	4	*Populus* spp. I (NE); *Populus* spp. II (NC); *Populus* spp. III (W); *Populus* spp. IV (SE)
	Robinia pseudoacacia	1	*R. pseudoacacia*
	Eucalyptus spp.	1	*Eucalyptus* spp.
	Hevea brasiliensis	1	*H. brasiliensis*
	Other hardwood	12	F.J.P.; C.S.P.; *Ulmus* spp.; *Schima superba*; *Juglans regia*; *Castanea mollissima*; *Quercus variabilis*; Other non-wood; Other hardwood I (NE + NC); Other hardwood II (NW); Other hardwood III (EC); Other hardwood IV (CS + SW)
	Other softwood	6	*Tilia tuan*; *Salix* spp.; Other softwood I (NE + NC); Other softwood II (NW); Other softwood III (SE); Other softwood IV (SW)

Note: (i) China is divided into six geographic regions, that is, North China (NC), Northeast (NE), East China (EC), Central South (CS), Southwest (SW), and Northwest (NW). NC includes Beijing, Tianjin, Hebei, Shanxi, and Inter Mongolia; NE includes Liaoning, Jilin, and Heilongjiang; EC includes Shanghai, Jiangsu, Zhejiang, Anhui, Fujian, Jiangxi, and Shandong; CS includes Henan, Hubei, Hunan, Guangdong, Guangxi, and Hainan; SW includes Chongqing, Sichuan, Guizhou, Yunnan, and Tibet; NW includes Shannxi, Gansu, Qinghai, Ningxia, and Xinjiang. NE, NC, and NW are merged to North (N); Northwest and Southwest are merged to West (W); East China and Central South are merged to Southeast (SE); and Southeast and Southwest are merged to South. The above forest classification and geographical division do not include Taiwan, Hong Kong, and Macau. (ii) Abbreviations in parentheses indicate geographic regions. (iii) P. K.T.D denotes *P. koraiensis*, *P. thunbergii*, and *P. densiflora*; F.J.P. means *Fraxinus mandshurica*, *Juglans mandshurica*, and *Phellodendron amurense*; C.S.P. represents *Cinnamomum* spp., *Sassafras* spp., and *Phoebe* spp.; and other non-wood includes the non-wood forests composed of other tree species except *Hevea brasiliensis*, *Juglans regia*, *Castanea mollissima*, and *Quercus variabilis*.

2.2.2. Model Development

Forest biomass is closely related to its volume, and the volume-derived biomass models have been extensively applied in previous studies [27,28,32,35,38]. According to Fang et al. [30], there is a linear correlation between forest biomass and volume for 21 forest types in China. This linearity was confirmed by a scatterplot depicting the relationship between forest biomass and volume data per hectare from the 52,700 sample plots. Moreover, linear models are advantageous for application across different scales and can avoid the scaling-up errors associated with overestimates of nonlinear models [37].

Additionally, both above- and below-ground biomass require consideration. For instance, reporting both kinds of biomass data in the Global Forest Resources Assessment by the FAO [3] is mandatory. The ratio of below-ground to above-ground biomass, known as the root-to-shoot ratio (RSR), varies across forest types [2]. Accurate estimation of below-ground biomass for various forest types depends on the reliable values of RSR, but these data are not available. After estimating total forest biomass, forest carbon storage is calculated by multiplying the biomass by the average carbon factor, usually 0.5 or 0.47 in previous studies [2,29,30]. Nonetheless, it should be recognized that carbon factors may differ among forest types [31].

Given the recursive nature of the relationship between total biomass and both above-ground biomass (or below-ground biomass) and carbon storage, this study employed simultaneous equations with error-in-variables and dummy variables [47–50] to develop a model system at three levels. The equations are as follows:

$$\begin{cases} B_T = \Sigma a_i S_i + (\Sigma b_i S_i)V + \varepsilon_1 \\ B_A = (\Sigma c_i S_i)\hat{B}_T + \varepsilon_2 \\ C = (\Sigma d_i S_i)\hat{B}_T + \varepsilon_3 \end{cases} \quad (1)$$

Here, B_T represents observed total biomass per hectare (t/ha), B_A indicates above-ground biomass (t/ha), and C refers to carbon storage (t/ha). V is the forest volume (m^3/ha) regarded as error-free-variable, $\hat{B}_T$ is estimated total biomass per hectare (t/ha) regarded as an error-in-variable in the second and third equations, and a_i, b_i, c_i, and d_i are parameters where i denotes the ith category, type, or sub-type. S_i is a dummy variable equal to 1 when data belong to the ith category, type, or sub-type, and 0 otherwise. ε_1, ε_2, and ε_3 are error items, assumed to follow a normal distribution with a mean of zero.

By dividing the first equation in model system (1) by V, we obtain a biomass conversion factor (BCF) model:

$$BCF = B_T/V = b_i + a_i/V \quad (2)$$

In this model, BCF combines basic wood density (WD), biomass expansion factor (BEF), and the ratio of shoot-to-root (RSR) as outlined in the IPCC Guidelines for national greenhouse gas inventories [2], where BCF = WD·BEF·(1 + RSR). The d_i parameter in the third equation in model system (1) aligns with the carbon factor (CF). From the c_i parameter in the second equation in model system (1), one can deduce the RSR:

$$RSR = B_B/B_A = (1 - c_i)/c_i \quad (3)$$

Like forest volume data, both forest biomass and carbon storage data display heteroscedasticity. This research advocates the use of the weighted regression method [47], with a weight function defined as $w = 1/V^{0.5}$, which was determined on compromise consideration of total relative error and average systematic error of the models [48]. Additionally, considering the recursive relationship between total biomass, above-ground biomass, and carbon storage, it is crucial to employ simultaneous equations with error-in-variables to fit accurately the model system (1) [50].

2.2.3. Model Evaluation

Six indices were employed to evaluate the models: coefficient of determination (R^2), standard error of the estimate (SEE), total relative error (TRE), average systematic error (ASE), mean prediction error (MPE), and mean percentage standard error (MPSE) [53,54], which were commonly used to evaluate the integrated performance of models [55,56]. TRE, ASE, MPE, and MPSE are defined as follows:

$$\text{TRE} = \sum(y_i - \hat{y}_i)/\sum\hat{y}_i \times 100 \quad (4)$$

$$\mathrm{ASE} = \sum (y_i - \hat{y}_i)/\hat{y}_i/n \times 100 \quad (5)$$

$$\mathrm{MPE} = t_\alpha \cdot (SEE/\overline{y})/\sqrt{n} \times 100 \quad (6)$$

$$\mathrm{MPSE} = \sum |(y_i - \hat{y}_i)/\hat{y}_i|/n \times 100 \quad (7)$$

In these equations, y_i represents observed values, $\hat{y}_i$ represents estimated values, $\overline{y}$ is the mean of observed values, n is the number of plots, and t_α is the t-value at the confidence level α. For the models developed, values of the six indices were computed and used for evaluation.

Practically, models are expected to have MPE values less than 3% or 5%, MPSE values less than 15% or 20%, TRE values within ±3%, and ASE values within ±5%. For extensive applicability assessments, TRE and ASE must also be independently verified using validation samples, with TRE ideally within ±3% or below the MPE value, and ASE within ±5%. If ASE exceeds ±5% but is within ±10%, the model has a moderate systematic error; if it exceeds ±10%, the systematic error is considered high.

3. Results

For this study, data from 35,120 plots were utilized to calibrate model system (1), applying simultaneous equations with error-in-variables and dummy variables. This encompassed one population, three forest categories, 20 forest types, and 74 forest sub-types. The parameter estimates and associated evaluation indices for the first equation in model system (1) are presented in Table 3. Notably, when assessing the SEE values for the second and third equations in model system (1) relative to the first equation, significant discrepancies emerged. However, for the other five evaluation indices, only minor differences were noted. Due to their negligible variance, the indices for the second and third equations in model system (1) have been excluded from Table 3 for succinctness.

A thorough analysis of the six evaluation indices in Table 3 revealed several insights into the model's performance. The TRE values were nearly zero, which suggests a strong agreement between the observed and estimated values across the models. The average values of the other five evaluation indices demonstrated a positive trend moving from one population to three forest categories and further to 20 forest types and 74 forest sub-types, except for a slight uptick in the average MPE values from 0.42% to 0.60%, 1.22%, and 2.11%, attributable to sample size variations. The average R^2 values increased progressively from 0.781 for the population average model to 0.853, 0.928, and 0.941 for levels I, II, and III models, respectively, while the average values of SEE, ASE, and MPSE decreased. For instance, the average MPSE values at the three levels fell from 22.38% for the population average model to 21.03%, 14.68%, and 12.96%, a reduction of approximately 6%, 34%, and 42% respectively.

Regarding the models for the three forest categories, the R^2 values exceeded 0.78, MPE values were below 1%, ASE values fell within ±10% (with two categories within ±5%), and MPSE values were around 20%. The coniferous biomass model did not perform so well due to the large variation in forest volume stock. Subsequently, data from 17,580 plots, classified as validation samples in Table 1, underwent independent validation. The outcomes affirmed that TRE values stayed within ±0.3% and ASE values within ±5%.

For the 20 forest-type models, the R^2 values were above 0.87, and MPE values were below 3% (with eight types under 1%). ASE values were within ±10% (18 types within ±5%), and MPSE values were under 20% (with the exception of the 'other coniferous' type and two types under 10%). Independent cross-validation confirmed that TRE values for most types were within ±3%, barring the third equation in model system (1) for the 'other coniferous' type at −3.16%; and ASE values were within ±5%, except for two types. The R^2 values and other evaluation indices, which were different for each forest type, depended on the size and structure of the modeling samples.

Table 3. The parameter estimates of three-level simultaneous models of forest biomass and carbon storage and evaluation indices of biomass model in China.

Level	Type	Parameter Estimate					Evaluation Indices of Biomass Model					
		a_i	b_i	c_i	RSR	d_i (CF)	R^2	SEE/t	TRE/%	ASE/%	MPE/%	MPSE/%
Population	Forest	1.4864	0.9792	0.7973	0.2542	0.4891	0.781	34.62	0.00	5.30	0.42	22.38
Level I	Coniferous	1.6355	0.8129	0.8069	0.2393	0.4994	0.787	35.69	−0.01	8.52	0.80	24.60
	Mixed	1.7434	1.0463	0.7956	0.2569	0.4868	0.884	25.39	−0.04	3.45	0.45	16.81
	Broadleaved	0.2986	1.1133	0.7959	0.2564	0.4838	0.887	23.06	−0.01	3.93	0.55	21.68
Level II	*Abies* spp.	5.5779	0.5937	0.8180	0.2225	0.4951	0.873	47.30	−0.06	2.20	2.21	17.30
	Picea spp.	1.6095	0.7157	0.7991	0.2514	0.4901	0.914	29.89	−0.01	5.50	1.23	15.65
	Larix spp.	1.0427	0.8228	0.7794	0.2830	0.4888	0.936	14.99	0.00	2.70	0.82	13.28
	Cunninghamia lanceolata	0.5904	0.7118	0.8109	0.2332	0.4969	0.936	12.06	0.00	4.36	0.87	14.04
	Cupressus spp.	0.4836	1.5336	0.8007	0.2489	0.5013	0.947	22.57	0.00	2.88	1.65	16.47
	Pinus massoniana	1.7401	0.9726	0.8297	0.2053	0.5162	0.954	11.20	0.02	0.91	0.67	10.01
	P. tabulaeformis	*0.0901*	1.1255	0.8098	0.2349	0.5130	0.966	16.18	0.00	0.05	1.66	15.60
	P. yunnanensis	0.9361	0.6499	0.8453	0.1830	0.5047	0.973	11.89	0.00	3.94	1.71	13.79
	Other coniferous	0.9884	0.8313	0.8041	0.2436	0.5009	0.874	20.90	−0.03	6.63	1.64	20.41
	Conifer mixed	0.5828	1.2964	0.8088	0.2364	0.5014	0.921	19.64	0.06	1.69	1.22	14.21
	Conifer–broadleaf mixed	0.5018	1.0149	0.7961	0.2561	0.4920	0.894	20.58	−0.01	3.37	0.89	15.80
	Broadleaf mixed	*0.0077*	0.9022	0.7938	0.2598	0.4835	0.916	22.67	−0.03	3.19	0.47	15.44
	Quercus spp.	0.6954	1.5388	0.7930	0.2610	0.4810	0.917	26.78	−0.01	3.27	0.85	17.44
	Betula spp.	0.3333	1.1740	0.7821	0.2786	0.4867	0.898	17.39	0.00	1.77	1.03	14.36
	Populus spp.	0.8226	0.8607	0.8246	0.2127	0.4725	0.933	12.21	0.00	1.86	0.78	12.83
	Robinia pseudoacacia	0.3439	1.2130	0.7832	0.2768	0.4838	0.934	10.75	−0.01	1.94	2.00	14.68
	Eucalyptus spp.	0.5703	0.9935	0.7793	0.2832	0.5238	0.962	8.98	0.00	1.32	1.18	8.88
	Hevea brasiliensis	4.0799	0.8272	0.7981	0.2530	0.4956	0.991	4.10	−0.01	2.48	0.53	5.96
	Other hardwood	2.6170	0.9515	0.7880	0.2690	0.4856	0.933	14.37	−0.01	2.95	1.17	17.67
	Other softwood	1.5064	1.1073	0.7954	0.2572	0.4873	0.896	18.66	−0.02	5.29	1.79	19.75
Level III	*Abies* spp. I	1.9730	0.7767	0.8017	0.2473	0.4955	0.989	13.10	0.00	0.56	1.02	5.09
	Abies spp. II	3.6239	0.5359	0.8267	0.2096	0.4950	0.971	23.16	0.01	1.52	1.36	8.97
	Picea spp. I	1.4994	0.7351	0.7918	0.2629	0.4900	0.899	31.11	0.00	5.75	1.42	16.25
	Picea spp. II	3.4253	0.6405	0.8282	0.2074	0.4903	0.978	17.05	−0.01	1.63	1.63	8.27
	Larix spp. I	0.6576	0.9237	0.7452	0.3419	0.4885	0.967	9.24	0.00	1.35	0.87	9.95
	Larix spp. II	0.4562	0.7999	0.7836	0.2762	0.4888	0.939	12.49	0.00	0.81	1.12	10.26
	Larix spp. III	2.3320	0.7342	0.8285	0.2070	0.4895	0.973	14.02	0.01	2.47	1.37	11.69

Table 3. *Cont.*

Level	Type	Parameter Estimate					Evaluation Indices of Biomass Model					
		a_i	b_i	c_i	RSR	d_i (CF)	R^2	SEE/t	TRE/%	ASE/%	MPE/%	MPSE/%
Level III	*Cunninghamia lanceolata* I	0.7850	0.6990	0.8100	0.2346	0.4965	0.920	14.32	0.00	4.60	1.40	16.21
	Cunninghamia lanceolata II	0.4758	0.7229	0.8120	0.2315	0.4974	0.954	8.85	0.00	4.52	1.29	11.56
	Cunninghamia lanceolata III	0.4392	0.7268	0.8115	0.2323	0.4973	0.953	10.36	0.00	2.99	1.56	12.57
	Cupressus spp. I	*0.0735*	1.7647	0.7997	0.2505	0.5020	0.947	9.14	0.00	0.89	2.92	11.70
	Cupressus spp. II	2.0093	1.6802	0.8006	0.2491	0.5030	0.956	28.59	0.00	2.14	2.54	12.57
	Cupressus spp. III	1.0093	1.3584	0.8012	0.2481	0.4995	0.968	13.95	0.00	2.07	1.55	10.56
	Pinus massoniana I	1.3985	0.9519	0.8303	0.2044	0.5157	0.947	11.84	0.02	1.10	1.29	11.00
	Pinus massoniana II	1.8000	1.0408	0.8278	0.2080	0.5166	0.953	10.28	0.02	0.23	1.12	10.08
	Pinus massoniana III	1.6417	0.9385	0.8309	0.2035	0.5163	0.979	8.03	0.01	0.38	0.75	7.13
	P. tabulaeformis	*0.0940*	1.1255	0.8098	0.2349	0.5130	0.916	16.18	0.00	0.03	1.66	15.60
	P. yunnanensis	0.9394	0.6499	0.8453	0.1830	0.5047	0.909	11.89	0.00	3.94	1.71	13.79
	P. sylvestris var. *mongolica*	0.5574	0.8918	0.7950	0.2579	0.5163	0.873	21.25	0.00	4.26	4.68	17.69
	P. K.T.D.	0.8327	0.8267	0.8028	0.2456	0.5065	0.864	23.09	0.05	9.86	5.22	24.67
	P. armandii	*0.5115*	0.9231	0.8021	0.2467	0.5175	0.981	7.78	0.00	1.02	1.84	8.62
Level III	*P. densata*	3.1127	0.6639	0.8267	0.2096	0.4904	0.970	10.63	0.00	0.97	1.86	8.42
	Foreign pine	0.7032	0.9988	0.7793	0.2832	0.4763	0.976	6.55	0.00	1.67	1.48	8.85
	Other pine	3.9524	0.8538	0.8331	0.2003	0.5070	0.869	21.56	0.05	1.61	3.99	22.33
	Cryptomeria spp.	0.9519	0.7296	0.7809	0.2806	0.5052	0.968	11.52	−0.01	4.84	2.89	10.82
	Other coniferous	0.8331	0.7273	0.8079	0.2378	0.4993	0.855	22.33	0.00	1.00	6.42	24.38
	Conifer mixed I	5.2106	0.7982	0.7963	0.2558	0.4944	0.929	23.92	0.06	4.28	2.41	17.35
	Conifer mixed II	2.4575	0.8529	0.8116	0.2321	0.5033	0.915	14.73	−0.02	1.76	1.76	13.81
	Conifer mixed III	1.2449	0.9207	0.8132	0.2297	0.5047	0.959	9.48	−0.01	0.82	1.73	10.35
	Conifer mixed IV	4.4595	0.8099	0.8159	0.2256	0.5044	0.906	24.82	0.05	1.71	2.97	13.66
	Conifer–broadleaf mixed I	1.8584	0.8430	0.7829	0.2773	0.4870	0.930	17.73	0.00	2.52	1.37	10.22
	Conifer–broadleaf mixed II	3.8893	1.1026	0.7938	0.2598	0.4934	0.877	21.66	0.01	0.89	2.35	16.44
	Conifer–broadleaf mixed III	1.7448	1.0257	0.7965	0.2555	0.4918	0.935	16.07	−0.01	2.12	1.42	14.12
	Conifer–broadleaf mixed IV	1.1497	1.0444	0.7974	0.2541	0.4950	0.940	10.91	−0.01	2.41	1.31	12.54
	Conifer-broadleaf mixed V	2.2794	0.8772	0.8121	0.2314	0.4945	0.923	19.06	−0.04	2.89	1.88	14.32
	Broadleaf mixed I	0.4864	0.9751	0.7927	0.2615	0.4798	0.909	19.82	0.00	1.37	0.69	11.40
	Broadleaf mixed II	*−0.0928*	1.3711	0.7894	0.2668	0.4838	0.927	16.58	0.00	0.75	1.85	14.87
	Broadleaf mixed III	*−0.0875*	1.3001	0.7965	0.2555	0.4849	0.954	16.91	0.00	1.28	1.02	11.90
	Broadleaf mixed IV	1.3414	1.2308	0.7905	0.2650	0.4833	0.952	17.96	−0.01	3.33	0.80	13.05
	Broadleaf mixed V	1.8432	1.1634	0.7911	0.2641	0.4865	0.960	15.28	−0.02	1.40	0.79	11.35
	Broadleaf mixed VI	0.9649	1.0453	0.8035	0.2446	0.4863	0.956	19.22	−0.03	3.09	1.06	12.15

Table 3. *Cont.*

Level	Type	Parameter Estimate					Evaluation Indices of Biomass Model					
		a_i	b_i	c_i	RSR	d_i (CF)	R^2	SEE/t	TRE/%	ASE/%	MPE/%	MPSE/%
	Quercus spp. I	−1.0068	1.1471	0.7818	0.2791	0.4795	0.970	13.96	0.00	1.13	0.80	9.89
	Quercus spp. II	*−0.0049*	1.5550	0.7821	0.2786	0.4808	0.946	15.61	0.00	−0.06	1.14	11.37
	Quercus spp. III	*−0.2365*	1.4851	0.7968	0.2550	0.4815	0.961	19.38	0.00	0.87	1.34	9.94
	Quercus spp. IV	1.0500	1.4135	0.7912	0.2639	0.4823	0.957	18.15	−0.02	2.81	1.77	14.07
	Quercus spp. V	*−0.1926*	1.1182	0.8217	0.2170	0.4820	0.969	23.36	0.01	4.84	1.54	11.61
	Betula spp. I	*0.1938*	0.9346	0.7613	0.3135	0.4862	0.984	6.21	0.00	−0.73	0.88	6.91
	Betula spp. II	1.5641	1.0444	0.7850	0.2739	0.4868	0.858	18.71	0.00	2.50	1.40	15.79
	Betula spp. III	0.7607	1.0042	0.7970	0.2547	0.4870	0.885	22.14	−0.01	3.63	2.82	12.73
	Populus spp. I	*0.1542*	0.7329	0.7968	0.2550	0.4731	0.982	7.14	0.00	1.33	1.17	11.25
	Populus spp. II	*−0.0628*	0.9791	0.8170	0.2240	0.4724	0.973	7.49	0.00	1.56	0.72	9.17
	Populus spp. III	*−0.3020*	0.8946	0.8195	0.2203	0.4729	0.956	11.94	0.01	0.81	1.97	13.26
	Populus spp. IV	*−0.3962*	0.9141	0.8601	0.1627	0.4721	0.907	12.13	0.01	1.31	1.50	14.09
	Robinia pseudoacacia	0.7447	1.5370	0.7832	0.2768	0.4838	0.934	10.74	0.00	1.75	2.01	14.68
	Eucalyptus spp.	0.3307	1.1740	0.7793	0.2832	0.5238	0.962	8.98	0.00	1.33	1.18	8.88
	Hevea brasiliensis	0.8497	0.8604	0.7981	0.2530	0.4956	0.991	4.09	−0.01	2.43	0.53	5.93
	F.J.P.	0.4219	0.9889	0.7929	0.2612	0.4822	0.908	14.30	0.00	4.38	3.49	14.22
Level III	C.S.P.	1.0519	1.1024	0.7934	0.2604	0.4894	0.975	8.00	−0.01	1.08	1.96	7.44
	Ulmus spp.	0.7656	1.0968	0.7844	0.2749	0.4839	0.938	10.41	0.00	4.89	2.73	16.44
	Schima superba	*0.4144*	1.2017	0.7728	0.2940	0.4746	0.942	15.88	0.00	1.67	3.81	14.17
	Juglans regia	0.3120	1.0375	0.7907	0.2647	0.4952	0.968	2.65	−0.02	1.59	2.76	11.32
	Castanea mollissima	*0.2104*	1.1672	0.7958	0.2566	0.4943	0.943	8.21	−0.01	1.72	3.46	13.91
	Quercus variabilis	*−0.2029*	1.5204	0.7823	0.2783	0.4945	0.963	10.46	0.00	−0.95	3.77	12.58
	Other non-wood	0.4858	1.0736	0.7882	0.2687	0.4944	0.964	5.32	−0.01	2.30	1.86	12.61
	Other hardwood I	0.4491	1.4058	0.7852	0.2736	0.4834	0.919	11.75	−0.01	3.80	4.08	15.92
	Other hardwood II	*−0.1049*	1.4276	0.7986	0.2522	0.4842	0.980	11.53	0.00	1.87	2.25	13.11
	Other hardwood III	1.0589	1.3343	0.7842	0.2752	0.4829	0.958	14.14	−0.01	5.09	2.74	15.51
	Other hardwood IV	0.4022	1.2749	0.7886	0.2681	0.4827	0.950	18.16	−0.01	4.25	3.67	15.88
	Tilia tuan	1.7288	0.8643	0.7949	0.2580	0.4617	0.879	19.92	−0.04	3.26	4.21	15.76
	Salix spp.	0.3393	0.9938	0.7948	0.2582	0.4929	0.932	9.15	−0.01	2.21	4.00	16.19
	Other softwood I	0.4991	0.9423	0.7900	0.2658	0.4888	0.933	14.43	−0.01	6.15	3.11	21.26
	Other softwood II	0.8547	1.3636	0.8010	0.2484	0.4938	0.898	19.83	−0.02	4.27	3.89	19.19
	Other softwood III	0.6294	0.9392	0.7895	0.2666	0.4937	0.963	9.45	−0.01	4.70	2.82	16.84
	Other softwood IV	1.3184	0.9434	0.8002	0.2497	0.4891	0.953	13.90	−0.04	3.00	2.53	10.55

Note: Only the parameter estimates in italics are not statistically significant, all others are highly significant ($p < 0.01$); the RSR values are calculated from Equation (3).

For the 74 forest sub-type models, the R^2 values surpassed 0.85, and MPE values remained under 5% (except for two sub-types), with 60 sub-types below 3%. ASE values were within ±10% (70 sub-types within ±5%), and MPSE values were under 20% (with four sub-types being the exception, and 55 sub-types under 15%). Independent cross-validation indicated that TRE values for most sub-types were within ±3%, with only six sub-types deviating, and ASE values for the majority were within ±5%, except for eight sub-types, and only two sub-types exceeding ±10%.

The model evaluations described above pertain especially to forest category, forest type, and forest sub-type, respectively, across three levels. If the evaluation indices target the entire population of forests, a comparable trend in the average values is observed, as seen in the three-level models. The distinction lies in the uniform sample size of the three dummy-variable simultaneous models, which results in a consistent decrease in MPE values from the categories to types and sub-types, all falling below 0.5% (refer to Table 4).

Table 4. The evaluation indices of three-level simultaneous models for the population.

Model	Level	R^2	SEE	TRE/%	ASE/%	MPE/%	MPSE/%
(1) Total biomass	Population	0.781	34.62	0.00	5.30	0.42	22.38
	Level I	0.857	28.04	−0.02	5.07	0.34	20.73
	Level II	0.928	19.88	−0.01	3.01	0.24	15.11
	Level III	0.955	15.80	0.00	2.15	0.19	12.42
(2) Above-ground biomass	Population	0.796	27.07	0.22	4.74	0.41	21.69
	Level I	0.864	22.07	−0.01	4.35	0.34	20.10
	Level II	0.933	15.55	0.00	2.31	0.24	14.69
	Level III	0.957	12.51	0.00	1.52	0.19	12.11
(3) Carbon storage	Population	0.786	16.72	0.05	5.48	0.42	22.36
	Level I	0.851	13.94	−0.07	5.32	0.35	21.11
	Level II	0.927	9.79	−0.05	3.06	0.24	15.15
	Level III	0.954	7.76	−0.04	2.14	0.19	12.43

As indicated by Table 4, the evaluation indices for forest biomass, above-ground biomass, and carbon storage models at each respective level exhibit minimal difference, barring the dimensionally variable SEE values, which, from a practical standpoint, can be disregarded. However, when comparing the indices across the three levels, it is evident that the level III models outperform others, followed by level II, with level I models lagging behind. Using forest carbon storage models as an example, the R^2 value improved from 0.851 at level I to 0.954 at level III, the MPE value diminished from 0.35% to 0.19%, and the MPSE value dropped from 21.11% to 12.43%, respectively, signifying a significant enhancement in model fit (as shown in Figure 1). The three-level model system has provided a quantitative framework for accurate evaluations of biomass and carbon storage on national, regional, and global scales. To diminish the uncertainty of forest biomass and carbon storage estimates, employing the level III models is highly recommended wherever feasible in application, especially for national estimation.

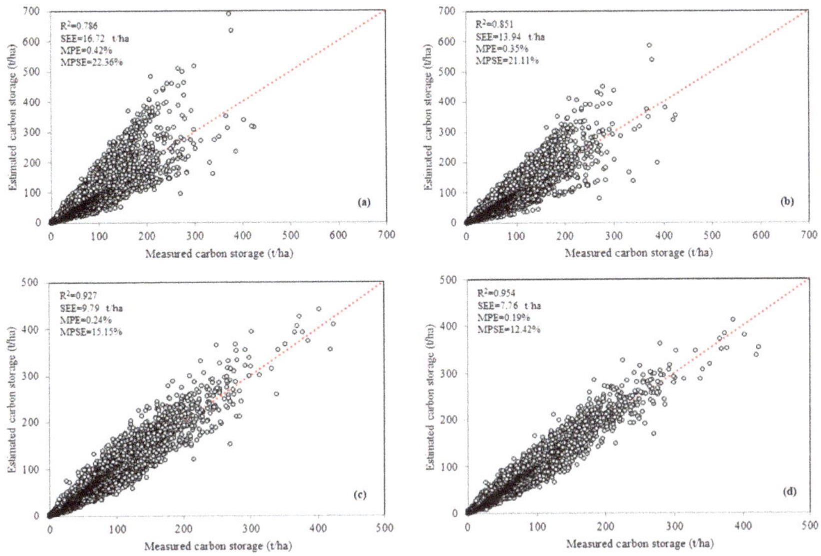

Figure 1. Comparison of fitting performance of three-level simultaneous forest carbon storage models: (**a**) population average model; (**b**) model at level I; (**c**) model at level II; (**d**) model at level III.

4. Discussion

4.1. Comparison with Related Models

In the introduction, we noted that, of the 21 biomass models developed by Fang et al. [30], only three models for larch (*Larix* spp.), Chinese fir (*Cunninghamia lanceolata*), and Chinese pine (*Pinus tabulaeformis*) were based on sample sizes larger than 30 plots. To provide a comparative analysis, we assessed the biomass models established by Fang et al. [30], Wang et al. [32], and Zhang et al. [41], using both validation samples and all samples from these three forest types, as shown in Table 5.

Table 5. Comparison of estimation results of biomass models between this study and other three sources.

Forest Type	Validation Samples		All Samples		Biomass Models	Source	Sample Size
	TRE/%	ASE/%	TRE/%	ASE/%			
Larix spp.	−10.13	−19.82	−10.41	−19.79	$B = 33.806 + 0.6096V$	Fang et al. [30]	34
	13.40	12.01	13.11	12.00	$B = V/(1.1111 + 0.0016V)$	Wang et al. [32]	39
	8.97	6.33	8.88	6.44	$B = 1.4091V^{0.8752}$	Zhang et al. [41]	241
	−0.29	3.13	−0.10	3.32	$B = 0.6986 + 0.8262V$	This study	1665
Cunninghamia lanceolata	6.10	−9.55	6.68	−8.75	$B = 22.5410 + 0.3999V$	Fang et al. [30]	56
	14.01	13.12	14.37	13.45	$B = V/(1.2917 + 0.0022V)$	Wang et al. [32]	70
	17.08	10.84	17.40	11.39	$B = 1.2877V^{0.8427}$	Zhang et al. [41]	88
	−0.28	3.75	−0.09	4.21	$B = 0.5743 + 0.7120V$	This study	2100
Pinus tabulaeformis	33.26	20.89	33.88	20.99	$B = 5.0928 + 0.7554V$	Fang et al. [30]	82
	38.56	33.33	39.52	33.28	$B = V/(1.0529 + 0.0020V)$	Wang et al. [32]	147
	24.27	14.11	24.99	14.14	$B = 1.7969V^{0.8416}$	Zhang et al. [41]	699
	−0.60	−0.47	−0.20	−0.58	$B = 0.2112 + 1.1235V$	This study	790

Table 5 reveals that the TRE values of three model sets exceed ±3%, with those for two models for Chinese pine being even higher than 30%; and the ASE values also surpass ± 5%, with the model by Wang et al. [32] for Chinese pine also higher than 30%. It is important to note that the performance of the three model sets generally improves with increased sample size. In general, Zhang et al.'s models [41] exhibit better performance than those Wang et al. [32] and Fang et al.'s models [30].

Examining the parameters from the models in our study shows that Fang et al.'s models [30] possess larger intercept parameters but smaller slope parameters (see Table 5). The first potential reason for these differences in parameter estimates might be the use of an improper estimation method. Using ordinary regression instead of weighted regression, especially in the presence of heteroscedasticity, may result in such distorted outcomes. Secondly, the sample size is also a significant factor. This is evident from the test results of the three forest types with more than 30 sample plots for modeling as indicated in Table 5, and their TRE values range from −10.41% to 33.88%. The modeling sample sizes for the other 18 forest types are all less than 30, and two types have even fewer than 10. As a result, the errors in forest biomass estimates are likely to be greater. Since the classification of forest types by Fang et al. [30] does not correspond exactly to that in this study, direct comparison is not possible. Eight forest types from the remaining 18 were selected for further validation—these include spruce and fir, cypress, *Pinus massonina*, *P. yunnanensis*, *P. sylvestris* var. *mongolica*, *P. armandii*, birch, poplar, and eucalyptus—which correspond to the forest type, sub-type, or their combinations in this study. The results showed that the TRE values ranged from −20.02% to 74.01%, and the ASE values from −19.32% to 77.62%, indicative of significant uncertainty.

To illustrate the differences between the three model sets more clearly, the validation results of the forest biomass model in this study and the three biomass models for *Pinus tabulaeformis* using all 1186 plots are compared in Figure 2. The models display substantial deviations, and the regression trend line (solid black) clearly diverges from the y = x line (dotted red), resulting in a non-random distribution of residuals and high TRE and ASE values. The validation results for the biomass models of the other two forest types, *Larix* spp. and *Cunninghamia lanceolata*, are similarly conclusive and are omitted here for brevity.

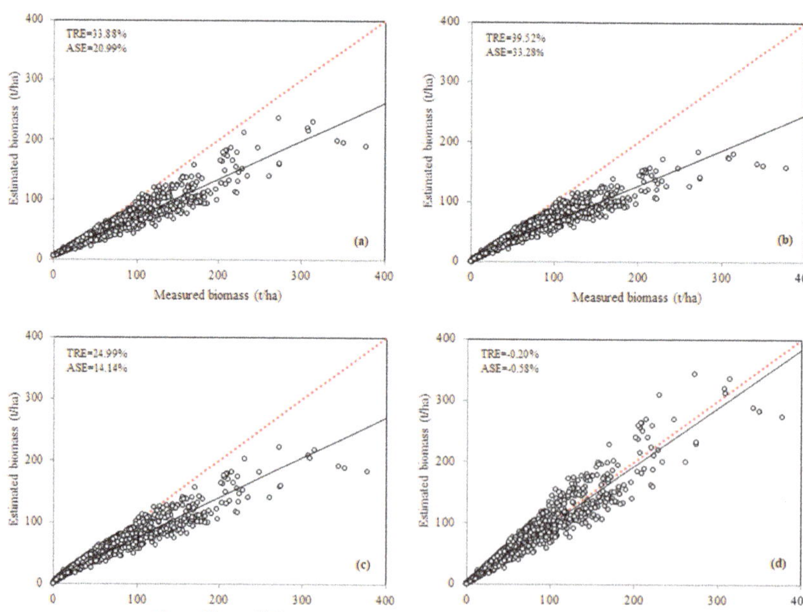

Figure 2. Comparison of fitting performance between the model in this study and the other three models for *Pinus tabulaeformis*: (**a**) model from [30]; (**b**) model from [32]; (**c**) model from [41]; (**d**) model used in this study. The red dotted line is $y = x$ line, the black solid line the fitted line, and the black circles are the observed values.

4.2. Correction of Negative Intercept Parameters

For the biomass models at level III in Table 3, the intercept parameters of 11 out of 74 forest sub-types are negative, leading to negative biomass estimates for young stands with a small volume or close to zero. Hence, the model parameters for these 11 sub-types required re-estimation, using simultaneous equations with error-in-variables and dummy variables for estimating two sub-types of broadleaf mixed forests, three sub-types of poplar forests, and four sub-types of oak forests, while the same approach without dummy variables was applied for the remaining two sub-types. If the intercept parameter remained negative, the weight function was modified to ensure a positive intercept. The updated parameter estimates and evaluation indices for the biomass model for the 11 forest sub-types are presented in Table 6.

Table 6. The parameter estimates of simultaneous forest biomass and carbon storage models and evaluation indices of biomass model for 11 forest sub-types.

Forest Sub-Type	Parameter Estimates						Evaluation Indices of Biomass Model				
	a_i	b_i	c_i	RSR	D_i(CF)	R^2	SEE/t	TRE/%	ASE/%	MPE/%	MPSE/%
Broadleaf mixed II	*0.3940*	1.3619	0.7893	0.2669	0.4838	0.927	16.60	−0.01	−0.82	1.86	14.74
Broadleaf mixed III	0.6511	1.2911	0.7964	0.2557	0.4849	0.955	16.90	0.01	−0.31	1.02	11.38
Quercus spp. I	*0.2090*	1.1363	0.7817	0.2793	0.4795	0.970	14.16	0.01	−2.87	0.81	8.46
Quercus spp. II	0.3647	1.5479	0.7820	0.2788	0.4808	0.946	15.61	0.01	−1.68	1.14	11.84
Quercus spp. III	*0.3070*	1.4795	0.7968	0.2550	0.4815	0.961	19.33	0.00	−0.67	1.34	10.01
Quercus spp. V	0.7385	1.1107	0.8216	0.2171	0.4820	0.969	23.29	0.00	1.98	1.53	11.01
Populus spp. II	0.3080	0.9727	0.8170	0.2240	0.4724	0.973	7.48	0.01	−0.92	0.72	10.08
Populus spp. III	*0.0941*	0.8891	0.8194	0.2204	0.4729	0.956	11.96	0.00	−1.23	1.97	12.92
Populus spp. IV	*0.4810*	0.9021	0.8600	0.1628	0.4721	0.908	12.10	0.01	−0.82	1.50	13.76
Quercus variabilis	*0.0492*	1.5131	0.7823	0.2783	0.4945	0.963	10.52	0.00	−4.88	3.79	12.73
Other hardwood II	*0.1409*	1.4235	0.7985	0.2523	0.4842	0.980	11.49	0.00	0.04	2.24	13.33

Note: Only the parameter estimates in italics are not statistically significant, all others are highly significant ($p < 0.01$); the RSR values are calculated from Equation (3).

In comparison with the original models in Table 3, the evaluation indices of the updated models for the 11 sub-types show minimal overall changes. Independent cross-validation confirmed that the TRE values fell within ±3% except for one sub-type of *Populus* spp. IV; the ASE values were within ±5% except for two sub-types, broadleaf mixed II and *Quercus variabilis*.

Thus, it is both straightforward and practical to develop simultaneous models of forest biomass and carbon storage for various types at once using simultaneous equations with error-in-variables and dummy variables. Nevertheless, if some model parameters are found to be atypical, separate re-estimation is necessary for adjustment and refinement.

5. Conclusions

Based on the measured data of 52,700 permanent plots from the ninth national forest inventory of China, a tiered volume-based forest biomass and carbon storage model system has been developed in this study according to one population, three forest categories (level I), 20 forest types (level II), and 74 forest sub-types (level III) through the use of weighted regression, dummy variable modeling, and simultaneous equations with error-in-variables. Additionally, the evaluation indices of the three-level models were examined. From the study results, the following conclusions can be drawn.

The performance of models generally improved as the modeling sample size increased. This study used almost all permanent plots in forests from the national forest inventory to develop a forest biomass and carbon storage model system, which is the first time to obtain the largest and most representative modeling samples.

The three-level models developed in this study will serve as a solid foundation for accurately assessing the status and changes of forest biomass and carbon storage at national, regional, and global scales; and the evaluation indices of the models would provide a scientific base for calculating the uncertainty in the estimates.

This study has provided several alternatives to select the models at different scales. To minimize uncertainty in forest biomass and carbon storage estimates, the use of level III models is recommended wherever feasible, especially for national or sub-national estimation.

Author Contributions: W.Z. (Weisheng Zeng) jointly conceived the study with W.Z. (Wentao Zou); X.C. and X.Y. designed the experiments and collected data; W.Z. (Weisheng Zeng) and W.Z. (Wentao Zou) developed the models and wrote the manuscript. All authors have read and agreed to the published version of the manuscript.

Funding: This research was funded by the National Key Research and Development Program of China (Grant No. 2022YFD2200501) and the Forest Resources Monitoring and Assessment Program of China (Grant No. 2130207).

Data Availability Statement: The data that support the findings of this study are available from the National Forestry and Grassland Administration (NFGA), but restrictions apply to the availability of these data, which were used for the current study with permission of NFGA, and so are not publicly available. Data are however available from the corresponding author upon reasonable request and with permission of NFGA.

Conflicts of Interest: The authors declare no conflicts of interest.

References

1. IUFRO (International Union of Forest Research Organizations). *International Guidelines for Forest Monitoring*; IUFRO Secretariat: Vienna, Austria, 1994.
2. IPCC (Intergovernmental Panel on Climate Change). *IPCC Guidelines for National Greenhouse Gas Inventories*; Institute for Global Environmental Strategies: Kanagawa, Japan, 2006.
3. FAO (Food and Agriculture Organization of the United Nations). *FRA 2025: Guidelines and Specifications*; FAO: Rome, Italy, 2023.
4. Molotoks, A.; Stehfest, E.; Doelman, J.; Albanito, F.; Fitton, N.; Dawson, T.P.; Smith, P. Global projections of future cropland expansion to 2050 and direct impacts on biodiversity and carbon storage. *Glob. Chang. Biol.* **2018**, *24*, 5895–5908. [CrossRef] [PubMed]
5. Walker, W.S.; Gorelik, S.R.; Cook-Patton, S.C.; Baccini, A.; Farina, M.K.; Solvik, K.K.; Ellis, P.W.; Sanderman, J.; Houghton, R.A.; Leavitt, S.M.; et al. The global potential for increased storage of carbon on land. *Proc. Natl. Acad. Sci. USA* **2022**, *119*, e2111312119. [CrossRef] [PubMed]
6. Somogyi, Z.; Cienciala, E.; Mäkipää, R.; Muukkonen, P.; Lehtonen, A.; Weiss, P. Indirect methods of large-scale forest biomass estimation. *Eur. J. For. Res.* **2007**, *126*, 197–207. [CrossRef]
7. Jagodziński, A.M.; Dyderski, M.K.; Gęsikiewicz, K.; Horodecki, P. Tree and stand level estimations of *Abies alba* Mill aboveground biomass. *Ann. Forest Sci.* **2019**, *76*, 56. [CrossRef]
8. Luo, Y.; Wang, X.; Ouyang, Z.; Lu, F.; Feng, L.; Tao, J. A review of biomass equations for China's tree species. *Earth Syst. Sci. Data* **2020**, *12*, 21–40. [CrossRef]
9. *LY/T 2260-2014*; Tree Biomass Models and Related Parameters to Carbon Accounting for Pinus tabulaeformis. China Standards Press: Beijing, China, 2015.
10. *LY/T 2261-2014*; Tree Biomass Models and Related Parameters to Carbon Accounting for Pinus elliottii. China Standards Press: Beijing, China, 2015.
11. *LY/T 2262-2014*; Tree Biomass Models and Related Parameters to Carbon Accounting for Pinus yunnanensis. China Standards Press: Beijing, China, 2015.
12. *LY/T 2263-2014*; Tree Biomass Models and Related Parameters to Carbon Accounting for Pinus massoniana. China Standards Press: Beijing, China, 2015.
13. *LY/T 2264-2014*; Tree Biomass Models and Related Parameters to Carbon Accounting for Cunninghamia Lanceolata. China Standards Press: Beijing, China, 2015.
14. *LY/T 2654-2016*; Tree Biomass Models and Related Parameters to Carbon Accounting for Larix. China Standards Press: Beijing, China, 2017.
15. *LY/T 2656-2016*; Tree Biomass Models and Related Parameters to Carbon Accounting for Abies. China Standards Press: Beijing, China, 2017.
16. *LY/T 2655-2016*; Tree Biomass Models and Related Parameters to Carbon Accounting for Picea. China Standards Press: Beijing, China, 2017.
17. *LY/T 2657-2016*; Tree Biomass Models and Related Parameters to Carbon Accounting for Cryptomeria. China Standards Press: Beijing, China, 2017.
18. *LY/T 2658-2016*; Tree Biomass Models and Related Parameters to Carbon Accounting for Quercus. China Standards Press: Beijing, China, 2017.

19. *LY/T 2659-2016*; Tree Biomass Models and Related Parameters to Carbon Accounting for Betula. China Standards Press: Beijing, China, 2017.
20. *LY/T 2660-2016*; Tree Biomass Models and Related Parameters to Carbon Accounting for Liquidambar formosana. China Standards Press: Beijing, China, 2017.
21. *LY/T 2661-2016*; Tree Biomass Models and Related Parameters to Carbon Accounting for Robinia pseudoacacia. China Standards Press: Beijing, China, 2017.
22. Shiver, B.D.; Brister, G.H. Tree and stand volume functions for *Eucalyptus saligna*. *For. Ecol. Manag.* **1992**, *47*, 211–223. [CrossRef]
23. Chamshama, S.A.O.; Mugasha, A.G.; Zahabu, E. Stand biomass and volume estimation for Miombo woodlands at Kitulangalo, Morogoro, Tanzania. *S. Afr. For. J.* **2004**, *200*, 59–69. [CrossRef]
24. Castedo-Dorado, F.; Gómez-García, E.; Diéguez-Aranda, U.; Barrio-Anta, M.; Crecente-Campo, F. Aboveground stand-level biomass estimation: A comparison of two methods for major forest species in northwest Spain. *Ann. Forest Sci.* **2012**, *69*, 735–746. [CrossRef]
25. Usoltsev, V.A.; Shobairi, S.O.R.; Chasovskikh, V.P. Triple harmonization of transcontinental allometric of *Picea* spp. and *Abies* spp. forest stand biomass. *Ecol. Environ. Conserv.* **2018**, *24*, 1966–1972.
26. Jagodziński, A.M.; Dyderski, M.K.; Gęsikiewicz, K.; Horodecki, P.; Cysewska, A.; Wierczyńska, S.; Maciejczyk, K. How do tree stand parameters affect young Scots pine biomass?—Allometric equations and biomass conversion and expansion factors. *For. Ecol. Manag.* **2018**, *409*, 74–83. [CrossRef]
27. Jagodziński, A.M.; Dyderski, M.K.; Gęsikiewicz, K.; Horodecki, P. Tree- and stand-level biomass estimation in a *Larix decidua* Mill. chronosequence. *Forests* **2018**, *9*, 587. [CrossRef]
28. Jagodziński, A.M.; Dyderski, M.K.; Gęsikiewicz, K.; Horodecki, P. Effects of stand features of aboveground biomass and biomass conversion and expansion factors based on a *Pinus sylvestris* L. chronosequence in western Poland. *Eur. J. For. Res.* **2019**, *138*, 673–683. [CrossRef]
29. Fang, J.Y.; Liu, G.H.; Xu, S.L. Biomass and net production of forest vegetation in China. *Acta Ecol. Sin.* **1996**, *16*, 497–508.
30. Fang, J.; Chen, A.; Peng, C.; Zhao, S.; Ci, L. Changes in forest biomass carbon storage in China between 1949 and 1998. *Science* **2001**, *292*, 2320–2322. [CrossRef] [PubMed]
31. Zeng, W.S.; Sun, X.N.; Wang, L.R.; Wang, W.; Pu, Y. Developing stand volume, biomass and carbon stock models for ten major forest types in forest region of northeastern China. *J. Beijing For. Uni.* **2021**, *43*, 1–8.
32. Wang, B.; Liu, M.C.; Zhang, B. Dynamics of net production of Chinese forest vegetation based on forest inventory data. *For. Resour. Manag.* **2009**, *1*, 35–42.
33. Hou, Y.-N.; Wu, H.-L.; Zeng, W.-X.; Xiang, W.-H. Conversion parameters for stand biomass estimation of four subtropical forests in southern China. In *DEStech Transactions on Environment Energy and Earth Science*; DEStech: Lancaster, PA, USA, 2017. [CrossRef]
34. Mei, G.Y.; Sun, Y.J.; Saeed, S. Models for predicting the biomass of *Cunninghamia lanceolata* trees and stands in southeastern China. *PLoS ONE* **2017**, *12*, e0169747.
35. Zhao, M.; Yang, J.; Zhao, N.; Liu, Y.; Wang, Y.; Wilson, J.P.; Yue, T. Estimation of China's forest stand biomass carbon sequestration based on the continuous biomass expansion factor model and seven forest inventories from 1977 to 2013. *For. Ecol. Manag.* **2019**, *448*, 528–534. [CrossRef]
36. Dong, L.H.; Zhang, L.J.; Li, F.R. Evaluation of stand biomass estimation methods for major forest types in the eastern Da Xing'an Mountain, northeast China. *Forests* **2019**, *10*, 715. [CrossRef]
37. Zhou, X.; Lei, X.; Peng, C.; Wang, W.; Zhou, C.; Liu, C.; Liu, Z. Correcting the overestimate of forest biomass carbon on the national scale. *Method Ecol. Evol.* **2016**, *7*, 447–455. [CrossRef]
38. Zhou, X.; Lei, X.; Liu, C.; Huang, H.; Zhou, C.; Peng, C. Re-estimating the changes and ranges of forest biomass carbon in China during the past 40 years. *For. Ecosyst.* **2019**, *6*, 51. [CrossRef]
39. Tang, X.; Zhao, X.; Bai, Y.; Tang, Z.; Wang, W.; Zhao, Y.; Wan, H.; Xie, Z.; Shi, X.; Wu, B.; et al. Carbon pools in China's terrestrial ecosystems: New estimates based on an intensive field survey. *Proc. Natl. Acad. Sci. USA* **2018**, *115*, 4021–4026. [CrossRef] [PubMed]
40. Zhang, Y.X.; Wang, X.J.; Pu, Y.; Zhang, J. Changes in forest resource carbon storage in China between 1949 and 2018. *J. Beijing For. Univ.* **2021**, *43*, 1–14.
41. Zhang, Y.X.; Wang, X.J. Study on forest volume-to-biomass modeling and carbon storage dynamics in China. *Sci. Sin. Vitae* **2021**, *51*, 199–214.
42. Zeng, W.S. Developing tree biomass models for eight major tree species in China. In *Biomass Volume Estimation and Valorization for Energy*; InTech: Zegreb, Croatia, 2017. [CrossRef]
43. *GB/T 43648-2024*; Tree Biomass Models and Related Parameters to Carbon Accounting for Major Tree Species. China Standards Press: Beijing, China, 2024.
44. Lambert, M.C.; Ung, C.H.; Raulier, F. Canadian national tree aboveground biomass models. *Can. J. For. Res.* **2005**, *35*, 1996–2018. [CrossRef]
45. Zianis, D.; Muukkonen, P.; Mäkipää, R.; Mencuccini, M. Biomass and stem volume equations for tree species in Europe. *Silva Fenn.* **2005**, *4*, 1–63. [CrossRef]
46. Ter-Mikaelian, M.T.; Korzukhin, M.D. Biomass equations for sixty-five north American tree species. *For. Ecol. Manag.* **1997**, *97*, 1–24. [CrossRef]

47. Zeng, W.S.; Tang, S.Z. Bias correction in logarithmic regression and comparison with weighted regression for non-linear models. *For. Res.* **2011**, *24*, 137–143.
48. Zeng, W.S. Comparison of different weight functions in weighted regression. *For. Resour. Manag.* **2013**, *5*, 55–61.
49. Wang, M.; Borders, B.E.; Zhao, D. An empirical comparison of two subject-specific approaches to dominant heights modeling the dummy variable method and the mixed model method. *For. Ecol. Manag.* **2008**, *255*, 2659–2669. [CrossRef]
50. Fu, L.; Lei, Y.; Wang, G.; Bi, H.; Tang, S.; Song, X. Comparison of seemingly unrelated regressions with error-in-variable models for developing a system of nonlinear additive biomass equations. *Trees* **2016**, *30*, 839–857. [CrossRef]
51. *GB/T 38590*; Technical Regulations for Continuous Forest Inventory. China Standards Press: Beijing, China, 2020.
52. Zeng, W.S.; Tang, S.Z.; Huang, G.S.; Zhang, M. Population classification and sample structure on modeling of single-tree biomass equations for national biomass estimation in China. *For. Resour. Manag.* **2010**, *3*, 16–23.
53. Parresol, B.R. Assessing tree and stand biomass: A review with examples and, critical comparisons. *For. Sci.* **1999**, *45*, 573–593. [CrossRef]
54. Zeng, W.S.; Tang, S.Z. Evaluation and precision analysis of tree biomass equations. *Sci. Silvae Sin.* **2011**, *47*, 106–113.
55. Fu, L.Y.; Zeng, W.S.; Tang, S.Z. Individual tree biomass models to estimate forest biomass for large spatial regions developed using four pine species in China. *For. Sci.* **2017**, *63*, 241–249.
56. Zeng, W.; Zhang, L.; Chen, X.; Cheng, Z.; Ma, K.; Li, Z. Construction of compatible and additive individual-tree biomass models for *Pinus tabulaeformis* in China. *Can. J. For. Res.* **2017**, *47*, 467–475. [CrossRef]

forests

MDPI

Article

Soil-Sensitive Weibull Distribution Models of *Larix principis-rupprechtii* Plantations across Northern China

Hong Guo, Xianzhao Liu * and Dan Liu

State Key Laboratory of Efficient Production of Forest Resources, Key Laboratory of Forest Management and Growth Modelling, Institute of Forest Resource Information Techniques, Chinese Academy of Forestry, Beijing 100091, China; hongguo@ifrit.ac.cn (H.G.); danliu@ifrit.ac.cn (D.L.)

* Correspondence: liuxianzhao@ifrit.ac.cn

Abstract: Tree diameter distribution models are important tools for forest management decision making. Soil variables affect tree growth and thus diameter distribution. However, few studies have been conducted on diameter distribution models describing the effects of soil. This study developed a soil-sensitive diameter distribution model based on 213 sample plots of *Larix principis-rupprechtii* plantations in northern China. The Weibull distribution model was modified by a compatible simultaneous system and the percentile method with the inclusion of soil variables. The most significant factors influencing the diameter distribution of *L. principis-rupprechtii* in terms of both scale and shape were stand characteristics and available K and alkali-hydrolysable N. The adjusted coefficient of determination for parameter γ significantly improved by 16.0%, while the root mean square error for parameter β decreased by 10.4%. The F test indicated a substantial difference between the models with and without soil variables. From the perspective of adjustable R^2 values, the Akaike information criterion, root mean square error, relative error index, and absolute error index, the inclusion of stand and soil factors in the tree diameter distribution model enhanced its performance compared to the model that did not consider soil factors. The soil-sensitive diameter distribution model is proven to be effective and accurate.

Keywords: Weibull distribution model; soil variables; percentile-based parameter recovery method; diameter structure

1. Introduction

The diameter structure of a stand is the most basic and significant component of the stand structure. Numerous studies have demonstrated a close relationship between changes in diameter structure, which describes the distribution status of trees of various sizes and diameters in forest stands according to their diameter order and variables such as tree height, stem shape, volume, species, crown, and biomass [1–3]. Therefore, research on the distribution of forest stand diameter structure can potentially serve as a foundation for methodologies and approaches related to forest management and tree measurement and tabulation. The distribution characteristics of stand diameter classes are described by a variety of probability density functions (PDFs) or models, including the β model, Johnson's sb model, gamma model, normal model, lognormal model, and Weibull model [4,5]. Weibull's three-parameter model, which includes a closed cumulative density formula and convenient parameter extraction, is a versatile tool for modeling the diameter distribution of *Larix principis-rupprechtii* plantations [6]. The fundamental variation in forest characteristics and ecological traits is likely to be captured by environmental variables such as soil conditions and climate, clarifying the biological linkages between the environment and tree growth. Recent studies have shown a significant impact of the environment on carbon and tree allotrophic growth, particularly the influence of soil factors [7,8]. Tree development and thus the distribution of tree diameter can be influenced by environmental variables. Several diameter distribution models, such as those by Sanquetta et al. (2014) and Guo et al. (2022) [6,9], describe the impacts of climatic characteristics. However, few studies have investigated how soil conditions influence the distribution of diameters. *Larix principis-rupprechtii*, as the primary afforestation tree in China, plays a significant role in wood production, carbon sequestration, and ecological services. According to the National Forestry and Grassland Administration in China (2019), arbor forests cover approximately 6.72% of the total area, with 29.19% of this area being planted. As indicated by Zang et al. (2016) and Lei et al. (2016) [10,11], the species under consideration is climate-sensitive. Due to this sensitivity, numerous studies have been conducted to develop growth and biomass models that are sensitive to climate factors, both at the individual tree level and at the stand level [12,13]. Nevertheless, there is a lack of understanding regarding the impact of soil variables on the diameter class distribution of *Larix principis-rupprechtii* plantations. The size structure of trees has been shown to be associated with the diversity of stand structure, which has been shown to have an impact on forest growth, biomass, productivity, and multifunctionality [13–17]. Forest managers employ silvicultural practices to manipulate stand density and preserve a desirable diameter distribution [18]. Additionally, the structure of stands plays a significant role in shaping the response of forests to climate and silviculture practices [15]. A lack of comprehensive understanding regarding the impact of changes in soil variables on the structure of larch forests can pose a significant challenge for forest managers in devising adaptive strategies to mitigate the potential consequences of these effects. Hence, the objectives of our study were twofold: (1) to establish a soil-sensitive diameter distribution model utilizing the Weibull distribution model by utilizing permanent sample plots from larch plantations located in the northern and northeastern regions of China and (2) to simulate the effects of soil on the diameter distribution. This study will help improve our understanding of the impact of soil on growth prediction.

2. Materials and Methods

2.1. Sample Plot Data

The data for the sample plots were obtained from the 6th National Forest Inventory of China, which was conducted between 1999 and 2003 in four provinces, namely, Beijing, Hebei, Inner Mongolia, and Shanxi Provinces (Figure 1). These plots are rectangular with an area from 0.06 ha to 0.08 ha in different provinces listed in Table 1. Only *Larix principis-rupprechtii* plantations that had not been harvested were chosen for this study. There were 213 plots in the dataset, with a total of 9980 trees. The plots that were examined were rectangular in shape and had areas that varied between 0.06 and 0.08 hectares in different provinces. The primary variables under investigation in the plots were tree diameter at breast height (DBH) greater than 5 cm, average age of the stand, average height of the trees, canopy closure, slope position, aspect, slope, elevation, soil thickness, and soil type. Other stand variables, such as the stand basal area (BA), quadratic mean diameter (QMD), density stems (S), minimum diameter (MinD), maximum diameter (MaxD), stand average age (SAA), average height (AH), and stand site index (SI), were derived from tree list measurements. The site indices were computed based on the model proposed by Zang et al. (2016) [10] in the literature. The summary statistics of the variables pertaining to the stand are presented in Table 1.

Table 1. Descriptive statistics of the stand variables.

Provinces	Plot Size	Statistic	SAA /Years	AH /m	QMD/cm	SI/m	BA/m^2/ha	S/ha
Beijing	0.0667 ha	Mean	24.56	7.02	10.86	26.47	9.29	773
		Standard Deviation	9.43	1.32	3.64	3.93	10.04	509
		Min	13.00	5.58	7.16	14.01	0.79	195
		Max	43.00	9.76	18.70	46.35	28.84	1694

Table 1. *Cont.*

Provinces	Plot Size	Statistic	SAA /Years	AH /m	QMD/cm	SI/m	BA/m²/ha	S/ha
Hebei	0.06 ha	Mean	17.86	6.30	9.39	19.25	7.57	1040
		Standard Deviation	5.27	1.27	3.49	3.30	5.82	610
		Min	9.00	4.77	5.65	9.70	0.75	217
		Max	31.00	9.85	20.00	33.42	24.11	2733
Inner Mongolia	0.06 ha	Mean	21.33	7.97	11.68	23.00	10.64	939
		Standard Deviation	5.96	1.80	3.39	3.33	6.31	496
		Min	13.00	5.89	8.32	14.01	1.9	350
		Max	33.00	11.17	18.09	35.57	18.84	2133
Shanxi	0.0667 ha	Mean	22.16	6.64	9.60	23.89	8.58	1128
		Standard Deviation	8.84	1.58	2.67	2.66	5.86	522
		Min	14.00	4.82	6.20	15.09	0.54	180
		Max	50.00	12.24	18.11	53.90	24.72	1979
Total		Mean	19	7.30	9.32	8.84	7.81	1076
		Standard Deviation	8	2.01	3.21	2.56	5.86	180
		Min	5	4.52	5.59	4.68	0.48	653
		Max	50	17.80	23.69	19.20	28.84	3820

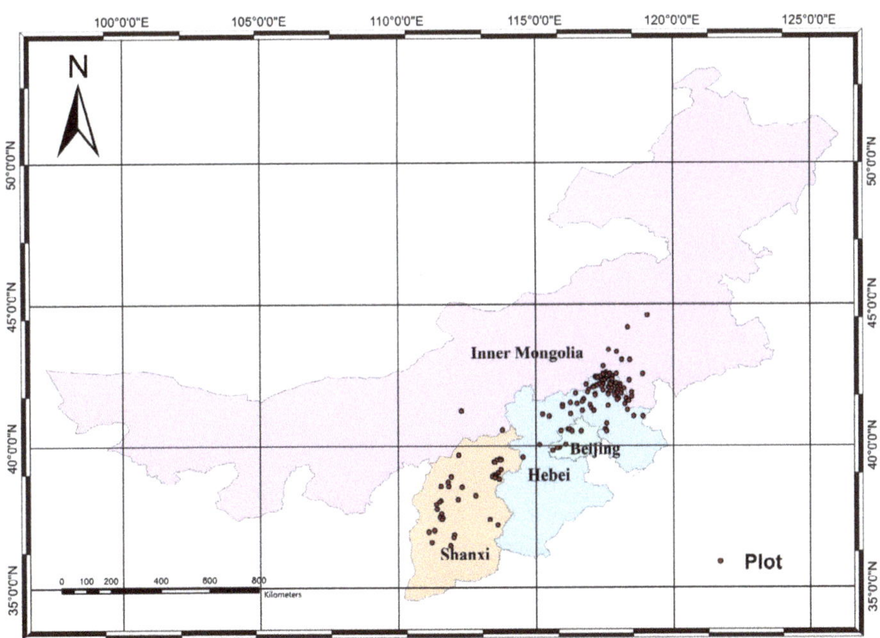

Figure 1. Location of the study area and spatial distribution of the sample plots.

2.2. Site and Soil Data

The soil data utilized in this study were derived from a comprehensive dataset of soil properties specific to China, which was collected and compiled by Shangguan and Dai (2014) [19]. This dataset served as the foundation for land surface modeling in the present research. The source data for this dataset were derived from the 1:1 million soil map of China and 8595 soil profiles. The dataset provided in raster format encompassed various parameters, such as pH value, soil organic matter, total N, total P, total K, alkali-hydrolysable N, available P, available K, cation exchange capacity, exchangeable Al3+, and others, totaling 16 indices (refer to Table 2). Each soil physical and chemical property in the database included data for eight different soil layer depths: 0–0.045 m, 0.045–0.091 m, 0.091–0.166 m, 0.166–0.289 m, 0.289–0.493 m, 0.493–0.829 m, 0.829–1.383 m, and 1.383–2.296 m. The 8 soil grid data points were weighted averages and registered with the basic geographic data of Shanxi province, Hebei province, and the Inner Mongolia Autonomous Region and were used as the soil environmental variables. In addition, altitude and soil thickness were also added as environmental variables.

Table 2. Description of soil and site variables for model development.

Category	Variable Unit	Description	Statistic		
			Maximum	Minimum	Mean
Soil chemical properties	TN (g/100 g)	Soil total nitrogen	0.48	0.04	0.19
	TK (g/100 g)	Soil total potassium	2.51	1.41	1.93
	TP (g/100 g)	Soil total phosphorus	0.13	0.04	0.06
	SOM (g/100 g)	Soil organic matter	8.90	0.55	3.69
	pH	pH value (H_2O)	8.39	6.07	6.88
	AN (mg/kg)	Alkali-hydrolysable nitrogen	284.15	24.78	140.64
	AK (mg/kg)	Available potassium	254.44	49.22	140.25
	AP (mg/kg)	Available phosphorus	13.81	2.19	6.66
	AL (Me/100 g)	Exchangeable Al^{3+}	3.21	0.03	0.37
	CEC (Me/100 g)	Cation exchange capacity (CEC)	27.90	6.39	16.84
Soil physical property	PDEP (cm)	Soil profile depth	147.85	43.86	85.62
	GRAV	Rock fragment	43.73	0.25	18.30
	BD (g/cm^3)	Bulk density	1.39	0.83	1.26
	CL	Clay	22.03	9.53	15.73
	SI	Silt	55.34	22.22	41.97
	SA	Sand	68.25	29.59	42.31
Site	LA (m)	Latitude	2383.00	690.00	1471.55
	SLD (cm)	Soil layer depth	100.00	1.00	24.69

2.3. Weibull Distribution Model and Parameter Estimation Methods

The three-parameter Weibull distribution model was selected because of its extensive utilization and flexibility, as shown by previous studies [20,21]. The probability density function (PDF) f(x) for the three-parameter Weibull distribution can be expressed as follows:

$$f(x) = (\frac{\gamma}{\beta})(\frac{x-\alpha}{\beta})^{\gamma-1}exp[-(\frac{x-\alpha}{\beta})^{\gamma}] \tag{1}$$

$$\alpha \leq x \leq \infty, \beta > 0, \gamma > 0$$

where x represents the measured tree diameter at breast height (DBH) and α, β, and γ are the position, scale, and shape parameters of the Weibull distribution, respectively.

This study compared the percentile method (PM) and maximum likelihood estimation method (MLE) for estimating PDF parameters. Both methods involve predicting the percentiles of the diameter distribution, which are dependent on the distribution parameters. However, PM has been found to have a stronger correlation with stand characteristics than with PDF parameters [22]. The probability proportional to size is a statistical method that directly represents the PDF parameters as a function of stand variables, as demonstrated by Liu et al. (2004) [23]. In this article, the Kolmogorov–Smirnov (KS) test [24,25] was employed to compare the suitability of the three-parameter Weibull distribution in fitting sample plots between PM and the MLE. The three-parameter Weibull distribution was fitted using maximum likelihood estimation [26] based on the tree list for each plot measurement (observation) in the moment method. The concept of PM can be defined as Equations (2)–(4) [23].

$$\alpha = \frac{n^{0.3333}\mu_0 - \mu_{50}}{n^{0.3333} - 1} \tag{2}$$

$$\gamma = \frac{2.343088}{\ln(\mu_{95} - \alpha) - \ln(\mu_{25} - \alpha)} \tag{3}$$

$$\beta = \frac{\alpha\Gamma(1+\frac{1}{\gamma})}{\Gamma(1+\frac{2}{\gamma})} + sqrt\{\frac{\alpha^2}{\Gamma^2(1+\frac{1}{\gamma})}[\Gamma^2(1+\frac{1}{\gamma}) - \Gamma(1+\frac{2}{\gamma})] + \frac{\overline{u_q^2}}{\Gamma(1+\frac{2}{\gamma})}]\} \tag{4}$$

In the given context, n represents the number of trees present in a plot, where μ_0, μ_{25}, μ_{50}, and μ_{95} correspond to the 0th, 25th, 50th, and 95th percentiles, respectively, in the distribution of trees. The gamma function Γ was utilized, and the mean square diameter u_q was also considered. If the estimated value of α is less than 0, α is forced to be set to 0.

2.4. Diameter Distribution Model with Soil Factors

First, the estimation of the three parameters of the Weibull distribution for each plot was conducted using the equations provided. Equations (2)–(4) were derived from the PM. Additionally, separate relationships were established between the parameters and stand variables, as well as between the parameters, the stand, and additional soil variables. Stand and soil variables were chosen through the use of SUR regression, which is a linear model that is employed to estimate sparse parameters and is particularly well suited for reducing the number of parameters [27]. A commonly employed method in regression analysis to identify multicollinearity is the variance inflation factor (VIF) [28]. The severity of multicollinearity increases with increasing VIF. Generally, a VIF exceeding 10 indicates the presence of a significant collinearity issue within the model. Moreover, to further elucidate the potential internal relationships among the three parameters of the Weibull distributions, we employed simultaneous equations to impose a constraint. In this study, the VIF was set at a threshold of 10 to examine the presence of multicollinearity. The calculations were conducted by the glmnet and Mass package in R4.4.0 [29,30].

$$\begin{cases} \alpha = f1(stand) \\ \beta = f2(stand) \\ \gamma = f3(stand) \end{cases} \tag{5}$$

$$\begin{cases} \alpha = f4(stand, soil) \\ \beta = f5(stand, soil) \\ \gamma = f6(stand, soil) \end{cases} \quad (6)$$

In the mathematical expression 5 and 6, the variable "stand" in this study refers to a set of stand variables, namely, BA, S, MaxD, MinD, QMD, and SI, as listed in Table 1. On the other hand, the term "soil" represents a collection of soil variables listed in Table 2.

2.5. Model Test

2.5.1. Kolmogorov–Smirnov Test for the Weibull Distribution

The Kolmogorov–Smirnov (KS) test, as proposed by Smirnov (1948) and further discussed by Reynolds et al. (1988) [25], was employed at a 95% confidence level to assess the suitability of fitting the diameter distribution of a sample plot using the three-parameter Weibull distribution through cumulative frequency (Equation (7)).

$$D = \underset{-\infty<\theta<\infty}{lim\ sup} |F^*(v) - F(v)| \quad (7)$$

where lim sup is the maximum value, v represents the measured tree diameter at breast height (DBH), and D is the KS statistic. $F^*(v)$ and $F(v)$ are the theoretical and realized probabilities of the diameter distribution of the plots, respectively. The original hypothesis is H0: the distributions of the two variables ($F^*(v)$ and $F(v)$) are consistent, or the data conform to the theoretical distribution. When the observed value D exceeds the critical value D (n, alpha), the null hypothesis H0 is rejected; otherwise, the null hypothesis H0 is accepted based on the assumption, where n represents the number of trees in a single plot and alpha denotes the specified test level, which is set at 0.05.

2.5.2. Model Evaluation for Regression

Leave-one-out cross-validation (LOOCV) was employed to assess the performance of the Weibull distribution parameter models (mathematical expression 5 and 6) [31]. Root mean square error (*RMSE*) and adjusted coefficient of determination (adjusted R^2) were calculated in Equations (10) and (11). The paper compared the differences between two models through the F-test in Equation (12). The Akaike information criterion (AIC) is a standard for measuring the goodness of fit of statistical models in Equation (13) [32]. E1 and E2 are the absolute error index and relative error index in Equations (14) and (15) respectively, that are specific to the *i*th plot measurement.

$$SSE = \sum_{i=1}^{N} (y_i - \hat{y}_i)^2 \quad (8)$$

$$SSR = \sum_{i=1}^{N} (y_i - \overline{y}_i)^2 \quad (9)$$

$$RMSE = \sqrt{\frac{SSE}{N}} \quad (10)$$

$$adjusted\ R^2 = 1 - \frac{SSE(N-1)}{(N-k-1)SSR} \quad (11)$$

$$F = \frac{SSR/(df_s - df_c)}{SSE/df_c} \quad (12)$$

$$AIC = e^{\left(\frac{2p}{N}\right)} \frac{SSE}{N} \quad (13)$$

$$\mathrm{E1} = \sum_{i}^{N} |\mathrm{N_{i,j}} - \hat{\mathrm{N}}_{i,j}| \quad (14)$$

$$\mathrm{E2} = \sum_{i}^{N} \left| \frac{\mathrm{N_{i,j}}}{\sum \mathrm{N_{i,j}}} - \frac{\hat{\mathrm{N}}_{i,j}}{\sum \hat{\mathrm{N}}_{i,j}} \right| \quad (15)$$

where SSEc and SSEs are the sum of square errors of the regression model (Equations (8) and (9); y_i and $\hat{y}_i$ are the observed and predicted numbers of trees for the ith plot, respectively; N is the total number of plots; p is the number of independent variables; dfc and dfs are the degrees of freedom for Equations (5) and (6), respectively; and $N_{i,j}$ and $\hat{N}_{i,j}$ are the observed and predicted numbers of trees per hectare within the jth diameter class (2 cm width), respectively.

3. Results

3.1. Parameter Estimation Methods for the Weibull Distribution Models

According to the Kolmogorov–Smirnov (KS) test, of a total of 213 samples, 208 samples were successfully fitted to parameters via the percentile method (PM), while 134 plots were successfully fitted to parameters via the maximum likelihood estimation method (MLE). Therefore, the use of PM is deemed more appropriate for modeling the diameter distribution of *Larix principis-rupprechtii* plantations. The parameter estimates obtained from PM exhibit the following ranges: $2.50 \leq \beta \leq 7.25$, $3.33 \leq \beta \leq 16.25$, and $1.28 \leq \gamma \leq 7.50$ (Table 3).

Table 3. Summary and statistics of diameter distribution parameters.

Variables	α	β	γ
Mean	2.89	7.49	3.99
Standard Deviation	0.90	2.99	1.19
Max	7.25	16.25	7.50
Min	2.50	3.33	1.28

3.2. Comparison of Models for Estimating Diameter Distribution Parameters with and without Soil Variables

The β and γ parameters of the Weibull distribution, obtained through the PM method, were utilized as dependent variables. Using the SUR method, two sets of parameter prediction equations for the Weibull distribution were derived: the traditional model, which includes only stand variables (Equation (16)), and the soil-sensitive model, which incorporates both stand and soil variables (Equation (17)). Stand variables, such as the maximum diameter (MaxD), minimum diameter (MinD), diameter growth (QMD), and stand density (SD), were included in the analysis. The soil and site variables that exhibited statistical significance were bulk specific gravity (BSG), altitude, available nitrogen, and DPEP. Tables 4 and 5 present the calculated values of the parameters β and γ obtained through Equations (16) and (17) using the seemingly unrelated regression (SUR) method. Compared to Equation (13), the adjusted R^2 value improved (16.0% for γ), while the RMSE decreased (3.5% for β and 10.4% for γ). Figures 2 and 3 demonstrate a higher level of concordance between the predicted and observed values for the parameter and γ when soil factors were considered. Therefore, the inclusion of soil variables in the models resulted in an increase in model performance. The severity of multicollinearity increased as the variance inflation factor (VIF) increased. Generally, when the variance inflation factor (VIF) exceeds 5, it indicates the presence of a significant collinearity issue within the model. According to Table 4, the variance inflation factor (VIF) values for all of the models ranged from 1.12 to 2.86, which are below the threshold of 5. There was a notable disparity observed, indicating that the models incorporating soil variables outperformed those without soil variables. This conclusion is supported by the fact that the Akaike information criterion (AIC) of Equation (17) was smaller than that of Equation (15) for both variables, γ. The β results indicate a significant improvement in the models, as evidenced by the F test, with the value decreasing from 100.80 to 49.25. The AIC for parameter β increased from −87.12 for the model without soil factors to −68.23 for the model with soil factors. The AIC for parameter γ increased from 503.09 for the model without soil factors to 523.82 for the model with soil factors.

$$\begin{cases} \alpha = 0.5 * MinD \\ \beta = f2(stand) = a_0 + a_1Dg + a_2MinD + a_3DE + a_4SLD \\ \gamma = f3(stand) = c_0 + c_1MaxD + c_2MinD + c_3LA + a_4SLD \end{cases} \quad (16)$$

$$\begin{cases} \alpha = 0.5 * MinD \\ \beta = f4(stand, soil) = a_0 + a_1Dg + a_2MinD + a_3DE + a_4SLD + b_1AP + b_2AL + b_3AK + b_4PDEP \\ \gamma = f5(stand, soil) = c_0 + c_1MaxD + c_2MinD + c_3LA + a_4SLD + d_1AL + d_2AK + d_3DPEP + d_4DPE + d_5AN + d_6TN \end{cases} \quad (17)$$

Table 4. Parameter estimates and fit statistics for Models (16) and (17).

Parameters	Type	Variables	Estimate	Standard Deviation	Error t Value	Pr	VIF
β	Stand	Intercept	2.37×10^{-3}	7.17×10^{-2}	0.033	0.97366	-
		MinD	-4.98×10^{-1}	1.15×10^{-2}	−43.309	$<2.00 \times 10^{-16}$	2.20
		QMD	1.05×10^{0}	6.34×10^{-3}	165.833	$<2.00 \times 10^{-16}$	2.39
		LA	-1.37×10^{-4}	4.44×10^{-5}	−3.083	0.00233	1.48
		Density	1.07×10^{-4}	2.58×10^{-5}	4.138	5.10×10^{-5}	1.12
		Soil depth	1.04×10^{-3}	5.21×10^{-4}	2.004	0.04641	1.22
	With soil	Intercept	-1.60×10^{-1}	9.62×10^{-2}	−1.661	0.09825	-
		MinD	-4.96×10^{-1}	1.13×10^{-2}	−43.969	$<2.00 \times 10^{-6}$	2.40
		QMD	1.05×10^{0}	6.42×10^{-3}	162.829	$<2.00 \times 10^{-6}$	2.75
		LA	-1.01×10^{-4}	4.54×10^{-5}	−2.215	0.02788	1.75
		Density	9.93×10^{-5}	2.53×10^{-5}	3.921	0.00012	1.21
		SLD	1.52×10^{-3}	5.18×10^{-4}	2.923	0.003856	1.36
		AP	2.68×10^{-2}	6.96×10^{-3}	3.847	0.00016	2.86
		AL	6.58×10^{-2}	1.97×10^{-2}	3.333	0.001022	1.34
		AK	-1.55×10^{-3}	4.32×10^{-4}	−3.584	0.000424	2.77
		PDEP	1.93×10^{-3}	6.89×10^{-4}	2.802	0.005577	1.15
γ	Stand	(Intercept)	3.298216	0.271389	12.153	$<2.00 \times 10^{-16}$	
		MaxD	−0.19526	0.013603	−14.354	$<2.00 \times 10^{-16}$	1.59
		MinD	0.446559	0.036948	12.086	$<2.00 \times 10^{-16}$	1.42
		LA	0.000775	0.000172	4.515	1.06×10^{-5}	1.38
		SLD	−0.00848	0.002084	−4.069	6.70×10^{-5}	1.22
	With soil	(Intercept)	2.163463	0.37838	5.718	3.82×10^{-8}	1.59
		(Intercept)	−0.19918	0.013451	−14.808	$<2.00 \times 10^{-16}$	1.76
		MaxD	0.47691	0.035632	13.384	$<2.00 \times 10^{-16}$	1.49
		MinD	0.000609	0.000168	3.623	0.000368	1.49
		LA	−0.00889	0.002054	−4.329	2.35×10^{-5}	1.33
		AL	0.208315	0.093252	2.234	0.026581	1.86
		AK	0.003616	0.001422	2.543	0.011721	1.87
		PDEP	0.005869	0.002973	1.974	0.049712	1.33
		AN	0.005541	0.001422	3.897	0.000132	2.84
		TN	−3.02752	1.124804	−2.692	0.007704	4.35

Table 5. Model valuation indices with leave-one-out cross-validation.

Parameters	Type	RMSE	Multiple R^2	AIC
β	Stand	0.2005	0.9955	−68.23
	With Soil	0.1864	0.9961	−87.12
γ	Stand	0.806	0.5472	523.83
	With Soil	0.750	0.6080	503.09

(Note: root mean square error and adjusted coefficient of determination).

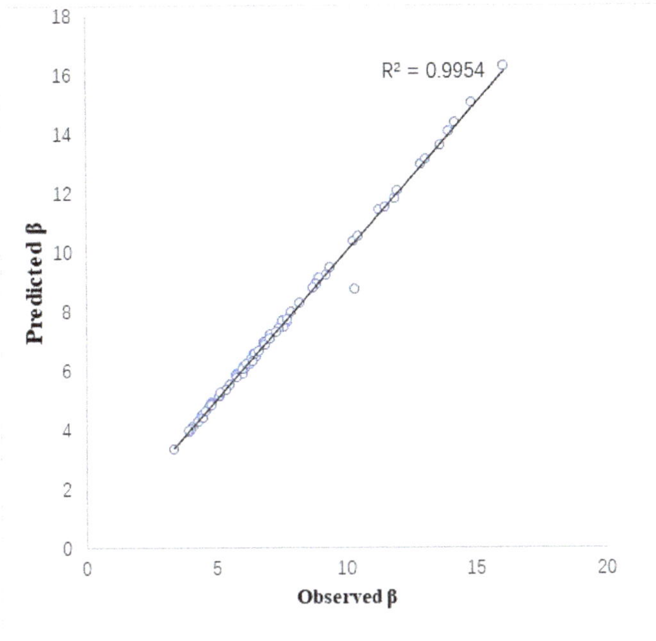

(a)

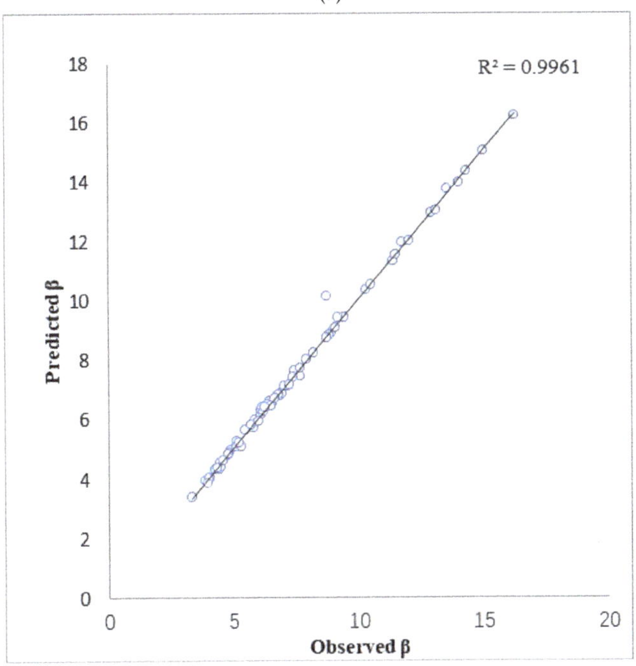

(b)

Figure 2. Observations and predictions of parameter β: (**a**) models without soil variables and (**b**) soil-sensitive models. The line represents the linear relationship between the predicted and observed values of the β parameter.

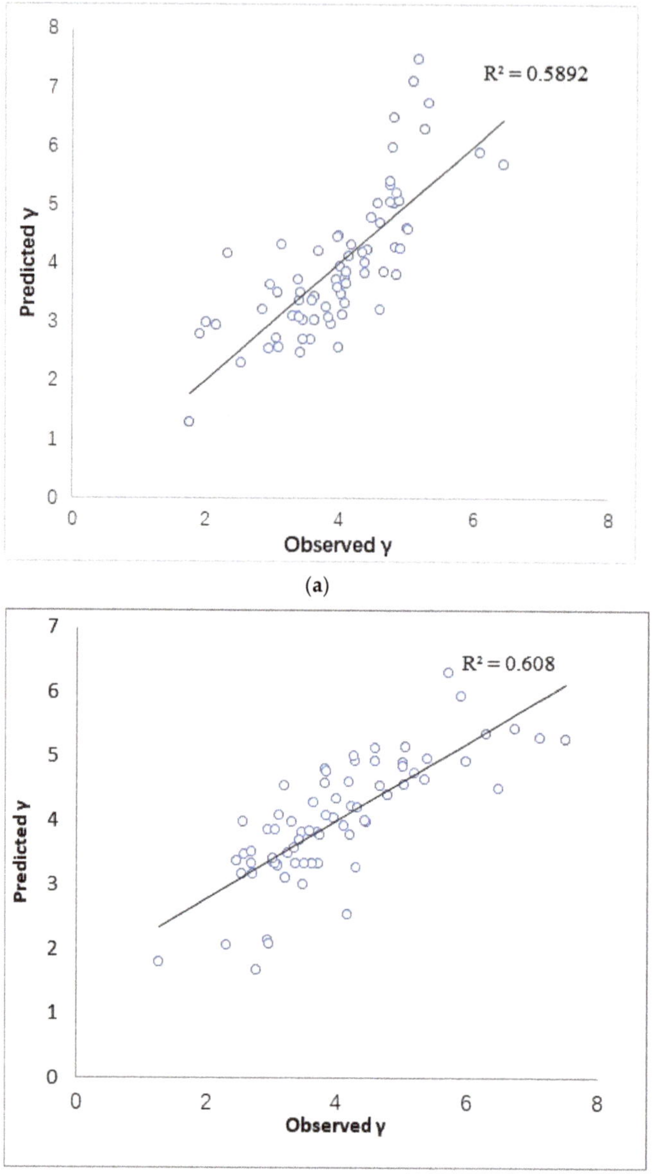

(**a**)

(**b**)

Figure 3. Observations and predictions of parameter γ: (**a**) models without soil variables and (**b**) soil-sensitive models. The line represents the linear relationship between the predicted and observed values of the γ parameter.

The absolute error index (E1) ranged from $33.3 \leq$ E1 (stand) ≤ 93.6 to $32.1 \leq$ E1 (soil) ≤ 73.4 (Table 6). The relative error (E) ranged from $4.08 \leq$ E2 (stand) ≤ 9.26 to $4.03 \leq$ E2 (soil) ≤ 8.4 (Table 6). Compared to Equation (15), the mean and maximum values of E1 and E2 decreased with the soil factors in the four provinces. In terms of the E1 index, the model that considered soil factors had a 3% to 21% decrease in the four provinces compared to the model that did not consider soil factors. Figures 2 and 3 show the fitting graphs of parameter β and parameter γ considering only forest factors and both soil factors and stand factors for predicted and measured values, respectively. Figure 2 shows slightly improved models when considering both soil factors and stand factors for parameter β because both models are much more accurate. Figure 3 shows a larger improved effect for models considering both soil factors and stand factors for parameter γ from scatter plots.

Table 6. Absolute error index E1 and relative error index E2 in different provinces.

	E1 (Stand)			E1 (Soil)			E2 (Stand) %			E2 (Soil) %		
Provinces	Mean	Min	Max	Mean	Min	Max	Mean	Min	Max	Mean	Min	Max
Hebei	33.3	0	145	32.1	0	136	4.08	0.13	48.8	4.03	1.05	43.3
Beijing	93.6	2	477	73.4	0	347	4.31	0.23	14.1	4.26	0.05	13.8
Shanxi	74.9	0	400	68.9	2	364	5.38	0.31	35.2	4.96	0.23	31.3
Inner Mongolia	57.4	8	202	48.9	8	175	9.2	0	43.2	8.4	0	38.3

4. Discussion

4.1. Attributes of the Diameter Distribution Model

Although many diameter class distribution models have been developed, soil variables are rarely introduced [9]. In this study, a soil-sensitive diameter distribution model was developed for *Larix principis-rupprechtii* plantations in northern China. When fitting with the PM method, 97.65% of the plots passed the KS test, while compared to the PPM method, only 62.91% of the plots passed the test. This result is the same as that of Guo et al. (2022) [6]. The PM method is more suitable for simulating the diameter distribution of *L. principis-rupprechtii* than the MLE method.

The distribution of tree diameter is influenced by intricate interactions among environmental, stand, and tree factors. Despite the existence of various diameter class distribution models [33], the incorporation of soil variables has been infrequently observed. In this research, a soil-sensitive diameter distribution model was developed using the SUR method. This method was employed to incorporate stand and soil variables into the model, aiming to improve the performance parameters for *Larix principis-rupprechtii* plantations in northern China. In our study, the model incorporating both stand and soil factors exhibited superior performance compared to the model that included only stand factors as predictors. Compared to Equation (13), the adjusted R^2 value improved (16.0% for γ), while the RMSE decreased (3.5% for β and 10.4% for γ). Compared with the model not considering soil factors, the model considering soil factors increased the AIC for both parameter β and parameter γ. The absolute error index (E) for both the mean value and maximum value decreased with increasing soil factors in the four provinces. From the perspective of adjustable R^2, AIC, RMSE, E1, and E2 values, the inclusion of stand and soil factors in the tree diameter distribution model enhanced its performance compared to that of the model that did not consider soil factors. Figure 4 shows the observed and fitting distribution with only stand factors and with the inclusion of stand and soil factors for four plots from Beijing, Heibei, Inner Mongolia, and Shanxi. We also can obtain the result that the fitting distribution with the inclusion of stand and soil factors is improved. The soil-sensitive diameter distribution model is proven to be effective and accurate.

The regression analysis examining the relationship between the scale parameter and the quadratic mean diameter in cork demonstrated a significant proportion of the variability in the β parameter ($R^2 > 0.99$). This was observed in both the models that included soil factors and those that did not. These findings are in line with previous studies that have demonstrated the positive impact of using the mean square diameter as an input variable [34,35]. Additionally, these studies reported a strong correlation between the mean square diameter and the observed outcomes [35,36].

A significant proportion (55%) of the plots exhibited γ values exceeding 3.6, with a mean value of 3.98. The distribution of diameters exhibited a mound-shaped pattern, with the majority of plots displaying negative asymmetry. These findings align with the typical attributes of the forests in the designated study area, which are primarily composed of stands of similar ages. Negative asymmetry is commonly observed in these forests due to low thinning and/or low competition, as noted by García Güemes et al. (2002) [36].

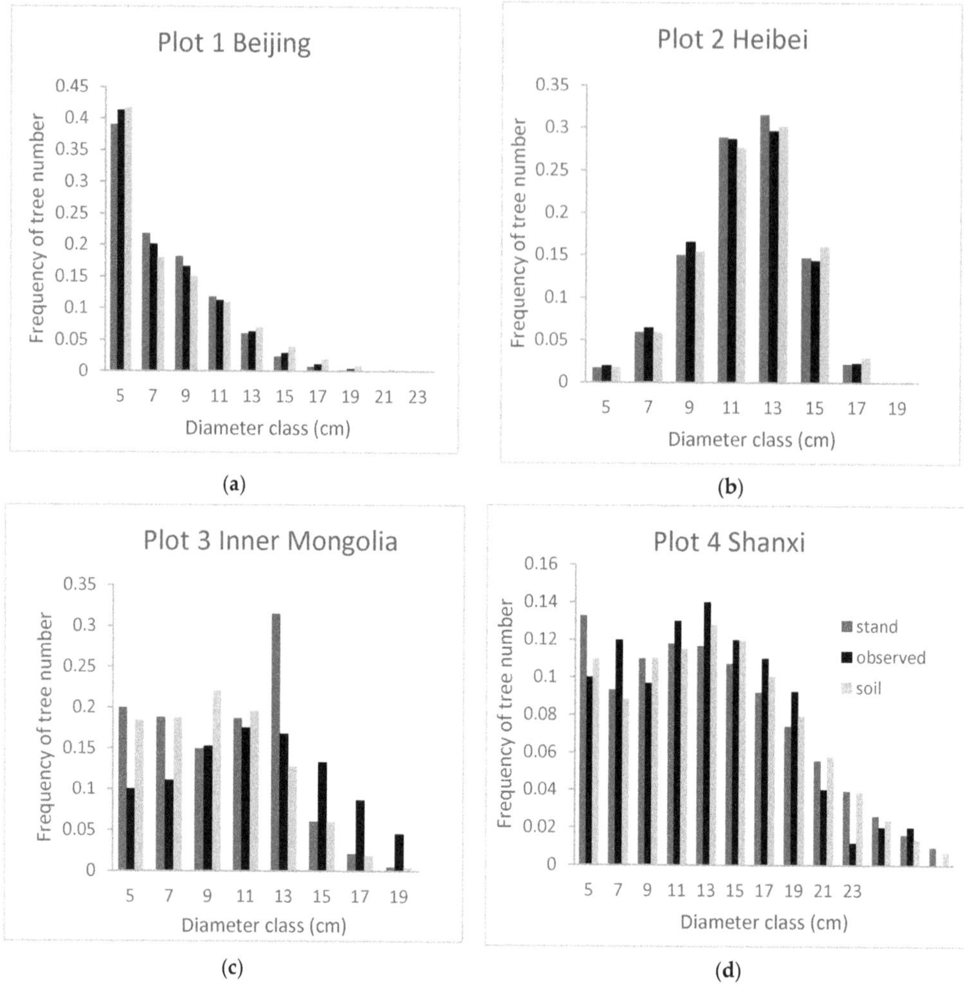

Figure 4. Observed and fitting diameter distribution in a plot: (**a**) Beijing, (**b**) Heibei, (**c**) Inner Mongolia, and (**d**) Shanxi.

4.2. Influence of Stand Factors and Soil Factors on the Parameters of the Weibull Distribution Model

Independent variables play a crucial role in influencing the parameters β and γ. Prior studies have highlighted the significance of various stand variables. The scale parameter β was effectively represented by the quadratic mean diameter under cork, resulting in a high adjusted R^2 value of 0.99. The shape parameter γ was estimated by incorporating the maximum diameter, minimum diameter, and plot elevation (R^2 adj = 0.40) [37]. Stand density is also a significant factor. The frequency and cumulative percentage of the diameter order distribution of the tree count suggest that low-density stands exhibit faster growth in diameter, with a greater number of trees in the larger diameter class. In this study, the scale parameter β was modeled based on the quadratic mean diameter and stand density, while the shape parameter γ was modeled using the maximum diameter, minimum diameter, and altitude. The results obtained from this analysis are in line with the findings reported in previous studies.

Tree growth has been found to be influenced by various soil factors, including topography, soil texture, soil moisture [38–43], and soil nutrient availability [44]. Soil, which serves as a crucial medium for tree growth, is subject to the influence of various soil-forming factors, such as topography, climatic variations, parent material, and living organisms (among others). The impact of nutrient variability on plant growth has been extensively investigated in grasslands, agricultural land systems, and certain forest ecosystems [44,45]. Numerous articles have indicated that soil properties exhibit a strong correlation with aboveground biomass, belowground biomass, and total basal area, suggesting a positive relationship between soil fertility and tree growth [46]. Although there is existing documentation on the impact of soil factors on tree growth, particularly in terms of tree diameter growth, there is limited knowledge regarding the effects of soil factors on tree diameter distribution. According to the Pr values of the soil factors, the minimum value was observed for AN. The results revealed the order of significance for the soil factors and β as AK, AP, AL, and PDEP. The results of the study revealed the order of significance for soil factors and γ as follows: AN, AK, AL, TN, and PDEP. The parameter γ exhibited significant correlations with AK, AL, and DPEP. The parameter β exhibited robust and statistically significant positive correlations with AP but strong and statistically significant negative correlations with AK. The parameter γ exhibited a robust and statistically significant positive correlation with AN. In general, the results of the study indicated that AK and AN are the most significant soil factors influencing the diameter distribution of *L. principis-rupprechtii* plantations.

4.3. Insufficient Research and Future Directions

Many techniques, including PM, MLE, the moment method (MOM), and hybrid methods, have been used in the field of dynamic prediction of stand diameter structure utilizing probability distributions to forecast these parameters in diverse tree species and forest types [47]. In regard to parameter estimation, using other techniques, such as MOM and the hybrid method, can produce very interesting outcomes.

Numerous studies have demonstrated that Weibull mixture distributions exhibit greater accuracy than the Weibull model. However, further research is required to develop robust analytical methods for Weibull mixture distributions. It is worth noting that this paper focuses solely on standard variables and soil factors, and it is possible that the prediction accuracy could be improved by incorporating climate factors.

5. Conclusions

PM is more suitable for predicting the parameters of the three-parameter Weibull diameter distribution function than the maximum likelihood estimation method (MLE) for the construction of *Larix principis-rupprechtii* plantations. Furthermore, compared to models that only consider stand factors, multiple regression models that consider the distribution parameters of both the stand and soil components are more accurate. The diameter distribution of *L. principis-rupprechtii* was found to be influenced primarily by stand density and soil composition, especially by AK and AN soil variables for both β and γ. One efficient way to estimate stand density, diameter distribution structure, and individual tree diameter is to use a three-parameter Weibull diameter distribution model that takes soil factors into account. As such, this model is a useful instrument for controlling the density and organization of *Larix principis-rupprechtii* plantations in China.

Author Contributions: We have a statement that X.L. had the idea for the article, H.G. performed the literature search and data analysis, and D.L. critically revised the work. All authors have read and agreed to the published version of the manuscript.

Funding: This research was supported by The National Key Research and Development Program of China (2023YFD220080302) and the National Natural Science Foundation of China (32271878).

Data Availability Statement: The datasets generated during the current study are available from the first author upon reasonable request.

Conflicts of Interest: The authors declare that they have no conflicts of interest.

References

1. Loetsch, F.; Zohrer, F.; Haller, K.E. *Forest Inventory*; BLV: Műnchen, Germany, 1973; p. 479.
2. Andrasev, S.; Bobinac, M.; Orlovic, S. Diameter structure models of Black Poplar selected clones in the section Aigeiros (Duby) obtained by the Weibull distribution. *Sumar. List* **2009**, *133*, 589–603.
3. Russell, M.B.; Burkhart, H.E.; Amateis, R.L.; Prisley, S.P. Regional Locale and Its Influence on the Prediction of Loblolly Pine Diameter Distributions. *South. J. Appl. For.* **2012**, *36*, 198–203. [CrossRef]

4. Burkhart, H.E.; Strub, M.R. A model for simulation of planted loblolly pine stands. In *Growth Models for Tree and Stand Simulation*; Fries, J., Ed.; Royal College of Forestry: Stockholm, Sweden, 1974; pp. 128–135.
5. Hafley, W.L.; Schreuder, H.T. Statistical distributions for fitting diameter and height data in even-aged stands. *Can. J. For. Res.* **1977**, *7*, 481–487. [CrossRef]
6. Guo, H.; Lei, X.; You, L.; Zeng, W.; Lang, P.; Lei, Y. Climate-sensitive diameter distribution models of larch plantations in north and northeast China. *For. Ecol. Manag.* **2022**, *506*, 119947. [CrossRef]
7. Urban, J.; Holušová, K.; Menšík, L.; Čermák, J.; Kantor, P. Tree allometry of Douglas fir and Norway spruce on a nutrient-poor and a nutrient-rich site. *Trees* **2013**, *27*, 97–110. [CrossRef]
8. Marshall, A.; Willcock, S.; Platts, P.; Lovett, J.; Balmford, A.; Burgess, N.; Latham, J.; Munishi, P.; Salter, R.; Shirima, D.; et al. Measuring and modelling above-ground carbon and tree allometry along a tropical elevation gradient. *Biol. Conserv.* **2012**, *154*, 20–33. [CrossRef]
9. Sanquetta, C.R.; Behling, A.; Corte, A.P.D.; Netto, S.P.; Rodrigues, A.L.; Simon, A.A. A Model Based on Environmental Factors for Diameter Distribution in Black Wattle in Brazil. *PLoS ONE* **2014**, *9*, e100093. [CrossRef]
10. Zang, H.; Lei, X.; Zeng, W. Height–diameter equations for larch plantations in northern and northeastern China: A comparison of the mixed-effects, quantile regression and generalized additive models. *For. Int. J. For. Res.* **2016**, *89*, 434–445. [CrossRef]
11. Lei, X.; Yu, L.; Hong, L. Climate-sensitive integrated stand growth model (CS-ISGM) of Changbai larch (*Larix olgensis*) plantations. *For. Ecol. Manag.* **2016**, *376*, 265–275. [CrossRef]
12. Fu, L.; Sun, W.; Wang, G. A climate-sensitive aboveground biomass model for three larch species in northeastern and northern China. *Trees-Struct. Funct.* **2017**, *31*, 557–573. [CrossRef]
13. He, X.; Lei, X.D.; Dong, L.H. How large is the difference in large-scale forest biomass estimations based on new climate-modified stand biomass models? *Ecol. Indic.* **2021**, *126*, 107569. [CrossRef]
14. Ali, A.; Yan, E.-R.; Chen, H.Y.H.; Chang, S.X.; Zhao, Y.-T.; Yang, X.-D.; Xu, M.-S. Stand structural diversity rather than species diversity enhances aboveground carbon storage in secondary subtropical forests in Eastern China. *Biogeosciences* **2016**, *13*, 4627–4635. [CrossRef]
15. Forrester, D.I. Linking forest growth with stand structure: Tree size inequality, tree growth or resource partitioning and the asymmetry of competition. *For. Ecol. Manag.* **2019**, *447*, 139–157. [CrossRef]
16. Ullah, F.; Gilani, H.; Sanaei, A.; Hussain, K.; Ali, A. Stand structure determines aboveground biomass across temperate forest types and species mixture along a local-scale elevational gradient. *For. Ecol. Manag.* **2021**, *486*, 118984. [CrossRef]
17. Sanaei, A.; Ali, A.; Yuan, Z.; Liu, S.; Lin, F.; Fang, S.; Ye, J.; Hao, Z.; Loreau, M.; Bai, E.; et al. Context-dependency of tree species diversity, trait composition and stand structural attributes regulate temperate forest multifunctionality. *Sci. Total Environ.* **2021**, *757*, 143724. [CrossRef] [PubMed]
18. Newton, P.F. Stand density management diagrams: Modelling approaches, variants, and exemplification of their potential utility in crop planning. *Can. J. For. Res.* **2021**, *51*, 236–256. [CrossRef]
19. Shangguan, W.; Dai, Y.; Duan, Q.; Liu, B.; Yuan, H. The global soil dataset for earth system modeling (2014). A Big Earth Data Platform for Three Poles. *J. Adv. Model. Earth Syst.* **2014**, *6*, 249–263. [CrossRef]
20. Bowling, E.H.; Burkhart, H.E.; Burk, T.E.; Beck, D.E.; Fortin, M.; Bédard, S.; DeBlois, J.; Meunier, S.; Packard, K.C.; Radtke, P.J.; et al. A stand-level multispecies growth model for Appalachian hardwoods. *Can. J. For. Res.* **1989**, *19*, 405–412. [CrossRef]
21. Zarnoch, S.J.; Feduccia, D.P.; Baldwin, V.C.; Dell, T.R. *Growth and Yield Predictions for Thinned and Unthinned Slash Pine Plantations on Cutover Sites in the West Gulf Region*; USDA Forest Service Southern Forest Experiment Station Research Paper SO1991; USDA: Washington, DC, USA, 1992; Volume 264.
22. Knowe, S.A. Basal Area and Diameter Distribution Models for Loblolly Pine Plantations with Hardwood Competition in the Piedmont and Upper Coastal Plain. *South. J. Appl. For.* **1992**, *16*, 93–98. [CrossRef]
23. Liu, C.; Zhang, S.Y.; Lei, Y.; Newton, P.F.; Zhang, L. Evaluation of three methods for predicting diameter distributions of black spruce (*Picea mariana*) plantations in central Canada. *Can. J. For. Res.* **2004**, *34*, 2424–2432. [CrossRef]
24. Smirnov, N. Table for Estimating the Goodness of Fit of Empirical Distributions. *Ann. Math. Stat.* **1948**, *19*, 279–281. [CrossRef]
25. Reynolds, M.R.; Burk, T.E.; Huang, W.-C. Goodness-of-fit tests and model selection procedures for diameter distribution models. *For. Sci.* **1988**, *34*, 373–399. [CrossRef]
26. Bailey, R.; Dell, T. Quantifying Diameter Distributions with the Weibull Function. *For. Sci.* **1973**, *19*, 97–104. [CrossRef]
27. Zellner, A. An efficient method of estimating seemingly unrelated regressions and tests for aggregation bias. *J. Am. Stat. Assoc.* **1962**, *57*, 348–368. [CrossRef]
28. Vu, D.H.; Muttaqi, K.M.; Agalgaonkar, A.P. A variance inflation factor and backward elimination based robust regression model for forecasting monthly electricity demand using climatic variables. *Appl. Energy* **2015**, *140*, 385–394. [CrossRef]
29. R Core Team. *R: A Language and Environment for Statistical Computing*; R Foundation for Statistical Computing: Vienna, Austria, 2021; Available online: https://www.R-project.org/ (accessed on 2 September 2024).
30. Henningsen, A.; Hamann, J.D. Systemfit: A Package for Estimating Systems of Simultaneous Equations in R. *J. Stat. Softw.* **2008**, *23*, 1–40. [CrossRef]
31. Gao, Z.; Wang, Q.; Hu, Z.; Luo, P.; Duan, G.; Sharma, R.P.; Ye, Q.; Gao, W.; Song, X.; Fu, L. Comparing independent climate-sensitive models of aboveground biomass and diameter growth with their compatible simultaneous model system for three larch species in China. *Int. J. Biomath.* **2019**, *12*, 19500530. [CrossRef]

32. Akaike, H. A new look at the statistical model identification. *IEEE Trans. Autom. Control* **1974**, *19*, 716–723. [CrossRef]
33. Newton, P.; Lei, Y.; Zhang, S. Stand-level diameter distribution yield model for black spruce plantations. *For. Ecol. Manag.* **2005**, *209*, 181–192. [CrossRef]
34. Kilkki, P.; Maltamo, M.; Mykkänen, R.; Päivinen, R. Use of the Weibull function in estimating the basal area dbh-distribution. *Silva Fenn.* **1989**, *23*, 311–318. [CrossRef]
35. Gorgoso, J.J.; Gonzalez, J.G.A.; Rojo, A.; Grandas-Arias, J.A. Modelling diameter distributions of Betula alba L. stands in northwest Spain with the two-parameter Weibull function. *For. Syst.* **2007**, *16*, 113–123. [CrossRef]
36. Güemes, C.G.; Cañadas, N.; Montero, G. Modelización de la distribución diamétrica de lasmasas de Pinus pinea L. de Valladolid (España) mediantela función de Weibull. *Investig. Agrar. Sist. Recur. For.* **2002**, *11*, 264–282.
37. Calzado, A.; Torres, E. Modelling diameter distributions of *Quercus suber* L. stands in "Los Alcornocales" Natural Park (Cádiz-Málaga, Spain) by using the two parameter Weibull functions. *For. Syst.* **2013**, *22*, 15–24. [CrossRef]
38. Boggess, W.R. Amount of throughfall and stemflow in a shortleaf pine plantation as related to rainfall in the open. *Trans. Ill. Acad. Sci.* **1956**, *48*, 55–61.
39. Bassett, J.R. Tree growth as affected by soil moisture availability. *Soil Sci. Soc. Am. J.* **1964**, *28*, 436–438. [CrossRef]
40. Hannah, P.R. Topography & soil relations for white and black oak in Southern Indiana. In *North Central Forest Experiment Station*; Forest Service; US Department of Agriculture: Washington, DC, USA, 1968.
41. Downs, J.M. Soil Water Regimes for Undisturbed Forest Communities in the Shawnee Hills, Southern Illinois. Ph.D. Thesis, Department of Forestry, Southern Illinois University, Carbondale, IL, USA, 1976.
42. Fralish, J.S. Forest site-community relationships in the Shawnee Hills region, southern Illinois. In Proceedings of the Central Hardwood Forest Conference, Proceedings of a Meeting, Carbondale, IL, USA, 17–19 October 1976; pp. 17–19.
43. Hills, S.; Fralish, J.S. The Effect of Site Environment on Forest Productivity in the Illinois. *Ecol. Appl.* **1994**, *4*, 134–143.
44. Mayoral, C.; van Breugel, M.; Turner, B.L.; Asner, G.P.; Vaughn, N.R.; Hall, J.S. Effect of microsite quality and species composition on tree growth: A semi-empirical modeling approach. *For. Ecol. Manag.* **2018**, *432*, 534–545. [CrossRef]
45. Lozano-García, B.; Parras-Alcántara, L.; Brevik, E.C. Impact of topographic aspect and vegetation (native and reforested areas) on soil organic carbon and nitrogen budgets in Mediterranean natural areas. *Sci. Total Environ.* **2016**, *544*, 963–970. [CrossRef]
46. Agbeshie, A.A.; Abugre, S. Soil properties and tree growth performance along a slope of a reclaimed land in the rain forest agroecological zone of Ghana. *Sci. Afr.* **2021**, *13*, e00951. [CrossRef]
47. Cao, Q.V. Predicting parameters of a Weibull function for modeling diameter distribution. *For. Sci.* **2003**, *50*, 682–685. [CrossRef]

forests

MDPI

Article

Optimizing Stand Spatial Structure at Different Development Stages in Mixed Hard Broadleaf Forests

Qi Sheng, Lingbo Dong and Zhaogang Liu *

Key Laboratory of Sustainable Forest Ecosystem Management—Ministry of Education, College of Forestry, Northeast Forestry University, Harbin 150040, China; harper@nefu.edu.cn (Q.S.); farrell0503@126.com (L.D.)
* Correspondence: lzg19700602@163.com; Tel.: +86-451-8219-0878

Abstract: Thinning plays a key role in regulating the stand spatial structure (SpS) to improve the development of stand quality, and the stand has different characteristics of stand structure (SS) at different growth and development stages (DSs), so it is most important to reasonably determine the stage of growth and development of the stand to optimize the stand structure. We applied the TWINSPAN two-way indicator species analysis method to classify the different development stages of mixed hard broadleaf forests. We provided a comprehensive stand spatial structure optimization model for three selected plots at different development stages, respectively, to optimize the SpS. The results demonstrated the classified DS of 29 mixed hard broadleaf plots for three forest stages: the establishment stage, competitive stage, and quality selection stage. We then applied the SpS optimization model to our three plots; the $Q(x)$ increased by 124.04%, 333.74%, and 116.83% when compared with those with no harvest, in which, upon the removal of 10% of the trees from the three plots, the maximum *RIP* values were all observed. Our results indicated that the SpS optimization model could regulate the SS for different growth stages and DSs.

Keywords: hard broadleaf forests; TWINSPAN classification; different development stages; spatial structure optimization; forest management

1. Introduction

The stage of forest growth and development is a guide to forest management. When the forest is in the primary stage, trees grow in groups or blocks at different densities, and the species composition is mostly dominated by pioneer species [1–3]. Forest management in the primary stage should aim to promote the growth of the forest and the formation of the microclimate environment within the forest [4,5]. There is an initial stage of tree growth, leading to an increase in stand biomass, stand density, and canopy cover; the initial stage is followed by an optimal stage, when dominant trees reach the upper levels, leading to tree maturity, characterized by an abundance of large trees during the maturity period; and the decline stage begins with the death of old trees, which reduces the number of old trees and increases the amount of dead tree, while gaps are formed, leading to regeneration in the gaps [6–8].

Vegetation age is the most straightforward method used for the identification of a stand's DS [9], but age alone is not a suitable forest stand parameter for use in predicting the different stand DSs because many factors influence the structural differences among sites within the same age class [8]. Instead, classification into DSs has been introduced for European primeval forests and is widely accepted as a surrogate for stand age [10,11]. Goodell et al. [12] used state forests in northern New York to classify different forest types into five structural stages, sapling, pole, mature, mosaic, and old-growth, by defining stand structural stage metrics (basal area weighted mean diameter at breast height (DBH); proportion of basal area of mature DBH > 60 cm) in conjunction with tree species composition and structural characteristics of the stand (stand density, diameter structure, and canopy density, etc.). In a study by Podlaski et al. [13], natural forests of classified fir (*Abies alba* Mill.) and beech (*Fagus sylvatica* L.) were divided into four stages based on changes in stand age distribution, SS, and volume. Zenner et al. [14], in their classification of the growth and DS of a primeval beech forest, found that the growth and DS dominated by regeneration and growth processes is characterized by steadily increasing biomass, which culminates in the optimum stage and then declines through the decay stage. According to this pattern of change, forests are divided into nine DSs: Gap (G), Regeneration (R), Establishment (E), Early Optimum (EO), Mid Optimum (MO), Late Optimum (LO), Terminal (T), Decay (D), and Plenter (P). Feldmann et al. [15] used temperate old-growth forests to propose a developmental stage index I_{SD} using two easily measurable stand structural parameters (stem density and basal area) and thus classified them into three stages, i.e., premature ($7 \leq$ DBH < 40 cm), mature ($40 \leq$ DBH < 70 cm), and over-mature (DBH ≥ 70 cm).

The SpS is an important factor influencing the growth of stands, and the growth of the stands will change the stands' structural characteristics [16]. Therefore, the study of SS is of important theoretical guidance and practical significance for forest management and optimal decision-making [17,18]. Many studies have proposed methods to quantify forest structure indicators, which can better describe forest structure. In general, SS can be divided into spatial and non-SpS; non-SpS measures are used to describe the average characteristics of a stand, do not require information on tree coordinates, and generally include diameter distribution, species composition, growth, and species diversity [19–21]. SpS measures describe the spatial distribution and characteristics of trees, including the degree of species separation in stands, competition between trees, distribution patterns of stands, and the vertical structure.

At present, the study of forest SpS is relatively mature, and how forests grow and develop under different management regimes is a sequence of spatial and temporal successions under both natural and perceived disturbances [22,23]. Forests at different stages of growing and developing respond to the same management measures in different ways; appropriate management measures can improve forest growth, seedling regeneration, etc.; on the contrary, irrational management can damage the forest structure, which will have a negative impact on tree growth and regeneration [24]. From the point of view of forest management, forest growth and development are processes of continuous change over time; different stages of forest growth and development with different forest structure characteristics and trees have different growth characteristics which require the adoption of different management plans [25].

In conclusion, it is practically important to formulate appropriate forest management planning according to the forest's different stages in scientific forest management. In this study, the following tasks were investigated: (1) the selection of the TWINSPAN two-way indicator species analysis method to classify the different DSs of mixed hard broadleaf forests, and (2) a proposed comprehensive SpS optimization model $Q(x)$, applied to three plots in different DSs, to determine the optimal cutting intensity in different DSs, which can effectively optimize the SpS.

2. Materials and Methods

2.1. Study Site

The study area is situated in the northern part of Tieli, located in the central region of Heilongjiang Province (46°28′40″–47°27′30″ N, 127°38′20″–129°24′10″ E), adjacent to the Lesser Khingan Mountains and bounded by the Songnen Plain to the west. The Lesser Khingan Mountains serve as a significant forestry industry base in China, boasting abundant forest vegetation with a forest cover of 72.6%. The eastern section of Yanan Forest Farm falls within the Changbai flora zone, characterized by over 2000 plant species and undisturbed forests exhibiting dense tree growth and ground coverage both above and below the forest canopy. The climate exhibits a temperate monsoon pattern with an annual minimum air temperature reaching −38.8 °C and an annual maximum air temperature peaking at 31.6 °C. Additionally, it experiences an average annual effective cumulative air temperature of 2685.8 °C along with approximately 2367.6 h of sunlight per year. Throughout all seasons, there are notable day-to-night temperature fluctuations observed on a wide annual scale basis. Snowfall accumulates to an average thickness of 31 cm each year while precipitation amounts reach approximately 714.4 mm annually. The frost-free period spans about 152 days, primarily concentrated during June–August due to warm and rainy summers, which provide favorable conditions for short-term forest growth (Figure 1). This area boasts ample broadleaved tree species, including *Quercus mongolica* Fisch, *Fraxinus mandshurica* Rupr., *Juglans mandshurica* Maxim., *Tilia tuan* Szyszyl., *Populus davidiana* Dode., *Acer pictum* Thunb., *Ulmus pumila* L., and *Betula platyphylla* Sukaczev. These trees play a crucial role in maintaining ecological balance throughout China.

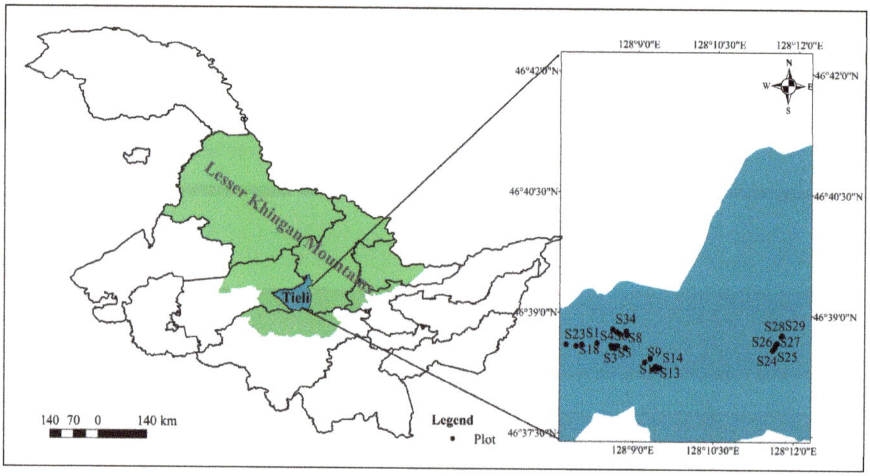

Figure 1. The locations of the studied forest in the Lesser Khingan Mountains in northeastern China and the distribution of the studied plots in the Yanan Forest Farm.

Study data were collected from 29 plots in mixed broadleaf forest surveyed at Yanan Forest Farm in 2023 (Table 1). The neighborhood grid method was used to divide each plot of 0.12 hm^2 into 10 m × 10 m grids, and all trees with a diameter at breast height (DBH) ≥ 5 cm were recorded for tree species, number, diameter at breast height (DBH), tree height (H), crown width, condition, and coordinates.

Table 1. Basic characteristics of the studied plots.

Plot	Mean Elevation (m)	Slope (°)	Mean DBH (cm)	Density (Tree/ha)	Number of Species
S1	321	<5	20.31	725	6
S2	341	<5	17.51	700	9
S3	338	<5	21.85	533	5
S4	451	<5	17.45	733	8
S5	339	<5	24.08	541	6
S6	378	<5	21.12	542	6
S8	296	<5	21.01	600	11
S9	343	<5	17.64	825	8
S10	329	<5	22.83	450	8
S11	331	<5	18.38	758	7
S12	271	<5	18.72	625	6
S13	252	<5	20.18	741	10
S14	337	<5	12.39	1266	10

Table 1. *Cont.*

Plot	Mean Elevation (m)	Slope (°)	Mean DBH (cm)	Density (Tree/ha)	Number of Species
S17	305	<5	20.57	625	6
S18	296	5~15	22.92	508	6
S23	323	<5	11.89	1383	6
S24	325	<5	18.26	641	10
S25	345	<5	21.47	500	10
S26	352	<5	19.94	616	9
S27	346	<5	22.62	566	9
S28	356	<5	17.26	775	8
S29	357	15~20	15.34	741	14
S30	376	5~15	18.08	958	10
S31	370	15~25	16.78	1050	10
S32	352	<5	17.68	775	11
S33	360	<5	17.44	991	11
S34	367	<5	16.65	750	10
S35	354	<5	14.99	1050	7
S36	354	<5	17.8	708	9

2.2. Classifying Development Stages of Mixed Hard Broadleaf Forests

We selected the TWINSPAN two-way indicator species analysis method to classify the different DSs of stands. The TWINSPAN two-way indicator species analysis method is modified based on indicator species analysis, which can complete the classification of samples and species at the same time [26]. The two-way indicator species analysis of the species plot matrix (TWINSPAN) method divides all samples and species into two main categories, 0 and 1, and then divides each of these two main categories into two categories until the required level of division is reached. The five levels of pseudo-species chosen per species were 0–2, 2–5, 5–10, 10–20, and >20, using the R software TwinspanR package 0.2.2 [27].

Our data from 29 plots were collated to group *Picea asperata*, *Abies fabri*, *Pinus koraiensis*, and *Larix gmelinii* as conifers, and others as soft broadleaf. Thus, 11 tree species (groups) were used for indicator species: *Fraxinus mandshurica*, *Juglans mandshurica*, *Phellodendron amurense*, *Acer pictum*, *Quercus mongolica* Fisch, *Tilia tuan*, *Ulmus pumila*, *Betula*, *Populus* L., *Conifers*, and other soft broadleafs. They were grouped into volume percentages by species to form a matrix (Equation (1)).

$$Y = \begin{pmatrix} B_{1\times m} \\ V_{n\times m} \end{pmatrix} \tag{1}$$

In the formula, $B_{1\times m}$ represent the $1\times m$ row vectors composed of tree species, $V_{n\times m}$ represents the $n\times m$ matrix composed of tree species volume, and n, m represent the number of plots and species trees, respectively.

To make a reasonable assessment of the DSs in mixed hard broadleafs, we should not only consider the stand structural condition but also the changes in the stand growth. We selected eight stand factors, namely average diameter at breast height (D_g), average height (H_t), per hectare tree number (N), basal area (BA), volume (V), percentage of three precious hard broadleaf species (*Fraxinus mandshurica*, *Juglans madshurica*, *Phellodendron amurensis*) (Yk), and diameter distribution (q), to calculate tree species diversity (H_1) and size diversity(H_2). H_1 and H_2 were calculated as follows.

$$H_1 = -\sum_{i=1}^{n} \frac{BA_i}{BA} ln(\frac{BA_i}{BA}) \tag{2}$$

$$H_2 = -\sum_{i=1}^{n} \frac{BA_j}{BA} ln(\frac{BA_j}{BA}) \tag{3}$$

In the formula, BA_i and BA_j represent the basal area of the trees of species group i and diameter class j, respectively.

2.3. Stand Spatial Structure Optimization

2.3.1. SS Parameters Selection

We selected a 3 m buffer area to mitigate edge effects. This implies that trees within the core area were considered as reference trees, and their corresponding parameters were calculated, while other trees in the buffer area were regarded as neighbor trees. Six indices were used to analyze the SpS: M_c reflects stand species diversity by measuring the dissimilarity of species between the reference tree and its four nearest neighbors; W indicates the horizontal distribution pattern of both the reference tree and its four nearest neighbors; U represents the proportion of neighboring trees larger than the reference tree among all neighboring trees; S is used as an indicator to describe the vertical SpS of the forest, referring to forest vertical structure diversity, in which the stand layer index is in accordance with the International Union of Forestry Research Organization's (IUFRO) standard for vertical stratification, which is based on the stand dominant height divided into three vertical layers, with a tree height less than or equal to 1/3 of the dominant height as the lower layer, between 1/3 and 2/3 dominant height as the middle layer, and greater than or equal to 2/3 dominant height as the upper layer; CI refers to the competitive status between the reference tree and its four nearest neighbors, which is based on diameter and distance; and K is a spatial indicator describing the level of light in a stand and is used to assess the adequacy of the space for individual tree growth [28–30] (Table 2).

Table 2. SpS parameters and index definition.

Index	Formula	Definition
Complete mingling index (M_c)	$M_{c_i} = \frac{1}{2}\left(D_i + \frac{C_i}{n_i}\right) \cdot M_i$	Where C_i is the number of different tree species in paired neighbours; C_i/n_i is the degree of isolation of the nearest neighbour to the reference tree; n_i is the number of nearest neighboring trees; D_i is the Simpson index of the SpS unit i, $D_i = 1 - \sum_{j=1}^{S_i} p_j^2$; S_i is the tree species of the SpS unit I; p_j is the proportion of trees of the jth species; M_i is a simple mingling index, $M_i = \frac{1}{n}\sum_{j=1}^{n} v_{ij}$, where $v_{ij} = 0$, if reference tree i and its neighbor tree j are the same white tree species, otherwise, $v_{ij} = 1$
Uniform angle index (W)	$W_i = \frac{1}{n}\sum\limits_{j=1}^{n} w_{ij}$	Where $w_{ij} = \begin{Bmatrix} 1; if\ \alpha_j < \alpha_0 = 72^\circ \\ 0; otherwise \end{Bmatrix}$; and α is the angle between two adjacent trees
Dominance index (U)	$U_i = \frac{1}{n}\sum\limits_{j=1}^{n} z_{ij}$	where z_{ij} takes the value 1 if the jth neighbor (d_j) is smaller than the reference tree $i(d_i)$, and the value 0, otherwise, $z_{ij} = \{1; d_j < d_i\}$
Stand layer index (S)	$S_i = \frac{1}{n}\sum\limits_{j=1}^{n} S_{ij}$	where $S_{ij} = 0$, indicating that the neighbour tree and the reference tree are not in the same vertical layer; otherwise, $S_{ij} = 1$
Hegyi competition index (CI)	$CI_i = \sum\limits_{i=1}^{n_i} \frac{d_j}{d_i \cdot L_{ij}}$	where d_i is the diameter of the reference tree, d_j is the diameter of the neighbouring trees, and L_{ij} is the Euclidean distance between them
Open degree (K)	$K_i = \frac{1}{n}\sum\limits_{j=1}^{n_i} \frac{D_{ij}}{H_{ij}}$	where H_j is the height of the neighboring tree j, and D_{ij} is the distance between them

2.3.2. SpS Optimization Model

The horizontal distribution, competition between trees, and tree height and growth space can be characterized using M_c, W, U, S, CI, and K. The quantitative SpS indices mentioned above can be quickly assessed and have clear biological significance. The determination of the optimal trees for harvesting can be achieved by evaluating the relationship between each reference tree and its four nearest reference trees [31]. However, it should be noted that these indices exhibit significant variations in both magnitude and measuring units. Thus, in this study, the SpS optimization model $Q(x)$ was proposed based on the multi-objective programming method. The maximum value $Q(x)$ indicates that the optimal SpS is closer to the natural and ideal spatial distribution; it shows that the tree species have the maximum mixing degree, minimal competition between trees, more adequate light conditions and space for tree growth, and the most stand vertical structure diversity.

The model is as follows:

$$Q(x) = \frac{\frac{1+Mc(x)}{\sigma_{M_c}} \cdot \frac{1+S(x)}{\sigma_S} \cdot \frac{1+K(x)}{\sigma_K}}{[1+CI(x)\cdot\sigma_C]\cdot[1+U(x)\cdot\sigma_U]\cdot[1+|W(x)-0.475|+|W(x)-0.517|]\cdot\sigma_W} \tag{4}$$

$$N = N_0 \tag{5}$$

$$D = D_0 \tag{6}$$

$$1.2 \leq q \leq 1.7 \tag{7}$$

$$M_c \geq M_{c0} \tag{8}$$

$$0.475 \leq W \leq 0.517 \tag{9}$$

$$U \leq U_0 \tag{10}$$

$$S \geq S_0 \tag{11}$$

$$CI \leq CI_0 \tag{12}$$

$$K \geq K_0 \tag{13}$$

where the q-value is a measure of diameter distribution commonly used in forest inventory management in the northeast of China, where a negative exponential function is used to fit the relationship between tree number and diameter class with the expression $q = exp(a \cdot d)$, where a is the estimated parameter and d is the diameter class. The q-value of natural heterogeneous forests varies between 1.2 and 1.7, which is a reasonable range for transformation [32,33]. N and N_0 are the numbers of diameter classes before and after optimization, and D and D_0 are the tree species before and after optimization, respectively. M_{c0}, U_0, S_0, CI_0, K_0, M_c, U, S, CI, K are the values of the complete mixing index, dominance index, stand layer index, Hegyi competition index, and open degree before and after optimization, respectively. W is the uniform angle index; when the distribution pattern of the trees is random, the W values range from 0.475 to 0.517 [34]. σ_{M_c}, σ_U, σ_w, σ_S, σ_{CI}, and σ_K are the standard deviations of the complete mixing index, uniform angle index, dominance index, stand layer index, Hegyi competition index, and open degree.

Equation (4) is the objective function for optimizing SpS. As the q-value increases, the existing SS approaches the optimum SS. Equation (5) is diameter distribution constraint. Equation (6) is the tree species constraint. Equation (7) is the q-value constraint which controls the optimized stand q-values between 1.2 and 1.7. Equations (8)–(13) are the SpS constraints.

2.3.3. Optimization Simulation

In this study, we selected 3 plots (S31, S23, and S28) at different DSs (forest establishment stage, competitive growth stage, and quality selection stage), respectively, to optimize the stand SpS. Since the model contains a large number of integer variables, solving it by the exhaustive method takes a lot of time. Therefore, it is acceptable to use the Monte Carlo algorithm for the simulation of interlopes to obtain a suboptimal solution to the problem [35]. The Monte Carlo algorithm is based on the principle of random sampling. It uses a random number function provided to perform a fast random sampling of feasible points in the constrained optimization problem.

Therefore, the tree removal intensity is set to 10%, 20%, and 30%, and the initial number of simulations is set to 5000. All objective function values that satisfy the constraints are defined as valid simulations, and the simulation with the largest objective function value is identified among these simulations. We consider this thinning simulation to be an approximate solution to the optimal solution of this problem.

3. Results

3.1. Classifying Development Stages of Mixed Hard Broadleaf Forests

3.1.1. Twinspan Classification Results

Figure 2 shows the 29 plots of the TWINSPAN classification process and their results. The 29 mixed broadleaf plots were divided into three categories by two divisions. With the exclusion of the misclassified plots and the combination of the information on the species composition of the plots, the stand characteristics of each class were as follows: there were 10 plots accounting for 34.5% in Type I, and the main tree species are *Fraxinus mandshurica, Juglans madshurica,* and *Phellodendron amurensis,* with volume composition ratios of 24.9%, 32.4%, and 15.6%, respectively; there were 14 plots accounting for 48.3% in Type II, where the main tree species are *Juglans madshurica, Ulmus pumila,* and *Fraxinus mandshurica* with volume composition ratios of 26.8%, 22.8%, and 16.1%, respectively; there were 5 plots accounting for 17.2% in Type III, where the main tree species are *Fraxinus mandshurica, Juglans madshurica, Phellodendron amurensis,* and *Ulmus pumila* with a total volume composition of 47.1%, in which the three precious hard broadleaf species (*Fraxinus mandshurica, Juglans madshurica, Phellodendron amurensis*) accounted for 29.9%, and *Ulmus pumila* accounted for 17.2%.

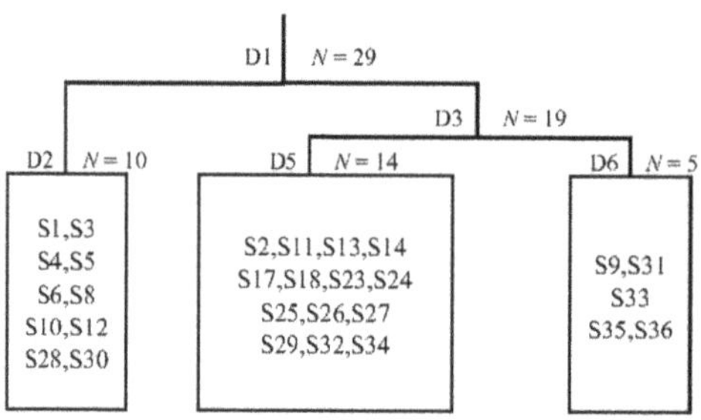

Figure 2. Plot TWINSPAN classification results.

3.1.2. Stand Growth Analysis for TWINSPAN Classification

Figure 3 shows the 29 plots' stand growth index change for TWINSPAN classification, where dbh, H_t, base area, number, v, q, H_1 and H_2 were not significant in different classifications. YK was significant in different classifications ($p < 0.05$); the dbh, H_t, base area, v, and YK values are sorted by Type I > Type II > Type III. The number and q values are sorted by Type I < Type II < Type III; Type II has the greatest tree species diversity (H_1) and the least size diversity (H_2), and Type III has the greatest size diversity (H_2) and the least tree species diversity (H_1). Therefore, Type III is in the early stage of development and Type I is in the late stage of development. As the stand grows and develops, the pioneer tree species gradually decline and tree species diversity (H_1) decreases, and tree size richness increases. Taking a comprehensive view of the tree species composition, SS, and growth, our study classified the DSs of 29 mixed hard broadleaf plots by Type III (forest establishment stage)—Type II (competitive growth stage)—Type I (quality selection stage).

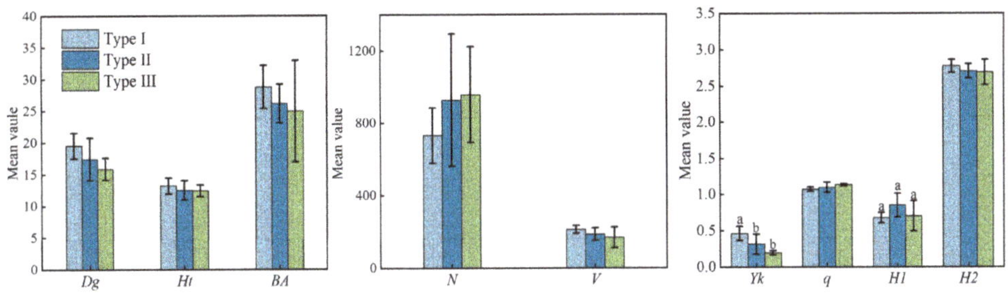

Figure 3. The stand factors of mixed hard broadleaf forest DS classification. (a b indicates level of significance in different classifications).

3.2. Stand Spatial Structure Optimization

Figure 4 shows the distribution of the number of plants in diameter steps at different growth and development stages. At the forest quality selection stage, the diameter distribution was the widest distribution, which was between 4 and 64 cm (S28). In S31 and S23, the diameter range was 4–40 cm, with obvious differences in diameter class distribution. Plot S31 and Plot S28 had 64% and 61% of the number of the total trees in the 4–16 diameter classes, respectively; Plot S23 had the largest number of trees in the 4–16 diameter classes, with 87% of the total number. In all three phases, the number of 4–24 cm diameter trees accounted for 95%, and the number of large-diameter trees accounted for about 2% of the total number.

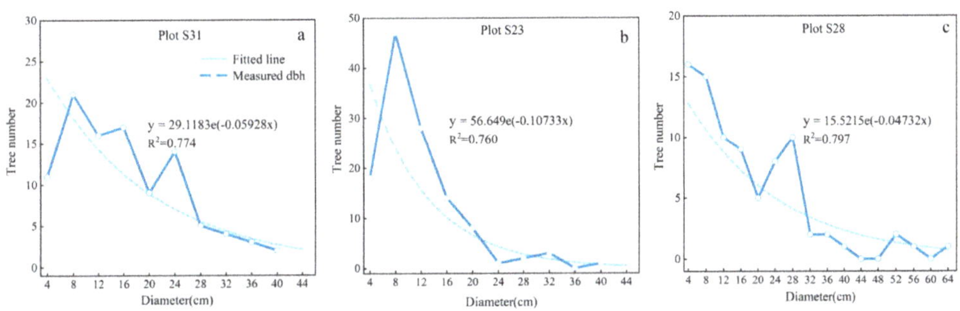

Figure 4. Diameter classes of the three plots from different DSs. ((**a**–**c**) represent Plot S31, Plot S23 and Plot S28).

In Plot S31, 8–24 cm trees were abundant, forming an asymmetric, left-sloping, multi-peaked curve. After the last peak in the 24 cm diameter class, the number of trees started to decrease with increasing diameter classes; Plot S23 showed that the diameter distribution of the unevenly aged stands is characterized by a single-peak inverted 'J' curve that peaks at an 8 cm diameter, it indicated that there were a lot of young trees and seedlings, and it could be continuously recruited, while large-diameter trees were few and had many deficiencies; in general, the number of trees in the stand decreased sharply as the diameter class increased from 6 cm upwards. Plot S28 was a multi-peaked mountainous curve with peaks at 8 cm and 28 cm, in which the middle- and small-diameter classes had more trees and large-diameter classes had fewer trees. When the diameter class was 20–28 cm, the number of trees increased, and the number of 8–20 cm and 28 cm trees decreased. When analyzed with the q-value, Plot S31 (1.267) had a slightly lower q-value than Plot S23 (1.536) and Plot S28 (1.227); however, Plot S28 was better fitted to the diameter of InN (R^2 = 0.797) than Plot S31 (R^2 = 0.774) and Plot S23 (R^2 = 0.760).

Table 3 shows that there is a certain difference in SS index optimization before and after different DS stands. It shows the statistical features of different spatial parameters for three different DS stands with 0%, 10%, 20%, and 30% of trees removed, respectively. The M_c value of the S23 plot is 0.400, which belongs to the medium mixing degree, the M_c value of the S28 plot is 0.455, which belongs to the medium mixing degree, and the M_c value of the S31 plot is greater than 0.5, which belongs to the strong mixing degree, greater than the S23 and S28 plots.

Table 3. The different SS-optimized variables for the three plots.

Plot	Intensity	Number (Tree)	M_c	S	CI	W	U	q	$Q(x)$	RIP
S31	0%	126	0.601	0.573	2.610	0.526	0.497	1.267	178.425	
	10%	12	0.661	0.603	2.220	0.516	0.500	1.260	607.397	124.04
	20%	24	0.685	0.631	2.124	0.497	0.500	1.208	677.869	11.60
	30%	36	0.667	0.618	1.946	0.503	0.506	1.339	733.350	8.18
S23	0%	166	0.400	0.386	3.043	0.523	0.486	1.536	57.312	
	10%	16	0.439	0.556	2.275	0.500	0.479	1.500	248.586	333.74
	20%	32	0.455	0.530	2.042	0.510	0.474	1.507	395.122	58.94
	30%	48	0.503	0.540	1.823	0.491	0.479	1.449	530.569	34.27
S28	0%	92	0.455	0.486	2.372	0.533	0.515	1.227	172.444	
	10%	10	0.469	0.593	2.096	0.513	0.513	1.212	373.914	116.83
	20%	20	0.465	0.598	1.969	0.511	0.511	1.228	407.176	8.89
	30%	30	0.470	0.560	1.972	0.5000	0.495	1.200	309.324	−24.03

The average U value of Plots S23, S28, and S31 are 0.484, 0.497, and 0.515, respectively, indicating that the stands are in homogeneous condition. The average W value of the different DS stands ranges from 0.523 to 0.533, which is greater than 0.517, indicating that the stands all have significant agglomeration distribution patterns. The average S values of Plots S23, S28, and S31 are 0.368, 0.468, and 0.573, respectively, indicating that Plot S31 has the highest vertical structural diversity. The average CI value (3.043) of Plot S23 reflects that its tree competition pressure is higher than that of Plots S31 (2.372) and S28 (2.610). The q-value of Plots S23, S28, and S31 are 1.536, 1.227, and 1.267, respectively, which are all significantly within the range of 1.2 and 1.7. The $Q(x)$ index is also a logical negative index; the $Q(x)$ values for plots S23, S28, and S31 are 57.312, 172.444, and 178.425, respectively; therefore, the lowest $Q(x)$ value was found in Plot S23.

Table 3 shows the changes in different DS stand structure indices with the increase in thinning intensity. When thinning intensity increases by 1%, the M_c, S, CI, W, U, and q in Plot S31 all increase, respectively; 0.099, 0.0523, 0.1756, 0.0193, 0.006, 0.0055. The trend of the increase or decrease in the SpS index before and after optimization is the same at different stages of the stands, but the magnitude of change is different. With the increase in thinning intensity, the spatial indices of M_c and S increased significantly, while CI decreased. The value of $Q(x)$ of the stand spatial structure optimization function in different stages increased with the increase in cutting intensity, and $Q(x)$ increased by 124.04, 11.60, and 8.18 for the S31 plot, respectively. The $Q(x)$ value increased with the optimized thinning intensity, which is equivalent to that when the thinning intensity increased by 1%, and the $Q(x)$ of the stand after different DS optimizations increased by 124.04%, 333.74%, and 116.83%, respectively. This result also shows that the relative increase ratio (RIP) of the optimization function $Q(x)$ between any two consecutive thinning intensifications decreases significantly as the selective thinning intensification increases, suggesting that the removal of 10% of trees may be the optimal intensity from the perspective of optimizing the forest SpS in all three plots.

For mixed hard broadleaf forests, the harvest possibilities of the three precious hard broadleaf species (*Fraxinus mandshurica*, *Juglans madshurica*, *Phellodendron amurensis*) were significantly higher than those of other species, and higher than in species when removing 30% of the trees, and more than twice as high as species when 10% of the trees were removed. The harvest of *Ulmus pumila* accounted for approximately 7.4%, *Tilia tuan* for approximately 14.2%, *Acer pictum* Thunb. for approximately 6.7%, and *Fraxinus mandshuric* for approximately 7.6% of the lower harvesting intensity in Plot S31; Figures 5–7 show the spatial distribution patterns of trees cut after the spatial optimization of different DS stands when the optimal cutting intensity is 10%, in which SpS is improved, reducing the competitive pressure among trees, adjusting the distribution pattern of trees, and providing space for tree growth.

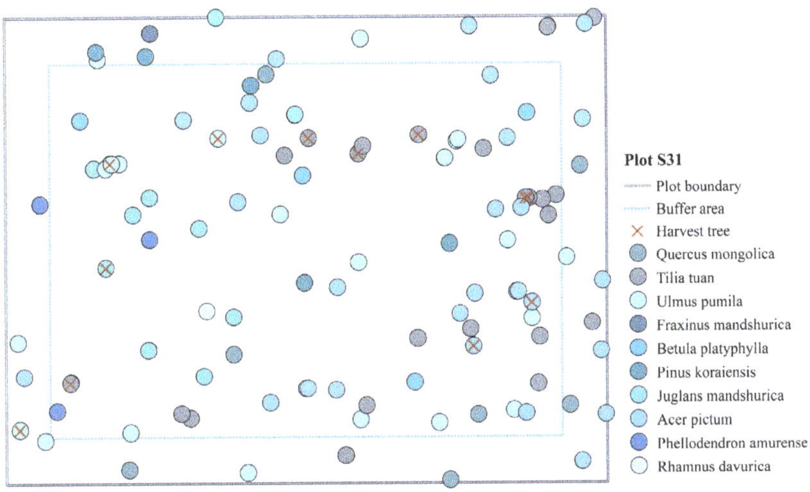

Figure 5. When the cutting intensity is 10%, the colors of the circles in Plot S31 indicate different tree species. × red indicates the distribution pattern of optimized harvested trees.

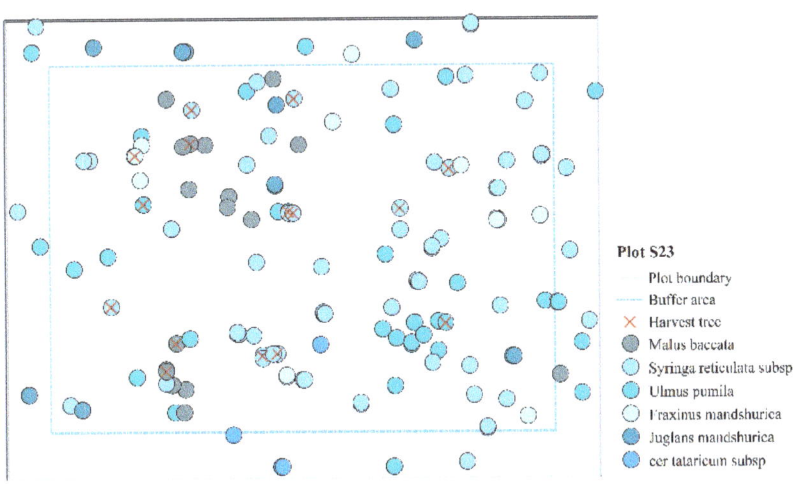

Figure 6. When the cutting intensity is 10%, the colors of the circles in Plot S23 indicate different tree species. × red indicates the distribution pattern of optimized harvested trees.

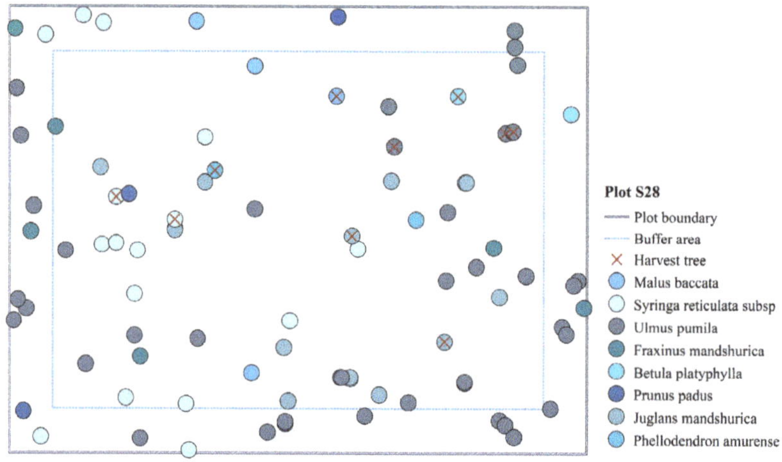

Figure 7. When the cutting intensity is 10%, the colors of the circles in Plot S28 indicate different tree species. × red indicates the distribution pattern of optimized harvested trees.

4. Discussion

Stand growth and DSs reflect natural and anthropogenic disturbances, and stands are usually located in each stage based on characteristics such as stand age, structure, and function. Their growth and DS are identified by the vertical and horizontal structure of the stand, canopy gaps, volume, number of dead trees, regeneration, and species composition of the stand [2,36]. Depending on habitat and site conditions, SS and the length of developmental stages vary considerably. Typically, forests have three main important developmental stages, including the primary stage (regeneration and growth), the optimum stage (maturity and senescence), and the decay stage (decomposition); each growth and development stage has processes and characteristics that create specific conditions for species response and the number of standing trees at different growth stages [3,37]. Stand types, tree ages, and stand densities vary, affecting changes in forest growth, biodiversity, soil fertility, soil nutrient levels, and plant nutrient uptake, which in turn affect forest productivity, diversity, and stability [38,39].

We selected the TWINSPAN two-way indicator species analysis method to classify the different DSs of stands. The obtained results indicated that by taking a comprehensive analysis of the tree species composition, SS, and growth through eight stand indicators (average diameter at breast height (D_g), average height (H_t), per hectare tree number(N), basal area (BA), volume (V), percentage of three precious hard broadleaf species (*Fraxinus mandshurica, Juglans madshurica, Phellodendron amurensis*) (Yk)), along with tree species diversity (H_1) and size diversity(H_2), our study classified the DSs of 29 mixed hard broadleaf plots by Type III (forest establishment stage)—Type II (competitive growth stage)—Type I (quality selection stage). In the process of forest growth and development, the pioneer species first occupied the main forest layer, and with the development of community succession, the pioneer species gradually declined and decreased, and the companion species invaded and then gradually occupied the main forest layer in the later stage of top pole species [40]. The broadleaved *Korean pine* forest is a typical horizontal zonal overstory community in the Lesser Khingan Mountains, in which the coniferous species are dominated by *Korean pine*, and some of the coniferous species, such as *Picea jezoensis*, *Picea koraiensis Nakai*, and *Abies nephrolepis*, etc., are accompanied by coniferous species, and the broadleaved species are mainly *Tilia amurensis*, *Acer pictum* Thunb., *Quercus mongolica* Fisch, *Betula costata* Trautv., *Ulmus pumila*, and three precious hard broadleaf species (*Fraxinus mandshurica, Juglans madshurica, Phellodendron amurensis*); due to differences in soil conditions, the succession process of secondary forests resulting from the disturbance of the broadleaved *Korean pine* forest will vary [41,42]. In the absence of the *Korean pine* seed source, the trees in the community will die and regenerate over many generations, and the succession will converge in the intermediate community consisting of coniferous, hard broadleaved, and soft broadleaved mixed forests composed of different species [43]. The intermediate stages in the development of community succession are the three stages of growth and development in this study.

Stand structure characteristics play an important factor in forest management optimization and evaluation, reflecting the tree's growth and basic characteristics in the stand. Different species compositions, diameter class distribution tree distribution patterns, and arrangements lead to different structural characteristics of DS stands [44,45]. SS is usually described in two directions, horizontal and vertical structure. Differences in the horizontal and vertical structure of different stands have implications for tree and understory plant growth, regeneration, biodiversity, stand ecological function, and wildlife habitat [46,47]. Therefore, how can we adjust the rationality and diversity of SSs through the stand optimization function, especially in the optimization of different DSs [23,48,49]? This study presented an SpS optimization model, which was applied to three mixed hard broadleaf stands of different DSs. Since the three plots were extracted from different DSs, there were obvious differences between their spatial and nonspatial structural characteristics. Differences between the three plots were also evident in the number of tree species and the distribution of diameter classes, which are traditional indicators of forest structure.

The $Q(x)$ value all increased and could effectively improve the SpS significantly. The $Q(x)$ increased by 124.01%, 333.74%, 116.83% compared with those with no harvest, the *RIP* decreased with the increase in thinning intensity in stands at different DSs. By comparing different optimal cutting intensities, it was found that the *RIP* value is the largest when the optimal cutting intensity is 10%. This means that improving SS characteristics is most effective when the optimization intensity is low at 10%. The State Forestry Bureau of China's forest management policies and regulations also coincide with this intensity [50]. Low-intensity harvest is the best optimization intensity from the perspective of SS optimization, though optimal harvesting intensity varies with the different optimization objectives (carbon sequestration and stocks, soil physical and chemical properties) [51,52].

The allocated harvest within this intensity was mainly focused on the trees with high competition, low health, and clumped distribution, while only a handful of trees were harvested to adjust to neighborhood-based structures (Figures 5–7). In SBFM management, the maximum thinning intensity is selected by considering the relationship between the central and adjacent trees in the structural unit, so as to adjust spatial distribution and species diversity [53,54]. Overall, the stand structure parameters were optimized by comparison. This optimized the post-stand mingling degree, increasing the uniform angle index, reduced competitive pressure among trees, and decreased densities. The change in the spatial distribution pattern of the stand can increase the light conditions, reduce the competition between neighboring trees, and promote the growth of regenerated trees.

5. Conclusions

We propose that the TWINSPAN two-way indicator species combination with stand growth indicators can more accurately classify the DS of mixed hard broadleaf forests. Meanwhile, we propose a stand SpS optimization model for three elected plots. Optimizing the SpS improved the $Q(x)$ by 124.04%, 333.74%, and 116.83%, and when the *RIP* reached the maximum value, it indicated that the cutting intensity was at the optimal intensity, and 10% is the optimal cutting intensity in this study. Moreover, this optimization model can be applied to the optimal management of stands at different stages to guide the optimization of natural forest structures under different growth and DSs.

Author Contributions: Conceptualization, L.D. and Z.L.; Methodology, Q.S. and L.D.; Software, Q.S.; Formal analysis, Q.S.; Data curation, Q.S.; Writing—original draft, Q.S.; Writing—review & editing, L.D. and Z.L.; Supervision, Z.L.; Project administration, Z.L.; Funding acquisition, Z.L. All authors have read and agreed to the published version of the manuscript.

Funding: This research was funded by the National Key Research and Development Program of China grant number [2022YFD2200502].

Data Availability Statement: Data is contained within the article.

Conflicts of Interest: The authors declare no conflict of interest.

References

1. Král, K.; McMahon, S.M.; Janík, D.; Adam, D.; Vrška, T. Patch mosaic of developmental stages in central European natural forests along vegetation gradient. *For. Ecol. Manag.* **2014**, *330*, 17–28. [CrossRef]
2. Gong, Z.; Kang, X.; Gu, L. Quantitative division of succession and spatial patterns among different stand developmental stages in Changbai Mountains. *J. Mt. Sci.* **2019**, *16*, 2063–2078. [CrossRef]
3. Dai, L.; Qi, L.; Wang, Q. Changes in forest structure and composition on Changbai Mountain in Northeast China. *Ann. For. Sci.* **2011**, *68*, 889–897. [CrossRef]
4. Collins, R.J.; Carson, W.P. The effects of environment and life stage on Quercus abundance in the eastern deciduous forest, USA: Are sapling densities most responsive to environmental gradients? *For. Ecol. Manag.* **2004**, *201*, 241–258. [CrossRef]
5. Hao, Z.; Zhang, J.; Song, B.; Ye, J.; Li, B. Vertical structure and spatial associations of dominant tree species in an old-growth temperate forest. *For. Ecol. Manag.* **2007**, *252*, 1–11. [CrossRef]
6. Král, K.; Vrška, T.; Hort, L.; Adam, D.; Šamonil, P. Developmental phases in a temperate natural spruce-fir-beech forest: Determination by a supervised classification method. *Eur. J. For. Res.* **2009**, *129*, 339–351. [CrossRef]
7. Glatthorn, J.; Feldmann, E.; Tabaku, V.; Leuschner, C.; Meyer, P. Classifying development stages of primeval European beech forests: Is clustering a useful tool? *BMC Ecol.* **2018**, *18*, 47. [CrossRef]
8. Huber, M.O. Statistical models to identify stand development stages by means of stand characteristics. *Can. J. For. Res.* **2011**, *41*, 111–123. [CrossRef]
9. Uhl, C.; Buschbacher, R.; Serrao, E.A.S. Abandoned Pastures in Eastern Amazonia. I. Patterns of Plant Succession. *J. Ecol.* **1988**, *76*, 663. [CrossRef]
10. Leibundgut, H. Uber Zweck und Methodik der Strukturund Zuwachsanalyse von Urw€aldern. *Schweiz. Z. Forstwes.* **1959**, *110*, 111–124.
11. Leibundgut, H. Uber die Dynamik europaischer Urwalder. *Allg. Forst Z.* **1978**, *33*, 686–690.
12. Goodell, L.; Faber-Langendoen, D. Development of stand structural stage indices to characterize forest condition in Upstate New York. *For. Ecol. Manag.* **2007**, *249*, 158–170. [CrossRef]
13. Podlaski, R. Suitability of the selected statistical distributions for fitting diameter data in distinguished development stages and phases of near-natural mixed forests in the Świętokrzyski National Park (Poland). *For. Ecol. Manag.* **2006**, *236*, 393–402. [CrossRef]
14. Zenner, E.K.; Peck, J.E.; Hobi, M.L.; Commarmot, B. Validation of a classification protocol: Meeting the prospect requirement and ensuring distinctiveness when assigning forest development phases. *Appl. Veg. Sci.* **2016**, *19*, 541–552. [CrossRef]
15. Feldmann, E.; Glatthorn, J.; Hauck, M.; Leuschner, C. A novel empirical approach for determining the extension of forest development stages in temperate old-growth forests. *Eur. J. For. Res.* **2018**, *137*, 321–335. [CrossRef]
16. Gao, T.; Hedblom, M.; Emilsson, T.; Nielsen, A.B. The role of forest stand structure as biodiversity indicator. *For. Ecol. Manag.* **2014**, *330*, 82–93. [CrossRef]
17. Ali, A. Forest stand structure and functioning: Current knowledge and future challenges. *Ecol. Indic.* **2019**, *98*, 665–677. [CrossRef]
18. Baran, J.; Pielech, R.; Kauzal, P.; Kukla, W.; Bodziarczyk, J. Influence of forest management on stand structure in ravine forests. *For. Ecol. Manag.* **2020**, *463*, 118018. [CrossRef]
19. Hui, G.; Zhang, G.; Zhao, Z.; Yang, A. Methods of Forest Structure Research: A Review. *Curr. For. Rep.* **2019**, *5*, 142–154. [CrossRef]
20. Aguirre, O.; Hui, G.; von Gadow, K.; Jiménez, J. An analysis of spatial forest structure using neighbourhood-based variables. *For. Ecol. Manag.* **2003**, *183*, 137–145. [CrossRef]
21. Gadow, K.V.; Zhang, C.Y.; Wehenkel, C.; Pommerening, A.; Corral-Rivas, J.; Korol, M.; Myklush, S.; Hui, G.Y.; Kiviste, A.; Zhao, X.H. Forest Structure and Diversity. In *Continuous Cover Forestry*; Springer: Dordrecht, The Netherlands, 2011; pp. 29–83. [CrossRef]
22. Pastorella, F.; Paletto, A. Stand structure indices as tools to support forest management: An application in Trentino forests (Italy). *J. For. Sci.* **2013**, *59*, 159–168. [CrossRef]
23. Ye, S.; Zheng, Z.; Diao, Z.; Ding, G.; Bao, Y.; Liu, Y.; Gao, G. Effects of Thinning on the Spatial Structure of *Larix principis-rupprechtii* Plantation. *Sustainability* **2018**, *10*, 1250. [CrossRef]
24. Lopes, J.F.B.; Lopes, F.B.; Araújo, I.C.D.S.; Pereira, E.C.B.; Brandão, M.L.S.M.; Feitosa, E.D.O.; Luna, N.R.D.S.; Sousa, G.G.D.; Amorim, A.V.; Iwata, B.D.F.; et al. How Forest Management with Clear-Cutting Affects the Regeneration, Diversity and Structure of a Seasonally Dry Tropical Forest in Brazil. *Forests* **2023**, *14*, 1870. [CrossRef]
25. Dittrich, S.; Hauck, M.; Jacob, M.; Rommerskirchen, A.; Leuschner, C. Response of ground vegetation and epiphyte diversity to natural age dynamics in a Central European Mountain spruce forest. *J. Veg. Sci.* **2012**, *24*, 675–687. [CrossRef]
26. Hill, M.O. TWINSPAN: A fortran program of arranging multivariate data in an ordered two way table by classification of individual and attributes. In *Section of Ecology and Systematics*; Cornell University: Ithaca, NY, USA, 1979.
27. Roleček, J.; Tichý, L.; Zelený, D.; Chytrý, M. Modified TWINSPAN classification in which the hierarchy respects cluster heterogeneity. *J. Veg. Sci.* **2009**, *20*, 596–602. [CrossRef]

28. Hu, Y.; Bo, H.Y.; Bo, X.H. *Structured Forest Management*; China Forestry Publishing: Beijing, China, 2007; pp. 56–60. ISBN 9787503847639.
29. Li, Y.; Hui, G.; Zhao, Z.; Hu, Y. The bivariate distribution characteristics of spatial structure in natural Korean pine broad-leaved forest. *J. Veg. Sci.* **2012**, *23*, 1180–1190. [CrossRef]
30. Hui, G.Y.; Von Gadow, K.; Hu, Y.B.; Chen, B.W. Characterizing Forest spatial distribution pattern with the mean value of uniform angle index. *Acta Ecol. Sin.* **2004**, *24*, 1225–1229. (In Chinese)
31. Pukkala, T.; Ketonen, T.; Pykäläinen, J. Predicting timber harvests from private forests—A utility maximisation approach. *For. Policy Econ.* **2003**, *5*, 285–296. [CrossRef]
32. Schwartz, J.W.; Nagel, L.M.; Webster, C.R. Effects of uneven-aged management on diameter distribution and species composition of northern hardwoods in Upper Michigan. *For. Ecol. Manag.* **2005**, *211*, 356–370. [CrossRef]
33. Gadow, K.V.; Hui, G. *Modelling Forest Development*; Forestry Sciences Series; Springer: Dordrecht, The Netherlands, 1999. [CrossRef]
34. Hui, G.Y.; Hu, Y.B.; Liu, R.H. Methods of Analyzing Stand Spatial Dominance in Forest Observational Studies. *Res. Inst. For.* **2019**, *2*, 12. (In Chinese)
35. Neto, T.; Constantino, M.; Martins, I.; Pedroso, J.P. A multi-objective Monte Carlo tree search for forest harvest scheduling. *Eur. J. Oper. Res.* **2020**, *282*, 1115–1126. [CrossRef]
36. Gu, L.; O'Hara, K.L.; Li, W.; Gong, Z. Spatial patterns and interspecific associations among trees at different stand development stages in the natural secondary forests on the Loess Plateau, China. *Ecol. Evol.* **2019**, *9*, 6410–6421. [CrossRef] [PubMed]
37. Gokturk, A.; Tiras, H. Stand structure and spatial distribution of trees at different developmental stages and stand layers in mixed stands in Artvin region, Turkey. *Appl. Ecol. Environ. Res.* **2020**, *18*, 6163–6179. [CrossRef]
38. Zhang, M.; Kang, X.; Meng, J.; Zhang, L. Distribution patterns and associations of dominant tree species in a mixed coniferous-broadleaf forest in the Changbai Mountains. *J. Mt. Sci.* **2015**, *12*, 659–670. [CrossRef]
39. Petritan, I.C.; Marzano, R.; Petritan, A.M.; Lingua, E. Overstory succession in a mixed *Quercus petraea—Fagus sylvatica* old growth forest revealed through the spatial pattern of competition and mortality. *For. Ecol. Manag.* **2014**, *326*, 9–17. [CrossRef]
40. Christensen, N.L.; Peet, R.K. Convergence During Secondary Forest Succession. *J. Ecol.* **1984**, *72*, 25. [CrossRef]
41. Zhou, C.F.; Zhang, H.R.; Lu, J.; Zhang, X.H. Disturbances and Succession Laws of Main Natural Secondary Forests in Northeast China. *For. Res.* **2021**, *34*, 175–183. [CrossRef]
42. Zhou, Y.L.; Zhao, G.Y. Types, distribution and succession patterns of natural secondary forests Lesser Khingan Mountains—Changbai Mountain forest area. *J. Northeast. For. Univ.* **1964**, 33–45. (In Chinese)
43. Lu, L.L. Study on Community Structure Characteristics of Different Succession Stages of Broad-leaved Korean Pine Forest in Changbai Mountain. Ph.D. Thesis, BeiHua University, Changchun, China, 2019.
44. Drössler, L.; Feldmann, E.; Glatthorn, J.; Annighöfer, P.; Kucbel, S.; Tabaku, V. What Happens after the Gap?—Size Distributions of Patches with Homogeneously Sized Trees in Natural and Managed Beech Forests in Europe. *Open J. For.* **2016**, *6*, 177–190. [CrossRef]
45. McElhinny, C.; Gibbons, P.; Brack, C.; Bauhus, J. Forest and woodland stand structural complexity: Its definition and measurement. *For. Ecol. Manag.* **2005**, *218*, 1–24. [CrossRef]
46. Wang, H.X.; Huang, S.X.; Zhang, S.S.; Peng, H.; Cao, K.F. Localized neighborhood species mingling is correlated with individual tree size inequality in natural forests in South China. *Ann. For. Sci.* **2021**, *78*, 102. [CrossRef]
47. Wang, Y.; Li, J.; Cao, X.; Liu, Z.; Lv, Y. The Multivariate Distribution of Stand Spatial Structure and Tree Size Indices Using Neighborhood-Based Variables in Coniferous and Broad Mixed Forest. *Forests* **2023**, *14*, 2228. [CrossRef]
48. Li, Y.; Hui, G.; Wang, H.; Zhang, G.; Ye, S. Selection priority for harvested trees according to stand structural indices. *iFor.-Biogeosci. For.* **2017**, *10*, 561–566. [CrossRef]
49. Li, Y.; Ye, S.; Hui, G.; Hu, Y.; Zhao, Z. Spatial structure of timber harvested according to structure-based forest management. *For. Ecol. Manag.* **2014**, *322*, 106–116. [CrossRef]
50. State Forestry Bureau. *Regulations for Forest Tending*; China Standards Press: Beijing, China, 2015. (In Chinese)
51. Dong, L.; Wei, H.; Liu, Z. Optimizing Forest Spatial Structure with Neighborhood-Based Indices: Four Case Studies from Northeast China. *Forests* **2020**, *11*, 413. [CrossRef]
52. Qiu, H.; Zhang, H.; Lei, K.; Hu, X.; Yang, T.; Jiang, X. A New Tree-Level Multi-Objective Forest Harvest Model (MO-PSO): Integrating Neighborhood Indices and PSO Algorithm to Improve the Optimization Effect of Spatial Structure. *Forests* **2023**, *14*, 441. [CrossRef]
53. Dong, L.; Bettinger, P.; Liu, Z. Optimizing neighborhood-based stand spatial structure: Four cases of boreal forests. *For. Ecol. Manag.* **2022**, *506*, 119965. [CrossRef]
54. Liu, H.; Dong, X.; Meng, Y.; Gao, T.; Mao, L.; Gao, R. A novel model to evaluate spatial structure in thinned conifer-broadleaved mixed natural forests. *J. For. Res.* **2023**, *34*, 1881–1898. [CrossRef]

forests

MDPI

Article

Carbon Stock Estimation of Poplar Plantations Based on Additive Biomass Models

Minglong Yin [1], Ting Gao [1], Yuhao Zhao [1], Ruiqiang Ni [1], Peijin Zheng [2], Yuyao Zhao [3], Jinshan Zhang [3], Kun Li [1,*] and Chuanrong Li [4,*]

1 Mountain Tai Forest Ecosystem Research Station of State Forestry Administration, Research Center for Forest Carbon Neutrality Engineering of Shandong Higher Education Institutions, Tai'an 271018, China; 2023010030@sdau.edu.cn (M.Y.); 2022110213@sdau.edu.cn (T.G.); 2022110215@sdau.edu.cn (Y.Z.); wind0309@163.com (R.N.)

2 Wulian Natural Resources and Planning Bureau, Rizhao 262300, China; wlxdclyk@rz.shandong.cn

3 Forest Construction and Development Service Center of Wangzhuang Sand Area of Lijin County, Dongying 257400, China; 13054657077@163.com (Y.Z.); 15666895396@163.com (J.Z.)

4 Key Laboratory of Ecological Protection and Security Control of the Lower Yellow River of Shandong Higher Education Institutions, Tai'an 271018, China

* Correspondence: kunli@sdau.edu.cn (K.L.); chrli@sdau.edu.cn (C.L.)

Abstract: Accurate estimation of biomass and carbon stocks in forest ecosystems is critical for understanding their roles in carbon sequestration and climate change mitigation. Currently, the development of stand biomass models and carbon stock estimation at the regional scale has emerged as a prominent research priority. In this study, 225 *Populus* spp. (poplar) trees in Shandong Province, China, were destructively sampled to obtain the biomass of their components. Two models (MS1 and MS2) were developed using allometric equations and the seemingly unrelated regression (SUR) method to ensure additive properties across tree components. The model evaluation employed the leave-one-out jackknife (LOO) method, considering statistics such as adjusted R-squared (R_a^2), root mean square error (RMSE), mean absolute percent error (MAPE), and mean absolute error (MAE). The results from our models demonstrated high accuracy, with MS2 slightly outperforming MS1 after incorporating tree height as an independent variable. The models reliably estimated component-specific biomass and carbon stocks, with distinct variations observed in the carbon content among foliage (47.14 ± 2.07%), branches (47.26 ± 2.48%), stems (47.67 ± 2.21%), and roots (46.37 ± 2.78%). Carbon stocks in poplar plantations increased with the diameter class, ranging from 5 to 35 cm and correspondingly from 3.670 to 172.491 Mg C ha^{-1}. As the diameter class increases, the carbon allocation strategy of poplars aligns with the CSR strategy, transitioning from prioritizing growth competition to emphasizing self-stabilization. Our research proposes a robust framework for assessing biomass and carbon stocks in poplar plantations, which is essential for evidence-based forest management strategies.

Keywords: carbon stocks; additive biomass models; poplar plantations; carbon content; allocation strategy

1. Introduction

The accurate determination of biomass is essential for understanding carbon stocks in forest ecosystems and assessing their contribution to climate change mitigation [1,2]. Biomass estimation provides critical information for forest management, carbon cycle modeling, the evaluation of carbon sequestration effectiveness [3,4], and carbon regulation mechanisms [5,6]. Over the years, researchers have made significant progress in developing biomass models that accurately estimate carbon content based on additive biomass modeling [7–9].

Additive biomass models have gained prominence because of their ability to estimate biomass at various organizational levels within a tree, including organs, compartments, and components [10–13]. These models account for the additive properties of biomass, where total biomass is the sum of the biomass from different parts of the tree, including foliage, branches, stems, and roots [14,15]. By incorporating species-specific allometric equations, these models facilitate accurate biomass estimation and offer insights into carbon allocation patterns within trees [4,9,12]. China's forests serve as a significant carbon sink globally [16,17], with plantations in Shandong Province playing a crucial role [18]. However, to date, there have been relatively few studies on accurate individual tree biomass models and carbon stocks in *Populus* spp. (poplar) plantations in Shandong Province.

Research on carbon stocks in various tree species has been instrumental in understanding carbon accumulation rates [8,19]. Methods for calculating carbon stocks in individual trees are categorized into direct and indirect approaches [12]. The method based on the biomass formulas of different components and their carbon content in individual trees is commonly employed [4,8,20]. When estimating forest carbon, an average carbon content of 50% or 45% is commonly utilized across all species and forest types at both regional and national scales [21]. However, some studies have utilized different carbon contents, as they found that carbon content is influenced by forest regions, tissue types, species, and tree sizes [8,20,22]. Additionally, environmental conditions, such as temperature, soil, and precipitation, significantly influence the carbon content in various plant organs [23,24]. Investigations focusing on organ-level carbon content have revealed variations in carbon allocation among various tree components, emphasizing the importance of understanding carbon distribution patterns and dynamics within trees [25]. Poplar trees, widely planted in Shandong Province, China, possess significant potential for carbon sequestration due to their rapid growth rate and high biomass production [26]. Understanding the carbon stocks and allocation characteristics of poplars is crucial for accurate carbon accounting and effective management strategies.

The objectives of this study are as follows: (i) to develop a precise additive biomass model tailored for poplar plantations in Shandong Province; (ii) to investigate carbon sequestration in poplar plantations across different diameter classes; and (iii) to analyze carbon allocation strategies in poplar trees.

2. Materials and Methods

2.1. Study Area

The study area is located in Shandong Province in east-central China (Figure 1). The region experiences a warm, temperate monsoon climate with rainy summers. The average annual temperature ranges from 11 to 14 °C, and the average annual sunshine duration varies between 2290 and 2890 h. Annual precipitation typically ranges from 550 to 950 mm. The second national soil census of China identified the primary soil types in Shandong Province as fluvo-aquic soils, skeletal soils, Brunisolic soils, and cinnamon soils. Collectively, these four soil types cover approximately 87.67% of the total soil area. Poplar is the predominant cultivated species in the plantation forests of Shandong Province. According to 2017 statistics, the cultivated area and stock volume of poplar accounted for 88.5% and 91.9% of the province's timber forests, respectively. According to the 2014–2018 forest inventory data of Shandong Province, we established 75 sample plots in 15 regions, taking into account the planted area and diameter class distribution. The poplar plantation area in the region containing each sample plot is at least 1% of the total poplar plantation area in Shandong Province. Stand biomass data were collected from 75 sample plots, each measuring 600 m^2 (20 m × 30 m). The community structure of these sample plots was similar, with poplar as the dominant tree species in the canopy, no shrubs in the understory, and the herbaceous primarily consisting of *Setaira viridis*, *Humulus scandens*, *Erigeron canadensis*, and *Digitaria sanguinalis*.

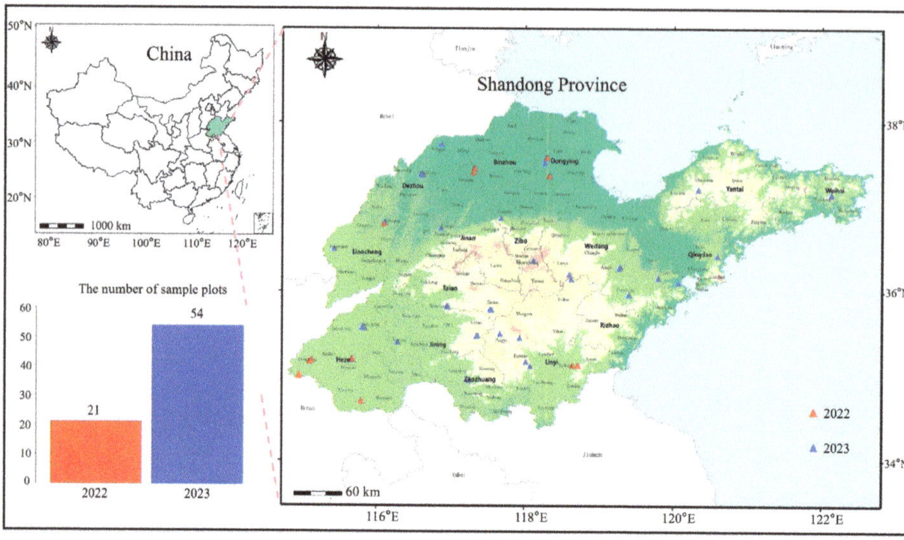

Figure 1. The locations and number of sample plots.

2.2. Measurement and Data Collection

To obtain representative tree samples, we considered the diameter at breast height (*DBH*) distributions during the selection process. Based on the average *DBH*, two to four sample trees were selected from each plot. Sample trees were selected with straight trunks, good growth, free of obvious pests and diseases, and representative of the growth of standing trees in the plots. A total of 225 trees ($DBH \geq 5$ cm) were surveyed (Table 1) and categorized into 16 diameter classes at 2 cm intervals [27] (Table S1). Additionally, due to the difficulty in obtaining sample trees with $DBH \geq 35$ cm during the field survey, these trees were assigned to the same diameter class (Table S1) based on previous studies [28]. Each diameter class represented the complete diameter range of the respective poplar species. Indicators for each diameter class were expressed as mean values. The *DBH* of sample trees, measured at 130 cm above ground, was recorded prior to conducting destructive sampling. Tree height (*H*) refers to the distance from the ground to the top of the crown and was measured after felling. The data were collected over a two-year period. Specifically, data for 60 poplars were collected between June and August 2022, while data for 165 poplars were collected during the same time frame in 2023 (Figure 1).

Table 1. Summary statistics of stand variables and biomass components for the sampled trees.

Statistics	*DBH*	*H*	B_f	B_b	B_s	B_r	B_t
Mean	19.7	17.3	14.85	38.61	148.04	54.13	255.64
Min	5.2	3.2	0.39	0.55	0.47	0.47	2.21
Max	59.1	31.1	147.28	602.70	1189.68	667.30	2607.00
SD	9.6	6.5	19.12	59.83	176.34	74.74	319.56

Note: *DBH* represents diameter at breast height (cm); *H* represents tree height (m); B_f, B_b, B_s, B_r and B_t represents foliage, branch, stem, root, and total biomass (kg), respectively. The same as below.

The total biomass was categorized into stem, branch, foliage, and root biomass. Stems of sample trees measuring 10 m or less in height were segmented into 1 m sections, whereas those exceeding 10 m in height were segmented into 2 m sections. The fresh weight of each stem section was recorded. The biomass of branches and leaves was estimated using the sample branch method [29,30]. The crowns of sample trees were divided into three layers, and the fresh weight of each layer was measured. Three sample branches were selected from each layer [31], and the fresh weights of their branches and leaves were measured to determine the ratio of branches to leaves for the sample trees. Roots are harvested by digging. The fresh weights of the stems, crowns, and roots were recorded in the field using an electronic scale with a capacity of 300 kg and an accuracy of ±0.1%.

Five hundred grams of branches, stems, foliage, and roots were collected separately as samples, weighed, and transported to the laboratory for carbon content analysis. The stems were sampled at one-quarter, one-half, and three-quarters of the tree's height, while the foliage and branches were sampled from each layer. Samples of stems and branches should be in the shape of a pie or a cylinder, including the bark [32]. The entire root systems were excavated, and the root stump, coarse roots, and small roots were weighed separately for fresh weight, excluding the fine roots (less than 2 mm) [33]. The fresh weights of the samples were measured using a platform scale with a capacity of 50 kg and an accuracy of 0.01 g.

In the laboratory, the subsamples were dried in an oven at 105 °C until a constant weight was achieved for weight determination and to estimate moisture content. The dry weights of branches, foliage, stems, and roots were determined by multiplying their respective moisture contents by their fresh weights [12]. The total dry weight of a tree was calculated by summing the dry weights of the branches, foliage, stems, and roots. The electronic scale used in the laboratory has an accuracy of 0.001 g.

2.3. Allometric Models

Biomass estimation serves as a foundation for estimating carbon stocks. Allometric models are commonly employed to estimate plant biomass [27,32,34]. The constant allometric ratio (CAR) model was employed to construct the biomass model for each component due to its simplicity, stability, and fewer variables [7]. In reference to previous studies on allometric equations for biomass [8,35], the following two models were utilized to fit the biomass of each component:

$$B = aDBH^b \tag{1}$$

$$B = aDBH^bH^c \tag{2}$$

where B is biomass (kg); DBH is the diameter at breast height (cm); H is the tree height (m); and a, b, and c represent the coefficients of the equation, respectively. The same as below.

The seemingly unrelated regression (SUR) method considers cross-equation constraints on structural parameters and accounts for cross-equation error correlations when analyzing total biomass and its individual components [14]. Additionally, SUR was applied to estimate coefficients of simultaneous equations because it ensures the additivity of these equations [10,11]. Therefore, in this study, the SUR method was used to fit the additive systems of allometric equations constructed based on Equations (1) and (2), as specified in the following forms:

$$\begin{aligned} B_f &= a_1DBH^{b_1} + \varepsilon_1 \\ B_b &= a_2DBH^{b_2} + \varepsilon_2 \\ B_s &= a_3DBH^{b_3} + \varepsilon_3 \\ B_r &= a_4DBH^{b_4} + \varepsilon_4 \\ B_t &= B_f + B_b + B_s + B_r + \varepsilon_5 \end{aligned} \tag{3}$$

$$\begin{aligned} B_f &= a_1DBH^{b_1}H^{c_1} + \varepsilon_1 \\ B_b &= a_2DBH^{b_2}H^{c_2} + \varepsilon_2 \\ B_s &= a_3DBH^{b_3}H^{c_3} + \varepsilon_3 \\ B_r &= a_4DBH^{b_4}H^{c_4} + \varepsilon_4 \\ B_t &= B_f + B_b + B_s + B_t + \varepsilon_5 \end{aligned} \tag{4}$$

where B_f, B_b, B_s, B_r, and B_t represent the foliage, branch, stem, root, and total biomass (kg), respectively; and a_i and b_i represent the coefficients of the equation, respectively (i = 1, . . ., 4 for foliage, branch, stem, and root, respectively). ε_j is the $n \times 1$ vector of residuals for the jth equation (j = 1, . . ., 5), and n is the number of observations.

Due to the heteroscedasticity of biological data [9], a weight function was defined and employed to eliminate the influence of heteroscedasticity during the model fitting process. Based on previous studies [8], $1/DBH^p$ was adopted as the weighting function for the elimination of heteroscedasticity in the biomass model, and p was confirmed for each component of the biomass model. Additionally, the Breusch–Pagan (BP) test was applied to assess the heteroscedasticity of the model based on p-value ($p > 0.05$).

2.4. Model Evaluation

The leave-one-out jackknife (LOO) method is widely employed to assess the accuracy and stability of models [14,34]. This method involves the repeated removal of one observation from the dataset, utilizing the remaining observations to re-estimate the model or estimator, thereby obtaining corresponding error estimates [36]. We utilized the LOO method to evaluate the performance of each component biomass prediction equation based on the adjusted coefficient of determination (R_a^2), root mean square error (RMSE), mean absolute percent error (MAPE), and mean absolute error (MAE) [9,34,35]. R_a^2, RMSE, and MAPE were used to assess the goodness of fit of the model, while MAE evaluated its predictive ability. The additive system biomass model with larger R_a^2 and smaller RMSE, MAPE, and MAE was selected as the optimal biomass model between Equations (3) (MS1) and (4) (MS2). The biomass of individual trees was calculated using the optimal additive system biomass model, and the total stand biomass was obtained by summing these values.

$$R_a^2 = 1 - \left(\frac{\sum_{i=1}^{n} (B_i - \hat{B}_i)^2}{\sum_{i=1}^{n} (B_i - \overline{B})^2} \right) \frac{n-1}{n-k-1} \tag{5}$$

$$\text{RMSE} = \sqrt{\frac{\sum_{i=1}^{n} (B_i - \hat{B}_i)^2}{n-k}} \tag{6}$$

$$\text{MAPE} = \frac{100}{n} \sum_{i=1}^{n} \left| \frac{B_i - \hat{B}_i}{B_i} \right| \tag{7}$$

$$\text{MAE} = \frac{\sum_{i=1}^{n} \left| B_i - \hat{B_{i,-i}} \right|}{n} \tag{8}$$

where B_i is the ith observed biomass value; $\hat{B}_i$ is the ith value predicted by the model using all the data; $\overline{B}$ is the mean of the observed value; $\hat{B_{i,-i}}$ is the ith value predicted by the model without ith observation; n is the number of observations for model fitting; and k represents the number of model coefficients.

2.5. Model Validation

In this study, 40% of the observed biomass data (Table 2) was randomly selected as test data to validate the predictive capability of the optimal additive biomass model. Graphical analysis was employed to compare the predicted and observed biomass components. Validation involved a statistical comparison of the fitted values with independent biomass observations using a t-test [9,34].

Table 2. Summary statistics of observed biomass data for model validation.

Statistics	*DBH*	*H*	B_f	B_b	B_s	B_r	B_t
Mean	19.5	17.2	14.83	38.85	142.25	53.49	249.42
Min	5.2	3.2	0.39	0.55	0.47	0.47	2.21
Max	59.1	31.1	147.28	602.74	1189.68	667.30	2607.00
SD	9.7	6.7	20.49	68.62	177.61	80.02	337.82

2.6. Comparison of Existing Biomass Models

In this study, we evaluated and compared the performance of an independent set of existing biomass equations using test data (Table 2). Although we did not find biomass models specifically applicable to poplar in Shandong, we were fortunate to find biomass models [37] from regions such as Beijing, Hebei, and others with similar climatic environments (Table 3). Aboveground biomass was defined as the sum of leaf, branch, and stem biomass, and total biomass was calculated as the sum of aboveground and belowground biomass in this study. We used the MAE as an indicator for each model to compare their predictive power. A smaller MAE value indicates better predictive performance [9,14].

Table 3. Parameters of existing models.

Source	Origin	Components	Model	*a*	*b*	*c*
[33]	Plantation	Above	$B = aDBH^bH^c$	0.047	2.125	0.592
		Root		0.024	2.311	0.085

2.7. Carbon Content and Carbon Stocks

Constant weight drying samples were collected to determine the carbon content of branches, foliage, stems, and roots using the potassium dichromate–sulfuric acid oxidation method [38]. The dry weight of each organ, multiplied by its respective carbon content, represents its carbon stock. The total carbon stock of individual trees is the sum of the carbon stocks from each organ, while the total carbon stock for the stand is derived from the sum of the carbon stocks of all individual trees in the sample plot (Figure 2). In this study, the biomass of each component predicted by the optimal additive biomass model was used to calculate its carbon stock.

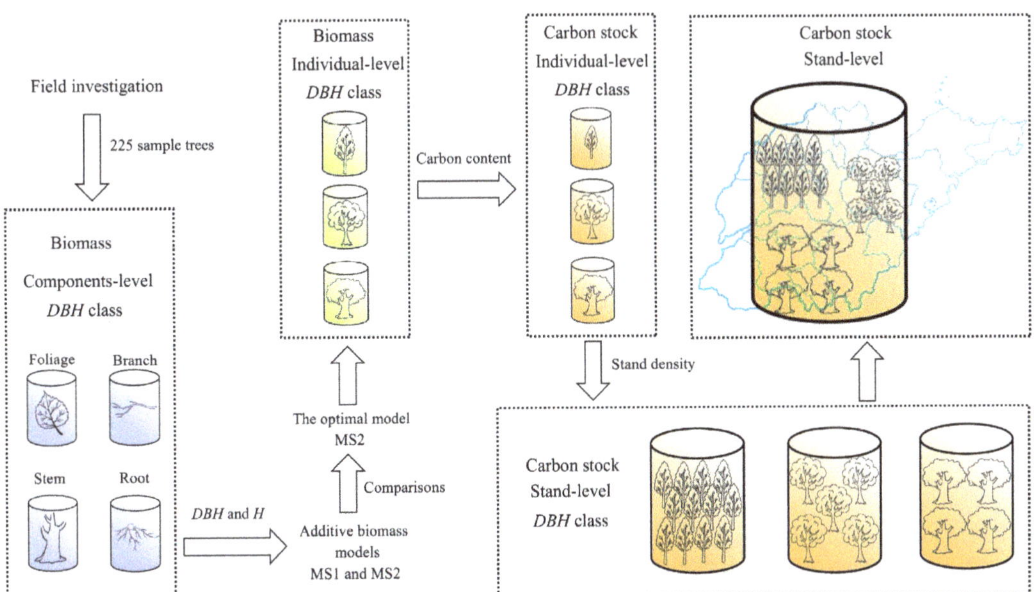

Figure 2. Procedure for estimating biomass and carbon stocks in poplar plantations. *DBH* represents the diameter at the breast height of sample trees—the same as below.

All statistics, the model fitting, heteroscedasticity test, normality test, and significance test of difference were performed in R studio software (version: 4.1.3). The additive system of nonlinear biomass equations was estimated by the function nlsystemfit in the systemfit package [39], while the BP test utilized the lmtest package. The Shapiro–Wilk test was used to check the normal distribution of the data, while the Kruskal–Wallis test was employed to assess the significance of differences. Additionally, the Duncan test was utilized to perform multiple comparisons of differences.

3. Results

3.1. Model Fitting for Biomass Models

The coefficient estimates, standard errors (SEs), model fitting statistics (R_a^2, MAPE, and RMSE), and weighting functions for MS1 and MS2 are presented in Table 4. The results indicated that the coefficient estimates for MS1 and MS2 were statistically significant at the 0.001 level, and the weighting function effectively eliminated the heteroscedasticity in each component equation (p-value > 0.05). As presented in Table 4, MS1 demonstrated a good fit with each component biomass, with R_a^2 ranging from 0.856 to 0.980, RMSE spanning from 7.789 to 54.380 kg, and MAPE ranging from 27.597 to 61.576. By adding the independent variable H to the additive system biomass model, MS2 enhanced the fit of the total biomass, stem biomass, root biomass, and foliage biomass data compared to MS1. However, this enhancement led to an increase in the fitting statistical index R_a^2 for branches in Equation (4), along with a slight increase in RMSE. A similar result was also observed in the MAPE for roots in the MS2. Additionally, MS1 and MS2 exhibited the highest conformity with total and stem biomass data, while relatively lower R_a^2 and larger RMSE were observed in the equations related to root, branch, and foliage biomass.

As illustrated in Figure 3, both MS1 and MS2 exhibited the highest MAE for the total biomass model and the lowest for the foliage biomass model. Compared to MS1, the MAE of all component biomass models in MS2 decreased, with a more substantial reduction in the MAE of stems (31.03%) and smaller reductions in foliage (0.58%) and branches (0.10%). With the exception of a slight increase in the RMSE for the branch biomass model, all other metrics in MS2 were superior to those in MS1 (Table 4). Furthermore, given that the MAE for the branch biomass model in MS2 was lower than that in MS1, MS2 was determined to be the appropriate model for predicting individual biomass.

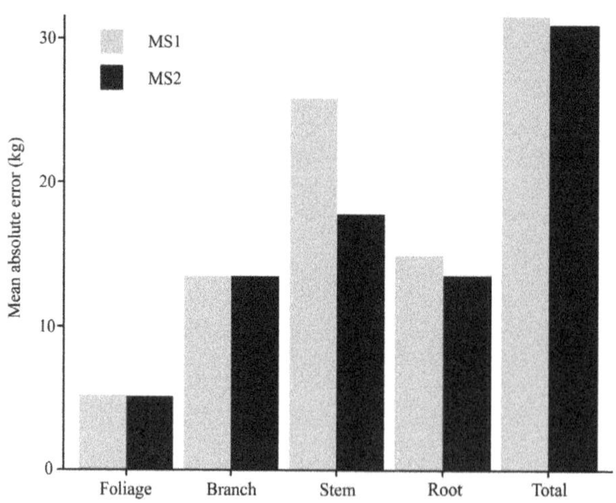

Figure 3. Mean absolute error (MAE) of the total and each component biomasses for the additive system biomass models MS1 and MS2.

Table 4. Model coefficient estimates, stand errors (SEs), model fitting statistics (R_a^2, RMSE, and MAPE), weight functions, and Breusch–Pagan (BP) test (p-value) of the additive system biomass models (Equations (3) and (4)) for each component biomass.

Models	Components	*a*		*b*		*c*		R_a^2	RMSE	MAPE (%)	Weight Function	BP Test
		Estimate	SE	Estimate	SE	Estimate	SE					(p-Value)
MS1	Foliage	0.018 ***	4.248×10^{-10}	2.184 ***	6.657×10^{-9}	-	-	0.856	7.789	61.576	$1/DBH^{1.06}$	0.397
	Branch	0.003 ***	2.682×10^{-11}	2.975 ***	2.538×10^{-9}	-	-	0.903	22.654	43.484	$1/DBH^{1.01}$	0.570
	Stem	0.212 ***	5.157×10^{-10}	2.127 ***	6.709×10^{-10}	-	-	0.962	39.370	70.562	$1/DBH^{1.17}$	0.557
	Root	0.010 ***	6.667×10^{-11}	2.718 ***	1.859×10^{-9}	-	-	0.924	22.475	33.011	$1/DBH^{1.03}$	0.545
	Total	-	-	-	-	-	-	0.980	54.380	27.597	$1/DBH^{1.12}$	0.626
MS2	Foliage	0.027 ***	1.901×10^{-3}	1.905 ***	1.550×10^{-2}	0.171 ***	2.655×10^{-2}	0.861	7.694	57.014	$1/DBH^{1.09}$	0.426
	Branch	0.012 ***	4.430×10^{-4}	2.947 ***	5.207×10^{-3}	−0.429 ***	1.173×10^{-2}	0.904	23.383	41.372	$1/DBH^{1.00}$	0.674
	Stem	0.028 ***	2.618×10^{-4}	2.074 ***	1.611×10^{-3}	0.704 ***	3.035×10^{-3}	0.982	30.049	23.821	$1/DBH^{1.07}$	0.518
	Root	0.046 ***	1.072×10^{-3}	2.645 ***	3.862×10^{-3}	−0.414 ***	7.861×10^{-3}	0.928	21.220	34.991	$1/DBH^{1.00}$	0.589
	Total	-	-	-	-	-	-	0.983	51.391	20.093	$1/DBH^{1.00}$	0.520

Note: R_a^2—adjust coefficient of determination; RMSE—root mean square error; MAPE—mean absolute percent error. Significance level: *** <0.001. The same as below.

3.2. Model Validation for the Optimal Model

The validity of the optimal additive biomass model (MS2) was evaluated using the test data. As shown in Figure 4, linear regression analysis revealed a strong correlation between predicted and observed biomass, with $R_a^2 > 0.85$ and p-values < 0.001 for all components. Furthermore, the paired samples t-test indicated no significant differences between the predicted and observed values for foliage ($t = -1.09$, p-value = 0.28), branch ($t = 1.36$, p-value = 0.18), stem ($t = -1.91$, p-value = 0.06), root ($t = 0.50$, p-value = 0.62), and total ($t = -0.47$, p-value = 0.64) biomass.

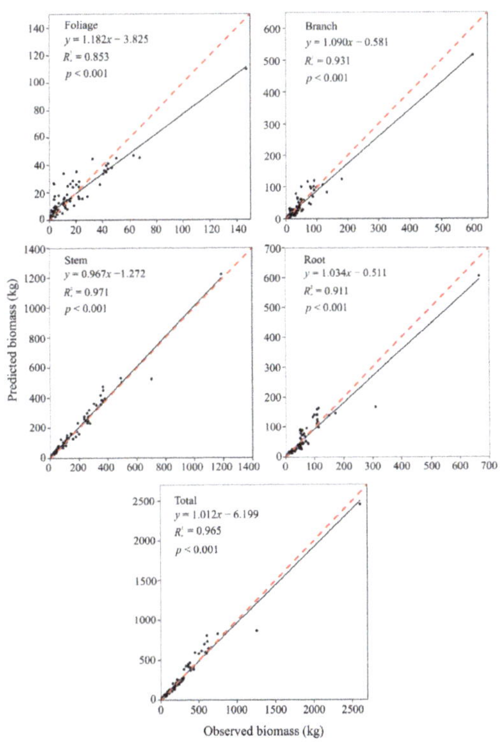

Figure 4. Scatter plots depicting the observed and predicted values for foliage, branches, stems, roots, and total biomass of the optimal model. The solid line is a liner fit, and the dashed line in red color indicates the 1:1 equivalence.

Table 5 presents the comparative results of the predictive abilities between the biomass model developed in this study and existing biomass models. MAEs for aboveground biomass, root biomass, and total biomass in the biomass models developed in this study were smaller than those in existing models, with reductions of 16.96%, 13.64%, and 7.74%, respectively.

Table 5. Comparison of biomass models.

Source	Origin	Components	MAE
This study	Plantation	Above	24.395
		Root	14.000
		Total	33.366
[33]	Plantation	Above	29.377
		Root	16.213
		Total	36.166

3.3. *Carbon Content in Various Components*

As shown in Table 6, the carbon content of roots was significantly different ($p < 0.001$) from that of foliage, branches, and stems. However, no significant differences ($p > 0.05$) were observed in carbon content among foliage, branches, and stems. There was no significant difference ($p > 0.05$) in the foliage carbon content of sample trees among different *DBH* classes, and a similar pattern was observed for branches, stems, and roots (Figure 5). Furthermore, we conducted linear regression analysis on various organs and their carbon content and found no significant linear relationship between carbon content and *DBH* across different organs (Figure 5). Consequently, the carbon stocks of different organs were calculated using the mean values of their carbon content.

Table 6. Means and standard deviation (SD) of carbon content in different components.

Components	N	Carbon Content (Mean ± SD)
Foliage	225	47.14 ± 2.07 a
Branch	225	47.26 ± 2.48 a
Stem	225	47.67 ± 2.21 a
Root	225	46.37 ± 2.78 b

Note: N is the number of observations. Different letters in the same column indicate significant differences at the 0.001 level.

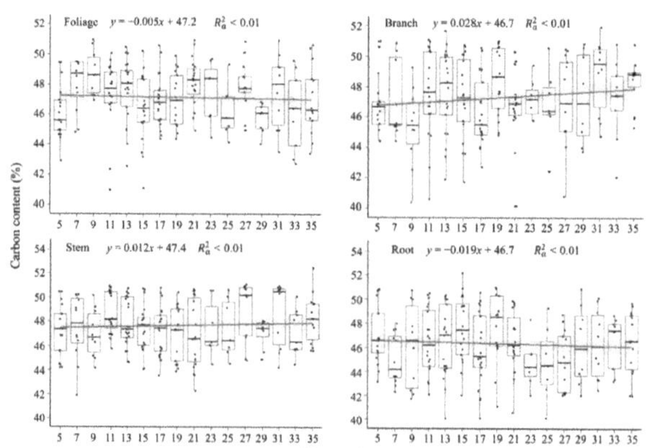

Figure 5. The carbon content of foliage, branches, stems, and roots. The red line is a linear fit, and the gray areas were 95% confidence intervals.

3.4. Carbon Stocks and Allocation in Standing Trees

The carbon stocks of standing trees were calculated based on their biomass and carbon content. As expected, carbon stocks increased with increasing *DBH*. The average carbon stock of poplar plantations in the 5 cm diameter class was the smallest, at 3.670 Mg C ha^{-1} (1 Mg = 10^6 g), whereas the 35 cm diameter class had the largest average carbon stock, at 172.491 Mg C ha^{-1}. The allocation of carbon stocks in poplars across various diameter classes is illustrated in Figure 6. The relative contribution of carbon stocks of each component to the total varied with *DBH*. The root–shoot ratio of poplars across different *DBH* classes ranged from 0.23 to 0.42. With the increase in *DBH*, the proportion of foliage in the total carbon stocks gradually decreased from 12.79% to 5.30%. Generally, the proportions of carbon stocks in branches and roots decreased and then increased, though the trends were not synchronized. The proportion of carbon stock in branches was minimal (11.35%) in the 9 cm diameter class and maximal (16.29%) in the diameter class greater than 35 cm. The proportion of root carbon stock was largest in the 5 cm diameter class (29.60%) and smallest in the 25 cm diameter class (18.30%), subsequently increasing with *DBH*. The stem constituted the major pool of carbon stocks in standing trees, ranging from 44.57% to 62.97%, with the proportion of carbon stocks in the stem peaking in the 25 cm diameter class. Since then, although stem carbon stock has increased, the marginal effect of resource investment on stem growth has diminished. The mean values of the carbon stock shares for foliage, branches, stems, and roots were 7.95%, 12.99%, 57.49%, and 21.57%, respectively.

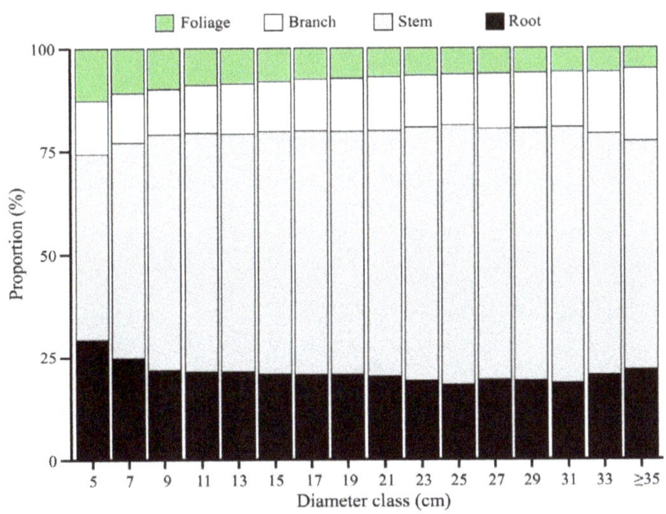

Figure 6. Carbon stocks allocation of poplars in different diameter classes.

4. Discussion

In recent years, research on the prediction of total and component biomasses of individual trees primarily utilizes an allometric growth model in the form of a power-law function [12,34,36,40]. Allometric models establish a relationship between the individual biomass of trees and various measurable parameters, such as *DBH*, *H*, and crown diameter [35,41]. The additive biomass model can reduce uncertainties between the sum of predictions for tree components obtained from allometric equations and the prediction for the entire tree [9,10]. We developed two additive biomass models: one based on *DBH* (MS1) and the other based on both *DBH* and *H* attributes (MS2). The additive biomass models presented in this study, MS1 and MS2, provided accurate predictions for all biomass components as well as total biomass for poplar plantations in Shandong Province.

Model MS1 exhibited a high degree of fit, as reflected by R_a^2 and RMSE, indicating reliable predictions of biomass within trees. Nevertheless, the introduction of *H* as an independent variable in model MS2 improved the fitting of total, stem, root, and foliage biomass while slightly increasing the RMSE of the branch component's prediction. This finding aligns with previous research that supports the inclusion of height as an essential predictor in forest biomass estimation models, enhancing model accuracy by incorporating an additional dimension of tree growth [27,34]. However, the slight increase in branch biomass RMSE when using MS2 may indicate a trade-off between model complexity and predictive precision for certain components [19]. Overall, model MS2 demonstrated superior prediction accuracy compared to model MS1.

Additionally, we observed that the *b* values in both MS1 and MS2 were greater than one, while the *c* values were less than one. This finding is consistent with the results of allometric models for deciduous broadleaf mixed forests and coniferous–broadleaf mixed forests in the Eastern Da Xing'an Mountains, Northeast China [14]. This can be explained by a negative correlation between *DBH* and *H* in allometric growth [27], indicating that poplar trees prioritize resources for diameter growth. Furthermore, the *b* values in the biomass equations for branches and roots in both MS1 and MS2 were greater than those for stems and foliage, indicating that the effect of *DBH* on the biomass of branches and roots was greater than on that of foliage and stems. In MS2, the absolute value of the *c* value in the stem biomass equation was greater than that of the *c* values in the equations for the other components. This indicates that *H* had a greater effect on stem biomass than on the biomass of the other components. Collectively, these findings show that both *DBH* and *H* influenced total biomass and component biomass, with *DBH* playing a predominant role [34].

A thorough investigation of carbon content among different components revealed that roots exhibited significantly different carbon percentages compared to other components, highlighting the intricate variability of carbon distribution within tree species [25]. Additionally, the carbon content of stems was the highest, followed by that of branches, foliage, and roots in this study; however, this finding needs to be verified with more data from different species. Dong et al. [8] found that the foliage carbon content in 10 broadleaf species in Northeast China was higher than in other tree tissues. This may be attributed to genetic differences among tree species and variations in individual habitats [20,42]. Since there was no significant difference in the carbon content of the same component across *DBH*, the average carbon contents of the various components were used to calculate their carbon stocks.

The findings indicated a direct correlation between *DBH* and carbon stocks, consistent with the general understanding that larger trees tend to store more carbon [7]. Stem carbon stocks represented the largest portion of the carbon stocks in poplars across all diameter classes, aligning with findings from 10 broadleaf species in Northeast China [8]. Furthermore, the carbon allocation analysis across different *DBH* classes revealed a trend of declining foliage carbon stock percentages with tree growth, while stems, branches, and roots exhibited dynamic patterns. Seedlings (5 cm *DBH* class) had a higher ratio of foliage biomass, enhancing their access to limited light [43], as well as a larger root biomass ratio, which increases stability and water absorption [44,45]. To capture more light, additional resources are allocated to stem and branch growth [25,34]. When the growth of stems ($DBH \geq 27$ cm) was constrained by genetic factors and environmental conditions [46,47], additional resources were redirected to roots and branches, enhancing tree stability and facilitating access to water and light [44]. These observations may reflect changes in tree architecture and resource allocation strategies associated with various growth stages [48]. At the seedling stage, poplars adopted the ruderal (R) strategy [49], prioritizing resources to foliage and roots to rapidly occupy above- and below-ground space, aiming to gain a first-mover advantage. As *DBH* increases, poplars adopted the competitors' (C) strategy, prioritizing resources to stems to gain more aboveground space and enhance competitive growth. When the $DBH \geq 27$ cm, poplars adopted the stress–

tolerators' (S) strategy, allocating more resources to roots and branches to secure additional resources while maintaining stability. These findings indicated that single-dominant species plant communities, such as poplar plantations, adhere to CSR strategies throughout long-term dynamic development. This observation aligned with the findings of Zhang et al. [50] regarding plant community succession in long-term abandoned cropland.

The understanding of carbon allocation dynamics, as highlighted by the improved model MS2, underscores the necessity of adopting flexible management practices that recognize the varying carbon sequestration capabilities across different tree components and sizes. As suggested by Figure 6, stems predominantly contribute to carbon stocks, particularly in large diameter classes, which supports prioritizing practices that enhance stem growth, such as optimized spacing, thinning regimes, and utilization of harvest residues [51,52]. In this study, the stem carbon stock accounted for the largest proportion in the *DBH* class of 25 cm. Although the stem carbon stock continued to increase, the marginal effect of resource investment in stem growth decreased. Therefore, thinning poplar plantations with a $DBH > 25$ cm enhances their carbon sequestration potential. The biomass and carbon stocks of standing trees vary with forest type, size, and tree age [48,53]. Therefore, further field trials are necessary to determine how to effectively enhance the carbon sequestration of poplar plantations through management practices, such as thinning intensity and the utilization of harvest residues.

Variations in climate, soil, and other factors can influence the biomass of standing trees [48,54]. Consequently, the sample plots in this study were distributed across the primary poplar plantation areas in Shandong Province. In addition, certain regions, such as Rizhao, Qingdao, Yantai, and Weihai, had fewer sample plots, potentially impacting the model's applicability in these areas. According to the results of the 9th National Forest Resources Survey of China (2014–2018), the total area planted with poplar in the four aforementioned regions constitutes only 15.01%. Nonetheless, the optimal model is still applicable for estimating the carbon stock in poplar plantation forests throughout Shandong Province. Meanwhile, field measurements were taken from June to August, with both 2022 and 2023 experiencing low rainfall in June and high rainfall in July and August, which may affect the accuracy of the fresh weight of the sample trees, consequently impacting biomass models. Additionally, the sample trees selected for this study were poplars with $DBH \geq 5$ cm, so the biomass models developed in this study are not applicable for estimating the biomass of poplars with $DBH < 5$ cm. Typically, there is a negative correlation between the accuracy of the biomass model and its scale of applicability [28]. Therefore, applying the model from this study to regions with significant environmental differences from Shandong Province requires validation with existing data to assess its applicability in those areas.

5. Conclusions

This study enhances our understanding of carbon sequestration in poplar plantations by developing additive biomass models that are suitable for precise carbon stock estimation in Shandong Province, China. The developed models, MS1 and MS2, demonstrated high accuracy, with MS2 exhibiting an improved fit after incorporating *H* as an independent variable. These models reliably quantify the contributions of individual tree components—foliage, branches, stems, and roots—to the overall biomass and, consequently, the carbon stock.

Carbon stocks in poplar trees increased with increasing *DBH*. The findings also highlighted the variation in carbon content among different tree organs, offering a more straightforward approach to carbon accounting. Furthermore, there was no significant correlation between the carbon content of each organ and *DBH*. The carbon content of the stem was the highest, and its proportion of the total carbon stock was also the greatest, indicating that the role of the stem should be emphasized when evaluating the carbon stocks of poplar plantations. Additionally, poplars employed CSR strategies, shifting from prioritizing growth competition to prioritizing self-stabilization.

Although this study focused on poplar plantations in a specific region, the methodology and findings are applicable to broader contexts. Future research should concentrate on the long-term monitoring of carbon sink capacity in plant communities within sample plots, encompassing the carbon content of trees, shrubs, herbs, and soils. Additionally, the remote sensing data were adjusted using the model from this study to enhance their accuracy in estimating large-scale forest carbon sinks. Overall, this study establishes a robust foundation for estimating biomass and carbon stocks of poplar trees and has significant implications for assessing and utilizing the full carbon sequestration potential of poplar plantations.

Supplementary Materials: The following supporting information can be downloaded at: https://www.mdpi.com/article/10.3390/f15101829/s1, Table S1: Statistics of sample plots for poplar plantations.

Author Contributions: Conceptualization, K.L. and C.L.; methodology, K.L., M.Y. and C.L.; software, R.N. and M.Y.; validation, K.L., M.Y. and R.N.; formal analysis, R.N. and M.Y.; investigation, K.L., M.Y., T.G., Y.Z. (Yuhao Zhao) and C.L.; resources, P.Z., Y.Z. (Yuyao Zhao), J.Z. and K.L.; data curation, R.N. and M.Y.; writing—original draft preparation, M.Y. and T.G.; writing—review and editing, M.Y. and K.L.; visualization, R.N., Y.Z. (Yuhao Zhao) and M.Y.; supervision, K.L.; project administration, K.L. and C.L.; funding acquisition, K.L. and C.L. All authors have read and agreed to the published version of the manuscript.

Funding: We would like to thank the Shandong Provincial Forestry Protection and Development Service Center (No. SDGP370000000202402002176), the National Natural Science Foundation of China (No. 31570705), and Taishan Forest Pest Control and Quarantine Station, Tai'an City, China (No. 2022TSGS001-5-2) for the financial support of this study.

Data Availability Statement: Data are contained within the article and Supplementary Materials.

Acknowledgments: We sincerely acknowledge the support of the faculty and students from Shandong Agricultural University, Qingdao Agricultural University, Shandong University of Technology, and Linyi University in conducting the field investigation and processing samples.

Conflicts of Interest: The authors declare no competing financial interests.

References

1. Canadell, J.G.; Le Quere, C.; Raupach, M.R.; Field, C.B.; Buitenhuis, E.T.; Ciais, P.; Conway, T.J.; Gillett, N.P.; Houghton, R.A.; Marland, G. Contributions to accelerating atmospheric CO_2 growth from economic activity, carbon intensity, and efficiency of natural sinks. *Proc. Natl. Acad. Sci. USA* **2007**, *104*, 18866–18870. [CrossRef] [PubMed]
2. Speak, A.; Escobedo, F.J.; Russo, A.; Zerbe, S. Total urban tree carbon storage and waste management emissions estimated using a combination of LiDAR, field measurements and an end-of-life wood approach. *J. Clean. Prod.* **2020**, *256*, 120420. [CrossRef]
3. Mugasha, W.A.; Eid, T.; Bollandsås, O.M.; Malimbwi, R.E.; Chamshama, S.A.O.; Zahabu, E.; Katani, J.Z. Allometric models for prediction of above- and belowground biomass of trees in the miombo woodlands of Tanzania. *For. Ecol. Manag.* **2013**, *310*, 87–101. [CrossRef]
4. Oliveira, N.; Rodríguez-Soalleiro, R.; Pérez-Cruzado, C.; Cañellas, I.; Sixto, H.; Ceulemans, R. Above- and below-ground carbon accumulation and biomass allocation in poplar short rotation plantations under Mediterranean conditions. *For. Ecol. Manag.* **2018**, *428*, 57–65. [CrossRef]
5. Eslamipoor, R.; Sepehriar, A. Enhancing supply chain relationships in the circular economy: Strategies for a green centralized supply chain with deteriorating products. *J. Environ. Manag.* **2024**, *367*, 121738. [CrossRef]
6. Sepehriar, A.; Eslamipoor, R. An economical single-vendor single-buyer framework for carbon emission policies. *J. Bus. Econ.* **2024**, *94*, 927–945. [CrossRef]
7. Alvarez, E.; Duque, A.; Saldarriaga, J.; Cabrera, K.; de las Salas, G.; del Valle, I.; Lema, A.; Moreno, F.; Orrego, S.; Rodríguez, L. Tree above-ground biomass allometries for carbon stocks estimation in the natural forests of Colombia. *For. Ecol. Manag.* **2012**, *267*, 297–308. [CrossRef]
8. Dong, L.H.; Liu, Y.; Zhang, L.; Xie, L.; Li, F. Variation in carbon concentration and allometric equations for estimating tree carbon contents of 10 broadleaf species in natural forests in Northeast China. *Forests* **2019**, *10*, 928. [CrossRef]
9. Meng, S.; Jia, Q.; Liu, Q.; Zhou, G.; Wang, H.; Yu, J. Aboveground biomass allocation and additive allometric models for natural larix gmelinii in the western Daxing'anling Mountains, Northeastern China. *Forests* **2019**, *10*, 150. [CrossRef]
10. Parresol, B.R. Additivity of nonlinear biomass equations. *Can. J. For. Res.* **2001**, *31*, 865–878. [CrossRef]

11. Sanquetta, C.R.; Behling, A.; Corte, A.P.D.; Péllico Netto, S.; Schikowski, A.B.; do Amaral, M.K. Simultaneous estimation as alternative to independent modeling of tree biomass. *Ann. For. Sci.* **2015**, *72*, 1099–1112. [CrossRef]
12. Bayen, P.; Noulèkoun, F.; Bognounou, F.; Lykke, A.M.; Djomo, A.; Lamers, J.P.A.; Thiombiano, A. Models for estimating aboveground biomass of four dryland woody species in Burkina Faso, West Africa. *J. Arid Environ.* **2020**, *180*, 104205. [CrossRef]
13. Dimobe, K.; Mensah, S.; Goetze, D.; Ouedraogo, A.; Kuyahf, S.; Porembski, S.; Thiombiano, A. Aboveground biomass partitioning and additive models for Combretum glutinosum and Terminalia laxiflora in West Africa. *Biomass Bioenergy* **2018**, *115*, 151–159. [CrossRef]
14. Dong, L.; Zhang, L.; Li, F. Evaluation of stand biomass estimation methods for major forest types in the eastern Da Xing'an Mountains, Northeast China. *Forests* **2019**, *10*, 715. [CrossRef]
15. Waring, R.H.; Landsberg, J.J.; Williams, M. Net primary production of forests: A constant fraction of gross primary production? *Tree Physiol.* **1998**, *18*, 129–134. [CrossRef]
16. Fang, J.Y.; Guo, Z.D.; Hu, H.F.; Kato, T.; Muraoka, H.; Son, Y. Forest biomass carbon sinks in East Asia, with special reference to the relative contributions of forest expansion and forest growth. *Glob. Chang. Biol.* **2014**, *20*, 2019–2030. [CrossRef]
17. Zhang, C.H.; Ju, W.M.; Chen, J.M.; Wang, X.Q.; Yang, L.; Zheng, G. Disturbance-induced reduction of biomass carbon sinks of China's forests in recent years. *Environ. Res. Lett.* **2015**, *10*, 114021. [CrossRef]
18. Zhang, C.; Ju, W.; Wang, D.; Wang, X.; Wang, X. Biomass carbon stocks and economic value dynamics of forests in Shandong Province from 2004 to 2013. *Acta Ecol. Sin.* **2018**, *38*, 1739–1749.
19. Chave, J.; Condit, R.; Lao, S.; Caspersen, J.P.; Foster, R.B.; Hubbell, S.P. Spatial and temporal variation of biomass in a tropical forest: Results from a large census plot in Panama. *J. Ecol.* **2003**, *91*, 240–252. [CrossRef]
20. Dong, L.H.; Zhang, L.J.; Li, F.R. Allometry and partitioning of individual tree biomass and carbon of *Abies nephrolepis* Maxim in northeast China. *Scand. J. For. Res.* **2016**, *31*, 399–411. [CrossRef]
21. Neumann, M.; Moreno, A.; Mues, V.; Härkönen, S.; Mura, M.; Bouriaud, O.; Lang, M.; Achten, W.M.J.; Thivolle-Cazat, A.; Bronisz, K.; et al. Comparison of carbon estimation methods for European forests. *For. Ecol. Manag.* **2016**, *361*, 397–420. [CrossRef]
22. Elias, M.; Potvin, C. Assessing inter- and intra-specific variation in trunk carbon concentration for 32 neotropical tree species. *Can. J. For. Res.* **2003**, *36*, 1039–1045. [CrossRef]
23. Lambers, H.; Boogaard, R.v.d.; Veneklaas, E.J.; Villar, R. Effects of global environmental change on carbon partitioning in vegetative plants of Triticum aestivum and closely related Aegilops species. *Glob. Chang. Biol.* **1995**, *1*, 397–406. [CrossRef]
24. Xia, J.Z.; Chen, Y.; Yuan, W.P.; Wang, Y.P. The effects of multiple environmental factors on global carbon allocation. *Ecol. Process.* **2023**, *12*, 60–77. [CrossRef]
25. Litton, C.M.; Ryan, M.G.; Knight, D.H. Effects of tree density and stand age on carbon allocation patterns in postfire lodgepole pine. *Ecol. Appl.* **2004**, *14*, 460–475. [CrossRef]
26. Wang, Z.; Yan, W.; Peng, Y.; Wan, M.; Farooq, T.H.; Fan, W.; Lei, J.; Yuan, C.; Wang, W.; Qi, Y.; et al. Biomass production and carbon stocks in poplar-crop agroforestry chronosequence in subtropical central China. *Plants* **2023**, *12*, 2451. [CrossRef]
27. Blujdea, V.N.B.; Pilli, R.; Dutca, I.; Ciuvat, L.; Abrudan, I.V. Allometric biomass equations for young broadleaved trees in plantations in Romania. *For. Ecol. Manag.* **2012**, *264*, 172–184. [CrossRef]
28. Shi, L.; Liu, S.R. Methods of estimating forest biomass: A review. In *Biomass Volume Estimation and Valorization for Energy*; Jaya Shankar, T., Ed.; IntechOpen: Rijeka, Croatia, 2017; pp. 23–46.
29. Edwards, T.C., Jr.; Cutler, D.R.; Zimmermann, N.E.; Geiser, L.; Alegria, J. Model-based stratifications for enhancing the detection of rare ecological events. *Ecology* **2005**, *86*, 1081–1090. [CrossRef]
30. Gregoire, T.G.; Valentine, H.T.; Furnival, G.M. Sampling methods to estimate foliage and other characteristics of individual trees. *Ecology* **1995**, *76*, 1181–1194. [CrossRef]
31. Poudel, K.P.; Temesgen, H.; Gray, A.N. Evaluation of sampling strategies to estimate crown biomass. *For. Ecosyst.* **2015**, *2*, 14–25. [CrossRef]
32. Basuki, T.M.; van Laake, P.E.; Skidmore, A.K.; Hussin, Y.A. Allometric equations for estimating the above-ground biomass in tropical lowland Dipterocarp forests. *For. Ecol. Manag.* **2009**, *257*, 1684–1694. [CrossRef]
33. Zeng, W. Developing tree biomass models for eight major tree species in China. In *Biomass Volume Estimation and Valorization for Energy*; Jaya Shankar, T., Ed.; IntechOpen: Rijeka, Croatia, 2017; pp. 3–21.
34. Xu, Z.; Du, W.; Zhou, G.; Qin, L.; Meng, S.; Yu, J.; Sun, Z.; SiQing, B.; Liu, Q. Aboveground biomass allocation and additive allometric models of fifteen tree species in northeast China based on improved investigation methods. *For. Ecol. Manag.* **2022**, *505*, 119918. [CrossRef]
35. Zhao, D.; Kane, M.; Markewitz, D.; Teskey, R.; Clutter, M. Additive tree biomass equations for midrotation loblolly pine plantations. *For. Sci.* **2015**, *61*, 613–623. [CrossRef]
36. Zheng, C.; Mason, E.G.; Jia, L.; Wei, S.; Sun, C.; Duan, J. A single-tree additive biomass model of *Quercus variabilis* Blume forests in North China. *Trees* **2015**, *29*, 705–716. [CrossRef]
37. *GB/T 43648-2024*; Tree Biomass Models and Related Paraments to Carbon Accounting for Major Tree Species. Standardization Administration of China (SAC): Beijing, China, 2024.
38. Schouwenburg, J.C.V. Comparison of methods for the determination of calcium in plant material. *Neth. J. Agric. Sci.* **1961**, *9*, 181–187. [CrossRef]

39. Henningsen, A.; Hamann, J.D. systemfit: A package for estimating systems of simultaneous equations in R. *J. Stat. Softw.* **2007**, *23*, 1–40. [CrossRef]
40. Zhao, D.; Westfall, J.; Coulston, J.W.; Lynch, T.B.; Bullock, B.P.; Montes, C.R. Additive biomass equations for slash pine trees: Comparing three modeling approaches. *Can. J. For. Res.* **2019**, *49*, 27–40. [CrossRef]
41. Jucker, T.; Caspersen, J.; Chave, J.; Antin, C.; Barbier, N.; Bongers, F.; Dalponte, M.; van Ewijk, K.Y.; Forrester, D.I.; Haeni, M.; et al. Allometric equations for integrating remote sensing imagery into forest monitoring programmes. *Glob. Chang. Biol.* **2017**, *23*, 177–190. [CrossRef]
42. Zhang, Q.; Wang, C.; Wang, X.; Quan, X. Carbon concentration variability of 10 Chinese temperate tree species. *For. Ecol. Manag.* **2009**, *258*, 722–727. [CrossRef]
43. Zhou, G.; Liu, Q.; Xu, Z.; Du, W.; Yu, J.; Meng, S.; Zhou, H.; Qin, L.; Shah, S. How can the shade intolerant Korean pine survive under dense deciduous canopy? *For. Ecol. Manag.* **2020**, *457*, 117735. [CrossRef]
44. Poorter, H.; Nagel, O.W. The role of biomass allocation in the growth response of plants to different levels of light, CO_2, nutrients and water: A quantitative review. *Aust. J. Plant Physiol.* **2000**, *27*, 595–607.
45. Walters, M.B.; Reich, P.B. Seed size, nitrogen supply, and growth rate affect tree seedling survival in deep shade. *Ecology* **2000**, *81*, 1887–1901. [CrossRef]
46. Poorter, H.; Niklas, K.J.; Reich, P.B.; Oleksyn, J.; Poot, P.; Mommer, L. Biomass allocation to leaves, stems and roots: Meta-analyses of interspecific variation and environmental control. *New Phytol.* **2012**, *193*, 30–50. [CrossRef]
47. Howe, G.T.; Aitken, S.N.; Neale, D.B.; Jermstad, K.D.; Chen, T.H.H. From genotype to phenotype: Unraveling the complexities of cold adaptation in forest trees. *Can. J. Bot.* **2003**, *81*, 1247–1266. [CrossRef]
48. Collalti, A.; Todaro, L.; Rita, A. Growth and allocation of woody biomass in forest trees based on environmental conditions. *Forests* **2021**, *12*, 154. [CrossRef]
49. Grime, J.P. Vegetation classification by reference to strategies. *Nature* **1974**, *250*, 26–31. [CrossRef]
50. Zhang, Y.S.; Meiners, S.J.; Meng, Y.N.; Yao, Q.; Guo, K.; Guo, W.Y.; Li, S.P. Temporal dynamics of Grime's CSR strategies in plant communities during 60 years of succession. *Ecol. Lett.* **2024**, *27*, e14446. [CrossRef] [PubMed]
51. Bird, S.; Coulson, R.N.; Crossley, D.A. Impacts of silvicultural practices on soil and litter arthropod diversity in a Texas pine plantation. *For. Ecol. Manag.* **2000**, *131*, 65–80. [CrossRef]
52. Yu, Z.; Liu, S.; Li, H.; Liang, J.; Liu, W.; Piao, S.; Tian, H.; Zhou, G.; Lu, C.; You, W.; et al. Maximizing carbon sequestration potential in Chinese forests through optimal management. *Nat. Commun.* **2024**, *15*, 3154. [CrossRef]
53. Chave, J.; Andalo, C.; Brown, S.; Cairns, M.A.; Chambers, J.Q.; Eamus, D.; Fölster, H.; Fromard, F.; Higuchi, N.; Kira, T.; et al. Tree allometry and improved estimation of carbon stocks and balance in tropical forests. *Oecologia* **2005**, *145*, 87–99. [CrossRef]
54. Sixto, H.; Cañellas, I.; van Arendonk, J.; Ciria, P.; Camps, F.; Sánchez, M.; Sánchez-González, M. Growth potential of different species and genotypes for biomass production in short rotation in Mediterranean environments. *For. Ecol. Manag.* **2015**, *354*, 291–299. [CrossRef]

forests MDPI

Article

Tree Height–Diameter Model of Natural Coniferous and Broad-Leaved Mixed Forests Based on Random Forest Method and Nonlinear Mixed-Effects Method in Jilin Province, China

Qigang Xu [1], Fan Yang [2], Sheng Hu [3], Xiao He [4] and Yifeng Hong [1,*]

1 East China Academy of Inventory and Planning, National Forestry and Grassland Administration, Hangzhou 310000, China; adslxqg@126.com
2 Academy of Forestry Inventory and Planning, National Forestry and Grassland Administration, Beijing 100714, China; yang2170057@163.com
3 Industrial Development and Planning Institute, National Forestry and Grassland Administration, Beijing 100010, China; hs356442192@163.com
4 State Key Laboratory of Efficient Production of Forest Resources, Key Laboratory of Forest Management and Growth Modelling, National Forestry and Grassland Administration, Institute of Forest Resource Information Techniques, Chinese Academy of Forestry, Beijing 100091, China; hexiao@ifrit.ac.cn
* Correspondence: hongyifeng007@126.com

Abstract: Objective: The purpose of this article was to use the Random Forest method and nonlinear mixed-effects method to develop a model for determining tree height–diameter at breast height (DBH) for a natural coniferous and broad-leaved mixed forest in Jilin Province and to compare the advantages and disadvantages of the two methods to provide a basis for forest management practice. Method: Based on the Chinese national forest inventory data, the Random Forest method and nonlinear mixed-effects method were used to develop a tree height–DBH model for a natural coniferous and broad-leaved mixed forest in Jilin Province. Results: The Random Forest method performed well on both the fitting set and validation set, with an R^2 of 0.970, *MAE* of 0.605, and *RMSE* of 0.796 for the fitting set and R^2 of 0.801, *MAE* of 1.44 m, and *RMSE* of 1.881 m for the validation set. Compared with the nonlinear mixed-effects method, the Random Forest model improved R^2 by 33.83%, while the *MAE* and *RMSE* decreased by 67.74% and 66.44%, respectively, in the fitting set; the Random Forest model improved R^2 by 9.88%, while the *MAE* and *RMSE* decreased by 14.38% and 12.05%, respectively, in the validation set. Conclusions: The tree height–DBH model constructed based on the Random Forest method had higher prediction accuracy for a natural coniferous and broad-leaved mixed forest in Jilin Province and had stronger adaptability for higher-dimensional data, which can be used for tree height prediction in the study area.

Keywords: tree height–diameter model; Random Forest; nonlinear mixed-effects model; coniferous and broad-leaved forest

1. Introduction

The tree height–diameter (H-D) model is one of the most useful tools in forest management. While measuring tree height in overcrowded and dense forests is time-consuming and difficult, DBH data can be accurately measured for all trees in a plot. Therefore, a tree H-D model is usually developed to predict the missing total height measurements for the remaining trees [1].

One of the most important factors of forest structure is the relationship between tree DBH and height [2]. As people's interest in the structure of natural mixed forests was growing, the demand for determining the relationship between tree height and diameter in mixed forests also rose. More comprehensive modeling methods were applied to the research on the H-D relationship. Walter et al. studied the relationship between tree height and diameter in unevenly aged pure beech forests using the mixed-effects model method and introduced the concept of "pseudo-residuals" to test the model [3]. It was found that the mixed-effects statistical method could fit bivariate normal distribution data very well, and the fitted model was easy to explain. Especially for forestry data, the NLME (nonlinear mixed-effects) method often showed a strong predictive ability. However, it must also be pointed out that the tree-level nonlinear mixed-effects model needs data from several groups of plots to be capable of random effect calibration when making predictions [4].

Models generated based on more dimensions and a larger amount of data had better performance. However, the nonlinear mixed-effects model had high requirements for modeling data. It was necessary to consider whether the data conformed to a normal distribution and the collinearity of independent variables. When fitting the model, the selection of initial values was also a major challenge. Machine learning methods did not have such issues. They had lower requirements for data and could contain more variables. Machine learning methods were gradually becoming a hot topic in the field of forest modeling. Chen et al. used an integrated neural network to conduct tree-level H-D modeling for six main tree species in central Canada [5]. More than 30 potentially important stand structures, sites, and climate variables were added to the model. The results showed that the model developed by this method performed excellently. Zhang et al. used traditional methods, Bayesian methods without prior information, and Bayesian methods with prior information to compare and analyze the estimation effect of the tree height–diameter relationship [6]. The results showed that the fitting results of traditional methods, Bayesian methods without prior information, and Bayesian methods with prior information were similar. However, the credible interval of Bayesian methods was more concentrated than that of traditional methods, and the parameter interval with prior information was more than 59% smaller than that of traditional methods. Shen et al. used a multilayer neural network to construct a tree height–diameter at breast height model in Guangdong Province. The research results showed that the neural network model was more accurate than the mixed-effects model [7]. Ozcelik et al. used a multilayer neural network and a mixed-effects model to develop a tree height–diameter model for Crimean juniper trees in southwestern Turkey. The results showed that both the nonlinear mixed-effects regression and backpropagation neural network modeling methods could produce accurate results. Compared with traditional nonlinear regression, the root mean square error (*RMSE*) of both was reduced by more than 20%. It was also pointed out that the backpropagation neural network seemed to be the method with the best generalization ability. In addition, from a practical perspective, compared with the mixed-effects model, its advantage is that it does not need to calibrate prior information when making predictions [8].

Among the many machine learning algorithms, the Random Forest method (RF) is an integrated decision tree algorithm. It fits models in a data-driven manner, has relatively low requirements for variables, is not restricted by statistical assumptions, and can include more variables. Consequently, in the field of forest modeling, the Random Forest algorithm is also receiving increasing attention [9]. Ou et al. used four machine learning methods (Random Forest, Boosted Regression Trees, cubist, and Multivariate Adaptive Regression Splines) to develop the individual tree basal area increment (BAI) growth model for a mixed forest in Northeast China. They found that the Random Forest method was an effective and powerful modeling method for predicting the growth of individual trees' BAI [10]. Jevšenak et al. used the Random Forest method to develop a BAI growth model based on the data of the national forest inventory of Slovenia. The results showed that the Random Forest method can provide similar verification statistical results to those of the previous traditional methods used in research reports [11].

In the field of forest management in Jilin Province, the main method used is the traditional regression model. Methods with stronger predictive abilities are currently in demand. Based on the national forest resources inventory data of natural coniferous and broad-leaved mixed forests in Jilin Province, this article used a total of 363 plots and 1305 sets of observations. The Random Forest method and the nonlinear mixed-effects method were used to develop and compare the tree height–diameter model of natural mixed forests in Jilin Province in order to identify the model with the best performance and analyze the advantages and disadvantages of the two methods. Our research results can provide help for forest management in Jilin Province.

2. Materials and Methods

2.1. Modeling Data

We used the following variables to represent stand density and stand competition status: $rSDI$ [12], BAL_j, $BAL_j\_interspecies$, $BAL_j\_intraspecies$, and $BA\_proportion$. The variable descriptions are shown in Table 1.

Table 1. Descriptions of stand density and stand competition variables.

Variable	Description	
$rSDI$	$N(\frac{D_0}{D_g})^{1.605}$	(1)
BAL_j	$\frac{1}{S}\sum_{i=1}^{n}(1_{D_i \le D_j} BA_i)$	(2)
$BAL_j\_interspecies$	$\frac{1}{S}\sum_{i=1}^{n}(1_{D_i \ge D_j \& Species_i \ne Species_j} BA_i)$	(3)
$BAL_j\_intraspecies$	$\frac{1}{S}\sum_{i=1}^{n}(1_{D_i \ge D_j \& Species_i = Species_j} BA_i)$	(4)
$BA\_proportion$	$\frac{\sum_{i=1}^{n} BA_{i,i \in species}}{\sum_{j=1}^{N} BA_{j,j \in stand}}$	(5)

Note: D represents the diameter at breast height (DBH) of an individual tree in the stand; D_0 represents the standard DBH, which is equal to the number of trees in fully stocked stands with an average diameter of 10 cm in China [13]; D_g is the average stand diameter (by basal area); BA is the basal area of an individual tree in the stand.

The tree-level data for developing the tree height–diameter model came from the sample plot data of the national forest resources inventory (NFI) of natural coniferous and broad-leaved mixed forests in Jilin Province for 2014. In each plot, the tree heights of 2 to 5 individual trees representing the average tree height level of the plot were accurately measured. After removing outliers, there were 363 sample plots and 1305 sets of individual tree height–diameter observations. The data were divided into a fitting set (1044 observations) and a validation set (261 observations). The statistical data are shown in Table 2.

Table 2. Statistics of tree height (m) and DBH (cm).

Tree Variable	Fitting Data				Validation Data			
	Max	Min	Mean	Standard Deviation	Max	Min	Mean	Standard Deviation
D/cm	79.50	7.40	22.92	10.84	71.80	7.80	23.27	11.37
rSDI	344.14	25.03	154.80	49.00	289.45	25.03	158.15	46.10
BA_proportion	0.65	0.01	0.34	0.16	0.64	0.01	0.34	0.16
BAL/m^2	59.93	0.00	13.98	8.76	45.57	0.00	14.12	7.51
$BAL\_interspecies/m^2$	55.41	0.00	9.51	7.21	30.77	0.00	9.53	6.52
$BAL\_intraspecies/m^2$	25.67	0.00	4.48	4.05	25.45	0.00	4.58	4.05
Altitude/m	1860.00	175.00	811.66	290.97	1860.00	220.00	807.19	288.46
Soil thickness/cm	70.00	10.00	42.92	10.83	70.00	10.00	42.00	11.59
H/m	32.50	4.50	16.38	4.60	32.20	5.20	16.52	4.22

Climate data were sourced from ClimateAP (V3.00). ClimateAP is an application for dynamic local downscaling of historical and future climate data in the Asia–Pacific region [14]. Based on the latitude, longitude coordinates, and altitude information of the sample plots, the ClimateAP software could be used to extract seasonal and annual climate variables for each sample plot (the time interval extracted in this study was the average value from 1980 to 2010). The candidate climate factors are shown in Table 3.

Table 3. Descriptions of the candidate climatic variables.

Variable	Description
AHM	The humiture index
CMD	The Hargreaves moisture deficit index
DD_0	The number of days below 0 °C
DD_18	The number of days below 18 °C
DD18	The number of days above 18 °C
DD5	The number of days above 5 °C
EMT/°C	The extreme low temperature in the past 30 years
EXT/°C	The extreme high temperature in the past 30 years
EREF	The Hargreaves precipitation index
MAP/mm	The mean annual precipitation
MAT/°C	The mean annual temperature
MCMT/°C	The mean coldest month temperature
MWMT/°C	The mean warmest month temperature
NFFD	The number of frost days
PAS/mm	The snowfall from August of the previous year to July of the current year
TD/°C	The temperature difference between MWMT and MCMT

2.2. *Tree Height–Diameter Model Based on Nonlinear Mixed-Effects Method*

2.2.1. Selection of Climate Variables

Due to the large number of climate variables and their tendency to exhibit collinearity, a separate screening analysis of the climate variables was carried out. Principal component analysis (PCA) can be used as an exploratory method for assessing climate variability and is robust and reliable as an auxiliary technique when combined with other statistical techniques [15,16]. We first used the PCA method to analyze the data for all climate variables. Since the units of the climate variables were different, all variables were standardized before the PCA. The principal components that explained more than 80% of the variance were retained. For each principal component, variables with large loads were selected for further analysis. The variables that had a strong correlation with H and had the least multicollinearity among them were used as candidate options for modeling.

For competing relevant indicators (density index ($rSDI$), basal area proportion of the corresponding tree species of the sample tree in the sample plot ($BA\_proportion$), BAL, $BAL\_interspecies$, $BAL\_intraspecies$), and the two aspects of indicators representing the site conditions of the sample tree (*Altitude*, *Soil thickness*), the stepwise regression method was used for screening.

2.2.2. Base Model

The basic model was based on the research of Zang [17]. The Richard model was selected. The Richard model always performed well in simulating the relationship between tree height and diameter at breast height. The formula is as follows:

$$H = 1.3 + (a_0 \times (1 - exp(-b_0 \times D))^c + \varepsilon \quad (6)$$

where H represents the tree height; D represents the diameter at breast height; a_0, b_0 and c represent the parameter, ε represents the error term.

Due to the research object being a natural mixed forest, indicators related to the competition of sample trees, namely $rSDI$, $BA\_proportion$, BAL, $BAL\_interspecies$, $BAL\_intraspecies$, and two indicators representing the site conditions of the sample trees (*Altitude* and *Soil thickness*) were selected for stepwise regression screening and then entered into the model. Therefore, the model could be written as follows:

$$\mathrm{H} = f(\beta, D, Competition\ status, Site\ condition) + \varepsilon \quad (7)$$

where β is a vector of the fixed effect parameters, $Competition\ status$ is a variable group of the competition status of individual trees, and $Site\ condition$ is a variable group representing the site condition of individual trees. Other variables are as defined before.

2.2.3. Tree Species Data and Dummy Variable

The divided groups of *Pinus koraiensis* Sieb et Zucc., *Picea koraiensis* Nakai, *Larix gmelinii* (Rupr.) Kuzen., *Pinus sylvestris* var. *sylvestriformis* (Takenouchi) Cheng et C. D. Chu, and other coniferous tree species groups *Quercus mongolica* Fisch. ex Ledeb. group, *Betula platyphylla* Sukaczev group, *Fraxinus mandshurica* Rupr., *Juglans mandshurica* Maxim. and *Phellodendron amurense* Rupr. group, *Ulmus pumila* L. group, *Acer mono* Maxim. group, *Tilia tuan* Szyszyl. group, and *Populus* L. group and miscellaneous tree group, constituting a total of 14 tree species groups, were used to construct tree species dummy variables. After adding the tree species dummy variables, the model could be written as follows:

$$H = f(\beta, D, Competition\ status, Site\ condition, S_m\ ) + \varepsilon \tag{8}$$

where S_m is the dummy variable of the tree species group. Among them, $S_1 = S_2 = \ldots = S_{13} = 0$ represents the *Pinus koraiensis* group, S_1 = 1 represents the *Picea koraiensis* group, and 0 represents other tree species groups, while S_2 = 1 represents the *Larix gmelinii* group. They were defined in turn according to the order introduced above. Other variables are as defined before.

2.2.4. Nonlinear Mixed-Effects Climate-Sensitive Model

In order to quantify the impact of the climate on *H-D* allometry, by reparameterizing the parameters in the basic *H-D* model and adding the selected climate variables to the model, the model could be written as follows:

$$H = f(\beta, D, Competition\ status, Site\ condition, Climate\ , S_m) + \varepsilon \tag{9}$$

where *Climate* represents the vector of climate variables screened by PCA and correlation analysis. Other variables are as introduced previously.

Mixed-effects models have been proven to perform excellently in the field of forestry modeling. We chose to develop a mixed-effects model with the sample plot as a random effect. The new *H-D* mixed-effects model can be written as follows:

$$H_{ij} = f(\beta, D_{ij}, Competition\ status, Site\ condition, Climate, S_m + u_i) + \varepsilon_{ij} \tag{10}$$

$$u_i \sim N(0, \sigma^2_{plot})$$

where H_{ij} and D_{ij} are the tree height and DBH of the jth individual tree of the ith tree species, repectively. u_i is the random effect representing the sample plot. ε_{ij} is the random term. Other variables are as defined before.

When making model predictions, the corresponding parameter values of the random effect were calculated using the Empirical Best Linear Unbiased Prediction method (EBLUP). The formula is as follows:

$$\hat{u}_i = \hat{\Psi}\hat{Z}_i^T(\hat{Z}_i\hat{\Psi}\hat{Z}_i^T + \hat{R}_i)^{-1}e_i \tag{11}$$

where $\hat{u}_i$ is the estimated value of the random effect. $\hat{\Psi}$ is the q × q variance–covariance matrix representing the variation between groups. q is the number of random effects. $\hat{R}_i$ is the k × k variance–covariance matrix representing the variation within groups. k is the number of observations within a group. $\hat{Z}_i$ is the partial derivative matrix of the random effect. e_i is the residual vector between the measured value and the estimated value of the fixed effect model.

The data in this paper were the plot survey data of Jilin natural forest plots in 2014. Therefore, there was no time autocorrelation problem in the data. The formula of $\hat{R}_i$ is as follows:

$$R_i = \sigma^2 G_i^{0.5} I_i G_i^{0.5} \quad (12)$$

where σ^2 is the residual vector of the model. G_i is the design matrix for explaining heteroscedasticity. In this paper, the power function form was adopted, which was $var(\varepsilon_{ij}) = \hat{H}_{ij}^{2\gamma}$; I_i is the identity matrix.

The parameter estimation of the nonlinear mixed effect was based on the nlme module of R (Version 4.3.1) [18]. The algorithm was the default restricted maximum likelihood method.

2.3. Individual Tree H-D Model of Natural Coniferous and Broad-Leaved Mixed Forest Based on Machine Learning Methods

2.3.1. Random Forest

Random Forest is an ensemble learning algorithm that parallelizes individual decision trees. The Random Forest algorithm simultaneously adopts the ideas of resampling and combined prediction. The input variables of each sub-decision tree are randomly sampled from all the feature variables of the fitting set. There is no strong dependency between each individual decision tree. Each decision tree independently learns and makes predictions. Finally, through voting, the final classification result is reached according to the principle of the minority obeying the majority. Therefore, it is called Random Forest [9]. The Random Forest algorithm has the advantages of being insensitive to missing values and having an extremely strong model generalization ability. But at the same time, it also has the disadvantage of being prone to overfitting in high-dimensional data sets. However, this can be avoided by adjusting the optimal parameters.

2.3.2. Input Variable

Unlike NLME methods, the Random Forest algorithm had good adaptability to handling high-dimensional and collinear data. Moreover, the selected variable information was all effective information that had an impact on the tree height growth. Therefore, variable screening was not performed, and all variables were selected to develop a model: competition-related indicators (*rSDI, BA_proportion, BAL, BAL_interspecies, BAL_intraspecies*), indicators representing the site conditions of the sample trees (*Altitude* and *Soil thickness*), indicators representing climate conditions (*MAT, MWMT, MCMT, TD, MAP, AHM, DD_0, DD5, DD_18, DD18, NFFD, PAS, EMT, Eref, CMD*), and *DBH* as independent variables. The fitting and validation of the model were based on the Scikit-learn package in Python (Version 3.11.5) [19,20]. In this study, Random Forest algorithms performed one_hot encoding processing on discrete factor variables (tree species). That is, a factor variable of level N was expanded into N columns of attributes. Among these N column attributes of each sample observation value, a value of 1 indicated that the sample observation belonged to this category, and all other extended attributes were 0.

2.3.3. Parameter Tuning

In the process of parameter tuning of Random Forest, we used R_{cv}^2 of ten-fold cross-validation as an observation index to select the optimal parameters in the fitting set. The formula is as follows:

$$R_{cv}^2 = \frac{1}{k}\sum_{j=1}^{k}(1 - \frac{\sum_{i=1}^{n_j}(O_{ij} - P_{ij})^2}{\sum_{i=1}^{n_j}(O_{ij} - \overline{O}_j)}) \quad (13)$$

where k is the number of folds in cross-validation. In this paper, k was set to 10 folds. O_{ij} and P_{ij}, respectively, represent the ith observed value and model predicted value of the jth fold; $\overline{O}_j$ represents the average value of the observed values of the jth fold. n_j represents the sample number of the jth fold.

2.4. Model Evaluation

This paper used three indicators for model evaluation and inspection: the coefficient of determination (R^2), mean absolute error (MAE), and root mean square error ($RMSE$). The calculation formulas were as shown in Equations (14)–(16).

$$R^2 = 1 - \frac{\sum_{i=1}^{n}(O_i - P_i)^2}{\sum_{i=1}^{n}(O_i - \overline{O})^2} \quad (14)$$

$$MAE = \frac{1}{n}\sum_{i=1}^{n}|P_i - O_i| \quad (15)$$

$$RMSE = \sqrt{\frac{1}{n}\sum_{i=1}^{n}(O_i - P_i)^2} \quad (16)$$

where O_i represents the observed value in the input data, P_i represents the predicted value of the model, and n represents the sample size of the input data. Both the MAE and $RMSE$ were indicators for measuring the distance between the predicted values and observed values.

The workflow associated with this study is illustrated in Figure 1.

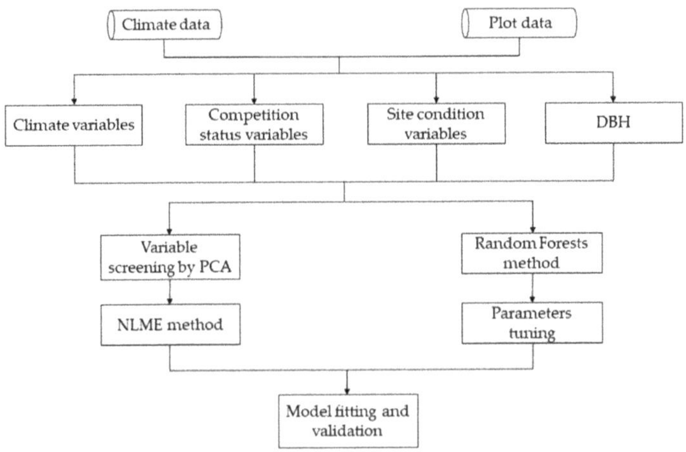

Figure 1. The workflow employed in this study to quantify the tree heights using NLME method and Random Forest method.

3. Results

3.1. Results of Nonlinear Mixed-Effects Model

3.1.1. Selection of Climate Variables

After standardizing the climate variables, principal component analysis was used. The results of the principal component analysis showed that the first two principal components explained 87.83% of the variation in all climate data.

The variable selection process of the principal component analysis was carried out according to the following principles: (1) Variables with an absolute value of load greater than 0.32 were selected. (2) If there was no variable with an absolute value of load greater than 0.32, the top three variables were selected. If there were variables with tied absolute values of load, all the tied variables were selected.

The results of each component load were shown in Table 4. For principal component 1, *DD5*, *MAT*, *MWMT*, and *DD_18* were selected for the next step. For principal component 2, *MCMT*, *TD*, and *EMT* were selected for the next step. The seven selected variables were subjected to Pearson correlation analysis with the dependent variable H. The analysis results are shown in Table 5. The results showed that the seven climate variables were all linearly significant with tree height, H, and there was collinearity among the seven climate variables. We used the variance inflation factor, VIF, to gradually eliminate variables with high collinearity. The final results are shown in Table 6. Therefore, finally, three climate variables, $MCMT$, TD, and EMT, were selected to represent the climate and added to the model.

Table 4. PCA results for the climate variables.

	Comp.1	Comp.2	Comp.3
MAT	0.279	0.107	0.15
MWMT	0.276		0.203
MCMT	0.214	0.447	−0.134
TD	0.203	−0.399	0.352
MAP	−0.228	0.212	0.473
AHM	0.258	−0.117	−0.347
DD_0	−0.256	−0.287	−0.102
DD5	0.281		0.175
DD_18	−0.276	−0.136	−0.13
DD18	0.267		0.224
NFFD	0.252	0.266	0.163
PAS	−0.256	0.132	0.299
EMT	0.196	0.455	−0.245
EXT	0.272	−0.174	
Eref	0.23	−0.261	0.137
CMD	0.232	−0.266	−0.373
Cumulative variation	76.15%	87.83%	95.22%

Table 5. Pearson correlation coefficient matrices between H and climatic variables.

	DD5	MAT	MWMT	DD_18	MCMT	TD	EMT	H
DD5	1.000	-	-	-	-	-	-	-
MAT	0.982 ***	1.000	-	-	-	-	-	-
MWMT	0.989 ***	0.957 ***	1.000	-	-	-	-	-
DD_18	−0.966 ***	−0.996 ***	−0.935 ***	1.000	-	-	-	-
MCMT	0.690 ***	0.800 ***	0.624 ***	−0.830 ***	1.000	-	-	-
TD	0.780 ***	0.662 ***	0.839 ***	−0.613 ***	0.099 ***	1.000	-	-
EMT	0.625 ***	0.700 ***	0.552 ***	−0.712 ***	0.883 ***	0.089 ***	1.000	-
H	−0.273 ***	−0.254 ***	−0.257 ***	0.238 ***	−0.180 ***	−0.201 ***	−0.232 ***	1.000

Note: *** $p < 0.001$.

Table 6. The VIFs of the final climatic factors.

Climate Variable	*MCMT*	*TD*	*EMT*
VIF	4.531	1.010	4.522

After stepwise regression to screen the competition indicators and site indicators, the indicators finally entering the model were diameter (D), stand density index ($rSDI$), basal area proportion ($BA\_proportion$), and interspecific BAL ($BAL\_interspecies$). After the PCA and correlation analysis to screen the climate variables, the climate variables that were finally entered into the model were $MCMT$, TD, and EMT. Then, the screened variables and tree species dummy variables were put into different positions of parameters a, b, and c of the basic model, respectively. The optimal model with the best coefficient significance performance was as follows:

$$H_{ij} = 1.3 + \left(a_0 + a_1 rSDI + a_2 BAL + a_3 MCMT + a_4 EMT + \sum_{m=1}^{13} f_m S_m + u_i\right)[1 - e^{-b_0 D_{ij}}]^{c_0} + \varepsilon_{ij} \quad (17)$$

where $a_0 \sim a_6$ are the parameters to be estimated. The remaining variables are as previously described.

3.1.2. Fitting and Test Results of the Final NLME Model

The model accuracy test results showed that the nonlinear mixed-effects model performed well in both the fitting set and the validation set. The *MAE* and *RMSE* of the fitting set were 1.056 m and 1.446 m, respectively, and the *MAE* and *RMSE* of the validation set were 1.420 m and 1.926 m, respectively (Table 7).

Table 7. The parameter estimation results and model validation results of final H-D NLME model.

	Parameters	Parameter Definition	Equation (17)
Fixed effects parameters	a		17.872 (0.000)
	b		0.035 (0.000)
	c		0.926 (0.000)
	a_1	*rSDI*	0.022 (0.000)
	a_2	*BAL*	−4.493 (0.005)
	a_3	*MCMT*	0.686 (0.052)
	a_4	*EMT*	−0.461 (0.025)
	f_1	*Picea asperata Mast.*	0.211 (0.609)
	f_2	*Larix gmelinii*	2.689 (0.000)
	f_3	*Pinus koraiensis*	0.122 (0.781)
	f_4	*Pinus sylvestris*	−2.296 (0.013)
	f_5	*Other coniferous tree species*	−3.829 (0.000)
	f_6	*Quercusmongolica*	−1.707 (0.002)
	f_7	*Birch*	2.157 (0.000)
	f_8	*Fraxinus mandshurica&Juglans mandshurica&Phellodendron amurense*	0.980 (0.106)
	f_9	*Elm*	−0.436 (0.488)
	f_{10}	*Acer pictumThunb.&Acer triflorum&Acer mandshuricum*	−0.103 (0.852)
	f_{11}	*linden*	0.029 (0.944)
	f_{12}	*poplar*	0.554 (0.428)
	f_{13}	*Miscellaneous wood*	−1.557 (0.060)
Variance components	σ_{plot}		0.490
	γ		0.445
Model performance			
AIC			4837.658
Fitting set R^2			0.901
Fitting set *MAE* (m)			1.056
Fitting set *RMSE* (m)			1.446
Validation set R^2			0.791
Validation set *MAE* (m)			1.420
Validation set *RMSE* (m)			1.926

3.2. Results of Random Forest Model

The Random Forest model was developed based on the scikit-learn module in Python [21]. For the three main parameters, n_estimator (the number of trees in the forest), max_depth (the number of splits that each decision tree was allowed to make) and max_features (the size of the random subsets of features to consider when splitting a node), the grid search method was used for parameter tuning. The parameter tuning process was as follows: iterative fitting with 10-fold cross-validation with the fitting set data as the object, taking the highest R^2_{cv} as the optimal parameter, and calculating step by step to obtain the optimal parameter combination. The final optimal parameters were n_sitimators = 370; max_depth = 17; and max_features = 15. Then, a Random Forest model was developed based on the optimal parameters.

The tree height–diameter model of the Jilin natural forest was developed by using Random Forest. The test results of the model on the fitting set and the validation set were as follows: The Random Forest method performed excellently on both the fitting set and the validation set. The R^2 of the fitting set was 0.970, the R^2 of the validation set was 0.801, the *MAE* was 1.44 m, and the *RMSE* was 1.881 m. Compared with the NLME model, in the training, the R^2 of Random Forest was increased by 33.83%, and the *MAE* and *RMSE* were decreased by 67.74% and 66.44%, respectively; in the performance of the Random Forest model on the validation set, R^2 was increased by 9.88%, and the *MAE* and *RMSE* were decreased by 14.38% and 12.05%, respectively (Table 8).

Table 8. Validation result of H-D model based on 3 machine learning methods and NLME method.

	Random Forest	NLME Method
Training set R^2	0.970	0.901
Training set *MAE* (m)	0.605	1.056
Fitting set *RMSE* (m)	0.796	1.446
Validation set R^2	0.801	0.791
Validation set *MAE* (m)	1.440	1.420
Validation set *RMSE* (m)	1.881	1.926

The residual plots of the finally fitted NLME model and the Random Forest model are shown in Figure 2. The results showed that the scatter points were randomly distributed on both sides of the 0-axis. No obvious heteroscedasticity trend was observed for the NLME model. There was an obvious trend in the lower graph, where the residuals in the Random Forest model were not uniformed distributed. The Random Forest model had a better predictive ability, and the scatter points were more closely distributed.

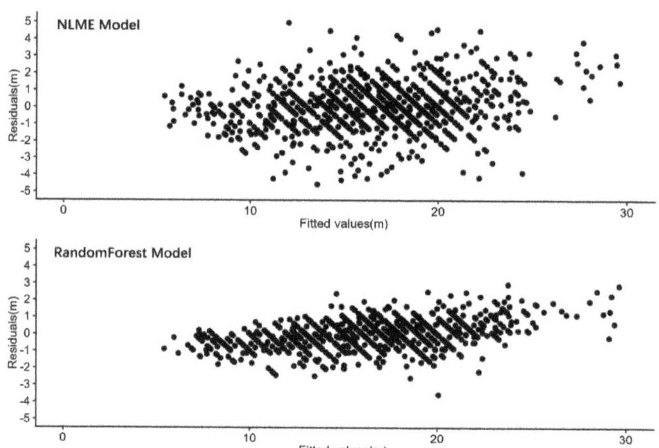

Figure 2. Residuals for tree height–diameter models (NLME model; Random Forest model).

4. Discussion

4.1. Nonlinear Mixed-Effects Model

The test results of using the nonlinear mixed-effects method to develop the tree height–diameter model for natural mixed forests in Jilin Province showed that the model had an excellent extrapolation ability. The nonlinear mixed-effects method had always performed well while developing the model. Sharma et al. used Czech national forest survey data to develop a nonlinear mixed-effects HDR (height–diameter ratio) model. The results showed that the R^2 of the HDR model for each tree species ranged from 0.8574 to 0.9605 [4]. Ciceu et al. used the nonlinear mixed-effects method to construct a tree height–diameter model for an unevenly aged mixed forest of Norway spruce in Romania. The research results showed that the addition of random effects increased the prediction accuracy of the tree height by 50 cm [22]. Meng et al. used the nonlinear mixed-effects method to construct a tree height–diameter model for the main dominant tree species in northern Ontario, Canada. After adding random effects, the fitting results and prediction accuracy of the models for all tree species were improved [23]. In forestry modeling, compared to methods that only used fixed effects, the mixed-effects method could incorporate more variable information, such as plot location information, into the model. And plot information contains many useful factors that cannot be included in fixed effects models. Therefore, compared with traditional regression methods, tree height–diameter models with random effects usually perform better in terms of model fitting performance and prediction accuracy.

4.2. The Influence of Temperature and Competition on the H-D Relationship

The results of the significance test of the model parameters showed that although the *p*-value of *MCMT* was at the 0.1 level, the rest of the coefficients were all significant at the 0.05 level. The coefficient of *rSDI* was positive, indicating that individual trees under high-density pressure will invest more resources in tree height growth to obtain more sunlight. Many studies have found that competition will have a significant impact on the allometric growth of trees. The greater the intensity of competition is, the thinner the individual trees in the stand will grow [24–28]. The coefficient of *MCMT* was positive, which meant that the higher the average temperature in the coldest month is, the faster the increase in the tree height-to-diameter ratio of individual trees in the study area is. Previous research results have shown that for the growth of individual trees, there was an optimal temperature for photosynthesis and growth, and the influence of the temperature on tree height growth was greater than that on diameter growth [26]. The study areas in this paper all belong to northern China, and the annual average temperature is relatively cold and has not reached the turning point of the optimal temperature. Therefore, for individual trees in the study area, an increase in temperature would make the tree height increase more rapidly.

4.3. Machine Learning Algorithms and Forestry Modeling

Machine learning methods have gradually become a hotspot in the field of forest modeling in recent years, because they often have better predictive capabilities than traditional regression methods. These methods can contain more variables. Ogana et al. [29] used three methods, the DLA (Deep Learning Algorithm), NLS, and NLME, to study the H-D relationship of complex tropical rainforest trees. It was found that the DLA model was superior to the NLS and NLME models. Compared with NLS and NLME, the error of estimating aboveground biomass by tree height predicted by using the DLA model was reduced by more than 30%. The deep learning network model can be regarded as an alternative to traditional nonlinear regression techniques. Qin et al. used the DLA method and the NLME method to construct a natural mixed forest crown model. It was found that the best DLA model can explain 69% of the crown variation. When all 22 input variables were used for modeling, the DLA model performed better than the NLME model [30].

However, Dantas et al. employed the ANN (Artificial Neural Network), SVM (Support Vector Machine), and NLME methods to construct a volume model for *eucalyptus* plantations in Minas Gerais, Brazil. They found that the nonlinear mixed-effects model performed the best [31]. A possible reason for this result was that the research object of the article was *eucalyptus* plantations, which had a relatively simple structure. The variables affecting the volume of *eucalyptus* were more explicit. Therefore, using the nonlinear mixed-effects method could already accurately express the variations in *eucalyptus* volume. The two machine learning methods, ANN and SVM, required more hyperparameters to be adjusted. It was more difficult to find the optimal parameters in the parameter space to express the variations in *eucalyptus* volume. Consequently, in this study, the nonlinear mixed-effects model showed the best performance.

Our research results showed that compared with the nonlinear mixed-effects method, in the performance of the fitting set of the Random Forest model, R^2 was increased by 33.83%, and the *MAE* and *RMSE* were decreased by 67.74% and 66.44%, respectively; in the performance of the validation set, R^2 was increased by 9.88%, and the *MAE* and *RMSE* were decreased by 14.38% and 12.05%, respectively. The Random Forest method showed excellent performance, which was similar to the results of previous studies. Yu Yang et al. used beta regression and the Random Forest algorithm to develop a crown ratio (CR) model. Their research results showed that the CR model developed based on the RF algorithm was superior to the model developed by beta regression. The algorithm idea based on the integrated system can improve the accuracy by itself. Coupled with the interpretation ability of multiple variables, the Random Forest algorithm has a stronger prediction ability [32].

5. Conclusions

This paper used the nonlinear mixed-effects method and the Random Forest method to construct a climate-sensitive tree height–diameter model for natural coniferous and broad-leaved mixed forests in Jilin Province. The parameter estimation and test results of the nonlinear mixed-effects method showed that temperature and competition were the key variables affecting the allometric increase in individual tree height–diameter. For individual trees in natural mixed forests in Jilin Province, an increase in temperature would make the increase in individual tree height more rapid. The model test results showed that compared with the mixed-effects method, the model constructed by the Random Forest algorithm had a higher prediction accuracy. The Random Forest method showed a strong generalization ability when constructing the tree height–diameter model and had low requirements for modeling data. In application scenarios with high requirements for prediction accuracy, it had advantages over traditional models. Our research results can provide decision-making support for forest management in Jilin Province.

Author Contributions: Methodology, data curation, formal analysis, writing—original draft preparation, and writing—review and editing, Q.X.; formal analysis, software, and writing—review and editing, F.Y., S.H. and X.H.; funding acquisition, conceptualization, and writing—review and editing, Y.H. All authors have read and agreed to the published version of the manuscript.

Funding: This research was funded by the National Key R&D Program of China (2022YFE0112700), the National Key R&D Program of China (2023YFD2201705-6), and the Research on the Path of Baishanzu National Park to Assist Regional Carbon Neutrality (2022JBGS08).

Data Availability Statement: Data are contained within the article.

Acknowledgments: The authors are grateful to the State Key Laboratory of Efficient Production of Forest Resources, Key Laboratory of Forest Management and Growth Modelling, National Forestry and Grassland Administration, Institute of Forest Resource Information Techniques, and Chinese Academy of Forestry for providing the data used in this study. We would also like to thank the Editors and anonymous reviewers for their valuable comments.

Conflicts of Interest: The authors declare no conflicts of interest.

References

1. Crecente-Campo, F.; Tomé, M.; Soares, P.; Diéguez-Aranda, U. A generalized nonlinear mixed-effects height–diameter model for Eucalyptus globulus L. in northwestern Spain. *For. Ecol. Manag.* **2010**, *259*, 943–952. [CrossRef]
2. Knoebel, B.R.; Burkhart, H.E. A bivariate distribution approach to modeling forest diameter distributions at two points in time. *Biometrics* **1991**, *47*, 241–253. [CrossRef]
3. Zucchini, W.; Schmidt, M.; von Gadow, K. A model for the diameter-height distribution in an uneven-aged beech forest and a method to assess the fit of such models. *Silva Fenn.* **2001**, *35*, 169–183. [CrossRef]
4. Sharma, R.P.; Vacek, Z.; Vacek, S.; Kučera, M. A nonlinear mixed-effects height-to-diameter ratio model for several tree species based on Czech national forest inventory data. *Forests* **2019**, *10*, 70. [CrossRef]
5. Chen, J.; Yang, H.; Man, R.; Wang, W.; Sharma, M.; Peng, C.; Parton, J.; Zhu, H.; Deng, Z. Using machine learning to synthesize spatiotemporal data for modelling DBH-height and DBH-height-age relationships in boreal forests. *For. Ecol. Manag.* **2020**, *466*, 118104. [CrossRef]
6. Zhang, X.; Duan, A.; Zhang, J.; Xiang, C. Estimating Tree Height-Diameter Models with the Bayesian Method. *Sci. World J.* **2014**, *2014*, 683691. [CrossRef]
7. Shen, J.; Hu, Z.; Sharma, R.P.; Wang, G.; Meng, X.; Wang, M.; Wang, Q.; Fu, L. Modeling height–diameter relationship for poplar plantations using combined-optimization multiple hidden layer back propagation neural network. *Forests* **2020**, *11*, 442. [CrossRef]

8. Özçelik, R.; Diamantopoulou, M.J.; Crecente-Campo, F.; Eler, U. Estimating Crimean juniper tree height using nonlinear regression and artificial neural network models. *For. Ecol. Manag.* **2013**, *306*, 52–60. [CrossRef]
9. Breiman, L. Random Forest. *Mach. Learn.* **2001**, *45*, 5–32. [CrossRef]
10. Ou, Q.; Lei, X.; Shen, C. Individual tree diameter growth models of larch–spruce–fir mixed forests based on machine learning algorithms. *Forests* **2019**, *10*, 187. [CrossRef]
11. Jevšenak, J.; Skudnik, M. A random forest model for basal area increment predictions from national forest inventory data. *For. Ecol. Manag.* **2021**, *479*, 118601. [CrossRef]
12. Reineke, L.H. Perfecting a stand-density index for even-aged forests. *J. Agric. Res.* **1933**, *46*, 627–638.
13. Meng, X.; She, G.; Li, F.; Wang, X. *Forest Mensuration*; China Forestry Publishing House: Beijing, China, 2006.
14. Wang, T.; Wang, G.; Innes, J.L.; Seely, B.; Chen, B. ClimateAP: An application for dynamic local downscaling of historical and future climate data in Asia Pacific. *Front. Agric. Sci. Eng.* **2017**, *4*, 448–458. [CrossRef]
15. Scolforo, J.R.S.; Maestri, R.; Ferraz Filho, A.C.; de Mello, J.M.; de Oliveira, A.D.; de Assis, A.L. Dominant height model for site classification of Eucalyptus grandis incorporating climatic variables. *Int. J. For. Res.* **2013**, *2013*, 139236.
16. Wold, S.; Esbensen, K.; Geladi, P. Principal component analysis. *Chemom. Intell. Lab. Syst.* **1987**, *2*, 37–52. [CrossRef]
17. Zang, H.; Lei, X.; Ma, W.; Zeng, W. Spatial heterogeneity of climate change effects on dominant height of larch plantations in northern and northeastern China. *Forests* **2016**, *7*, 151. [CrossRef]
18. Team, R.C. *R: A Language and Environment for Statistical Computing*; Foundation for Statistical Computing: Vienna, Austria, 2013; Available online: http://www.r-project.org (accessed on 12 January 2016).
19. Abadi, M.; Barham, P.; Chen, J.; Chen, Z.; Davis, A.; Dean, J.; Devin, M.; Ghemawat, S.; Irving, G.; Isard, M. {TensorFlow}: A system for {Large-Scale} machine learning. In Proceedings of the 12th USENIX Symposium on Operating Systems Design and Implementation (OSDI 16), Savannah, GA, USA, 2–4 November 2016; pp. 265–283.
20. Gulli, A.; Pal, S. *Deep Learning with Keras*; Packt Publishing Ltd.: Birmingham, UK, 2017.
21. Pedregosa, F.; Varoquaux, G.; Gramfort, A.; Michel, V.; Thirion, B.; Grisel, O.; Blondel, M.; Prettenhofer, P.; Weiss, R.; Dubourg, V. Scikit-learn: Machine learning in Python. *J. Mach. Learn. Res.* **2011**, *12*, 2825–2830.
22. Ciceu, A.; Garcia-Duro, J.; Seceleanu, I.; Badea, O. A generalized nonlinear mixed-effects height–diameter model for Norway spruce in mixed-uneven aged stands. *For. Ecol. Manag.* **2020**, *477*, 118507. [CrossRef]
23. Meng, S.X.; Huang, S.; Lieffers, V.J.; Nunifu, T.; Yang, Y. Wind speed and crown class influence the height–diameter relationship of lodgepole pine: Nonlinear mixed effects modeling. *For. Ecol. Manag.* **2008**, *256*, 570–577. [CrossRef]
24. Crecente-Campo, F.; Corral-Rivas, J.J.; Vargas-Larreta, B.; Wehenkel, C. Can random components explain differences in the height–diameter relationship in mixed uneven-aged stands? *Ann. For. Sci.* **2014**, *71*, 51–70. [CrossRef]
25. Forrester, D.I.; Benneter, A.; Bouriaud, O.; Bauhus, J. Diversity and competition influence tree allometric relationships–Developing functions for mixed-species forests. *J. Ecol.* **2017**, *105*, 761–774. [CrossRef]
26. Fortin, M.; Van Couwenberghe, R.; Perez, V.; Piedallu, C. Evidence of climate effects on the height-diameter relationships of tree species. *Ann. For. Sci.* **2019**, *76*, 1. [CrossRef]
27. Garber, S.M.; Temesgen, H.; Monleon, V.J.; Hann, D.W. Effects of height imputation strategies on stand volume estimation. *Can. J. For. Res.* **2009**, *39*, 681–690. [CrossRef]
28. Temesgen, H.; v Gadow, K. Generalized height–diameter models—An application for major tree species in complex stands of interior British Columbia. *Eur. J. For. Res.* **2004**, *123*, 45–51. [CrossRef]
29. Ogana, F.N.; Ercanli, I. Modelling height-diameter relationships in complex tropical rain forest ecosystems using deep learning algorithm. *J. For. Res.* **2022**, *33*, 883–898. [CrossRef]
30. Qin, Y.; Wu, B.; Lei, X.; Feng, L. Prediction of tree crown width in natural mixed forests using deep learning algorithm. *For. Ecosyst.* **2023**, *10*, 100109. [CrossRef]
31. Dantas, D.; Calegario, N.; Acerbi, F.W.; Carvalho, S.d.P.C.; Isaac, M.A.; Melo, E.d.A. Multilevel nonlinear mixed-effects model and machine learning for predicting the volume of Eucalyptus spp. trees. *Cerne* **2020**, *26*, 48–57. [CrossRef]
32. Yu, Y.; Zhou, Z.; Sharma, R.P.; Zhang, L.; Du, M.; Zhang, H. Comparing crown ratio models for spruce-fir broadleaved mixed forests using beta regression and random forest algorithm. *Comput. Electron. Agric.* **2024**, *225*, 109302. [CrossRef]

forests

MDPI

Article

LiDAR-Based Modeling of Individual Tree Height to Crown Base in *Picea crassifolia* Kom. in Northern China: Comparing Bayesian, Gaussian Process, and Random Forest Approaches

Zhaohui Yang [1,*], Hao Yang [1], Zeyu Zhou [2], Xiangxing Wan [3,4], Huiru Zhang [2] and Guangshuang Duan [5]

1 School of Forestry, Shanxi Agricultural University, Taiyuan 030031, China
2 Experimental Center of Forestry in North China, Chinese Academy of Forestry, Beijing 102300, China
3 China Aero Geophysical Survey and Remote Sensing Center for Natural Resources, Beijing 100083, China
4 Key Laboratory of Airborne Geophysics and Remote Sensing Geology, Ministry of Nature and Resources, Beijing 100083, China
5 School of Mathematics and Statistics, Xinyang Normal University, Xinyang 464000, China
* Correspondence: yangzhaohui@sxau.edu.cn; Tel.: +86-15501150306

Abstract: This study compared hierarchical Bayesian, mixed-effects Gaussian process regression, and random forest models for predicting height to crown base (*HCB*) in Qinghai spruce (*Picea crassifolia* Kom.) forests using LiDAR-derived data. Both modeling approaches were applied to a dataset of 510 trees from 16 plots in northern China. The models incorporated tree-level variables (height, diameter at breast height, crown projection area) and plot-level spatial competition indices. Model performance was evaluated using leave-one-plot-out cross-validation. The Gaussian mixed-effects process model (with an *RMSE* of 1.59 and *MAE* of 1.25) slightly outperformed the hierarchical Bayesian model and the random forest model. Both models identified LiDAR-derived tree height, *DBH*, and LiDAR-derived crown projection area as primary factors influencing *HCB*. The spatial competition index (*SCI*) emerged as the most effective random effect, with the lowest *AIC* and *BIC* values, highlighting the importance of local competition dynamics in *HCB* formation. Uncertainty analysis revealed consistent patterns across the predicted values, with an average relative uncertainty of 33.89% for the Gaussian process model. These findings provide valuable insights for forest management and suggest that incorporating spatial competition indices can enhance *HCB* predictions.

Keywords: height to crown base; LiDAR; hierarchical Bayesian model; Gaussian process regression; *Picea crassifolia* Kom.

1. Introduction

Height to crown base (*HCB*) is an important tree attribute that determines the quality class of sawn logs and the value of timber [1]. At the stand level, mean *HCB* is also an indicator of wood quality [2,3]. Theoretically, *HCB* is a critical parameter in constructing allometric growth models of trees [4] and plays an important role in the allometric equations of biomass components, such as branches, bark, and foliage; it holds considerable value in elucidating the intrinsic connections between crown and trunk growth in various processes, such as carbon allocation and nutrient transfer [5,6]. Practically, *HCB* serves as a crucial basis for designing harvesting plans and determining appropriate harvesting intensity [7,8]. The accurate prediction of *HCB* can maximize the usable timber volume, reduce harvesting residues, and enhance both wood utilization and economic benefits [9,10].

Moreover, *HCB* is closely linked to stand tending and fire management. The proper regulation of *HCB* can improve light conditions within the stand and promote tree growth through tending measures, such as pruning [11,12]. On the other hand, HCB significantly influences the risk and intensity of wildfires' spread [13,14]. Therefore, effective management of *HCB* is crucial for the prevention and control of forest fires [15,16].

Despite the significant importance of height to crown base (HCB) in forestry science and management, current research on *HCB* remains insufficient in several areas. Firstly, many existing *HCB* models are based on traditional statistical methods, such as linear regression, which often fail to capture the complexity and spatial heterogeneity of the processes involved in *HCB* formation. Advanced modeling approaches, including nonlinear simultaneous regression and the use of remote sensing data, such as LiDAR data, have shown promise in improving the accuracy of *HCB* predictions by accounting for these complexities [17]. Maltamo [18] developed *HCB* models by applying the linear mixed-effects (LME) modeling approach using LiDAR-retrieved percentiles of tree height, diameter at breast height (*DBH*), and field-measured tree height. However, given that *HCB* data often exhibit distinct nonlinear patterns, the use of linear functions to describe such relationships may be inadequate. Recognizing this limitation, Yang [17] established an *HCB* model based on LiDAR data using a nonlinear mixed-effects approach. Nevertheless, these studies have not fully addressed the inherent uncertainties associated with *HCB* modeling. Furthermore, the impact of spatial structure on *HCB* modeling has not been adequately accounted for in existing research. Spatial structure, including tree distribution patterns and competitive relationships, significantly influences trees' growth and morphological development. Incorporating these spatial elements into the model could potentially enhance the accuracy of *HCB* predictions substantially.

Enhanced predictive accuracy of *HCB* models is crucial for both practical forestry operations and theoretical research [19]. Accurate *HCB* models can not only optimize forest management practices, such as thinning and pruning, to improve stand productivity and health, but they can also deepen our understanding of the processes governing trees' growth and development, thereby providing important theoretical support for sustainable forest management.

Another crucial factor is data precision. This study employs data acquired through UAV-LiDAR technology, which provides more accurate measurements of tree height (H) and crown area compared to traditional ground-based techniques. UAV-LiDAR not only covers larger areas but also offers higher resolution and more precise three-dimensional forest structure information, thereby furnishing more reliable input data for *HCB* modeling. The use of high-precision UAV-LiDAR data is one of the major innovations of this study.

This study aims to systematically assess the uncertainties in *HCB* predictions using hierarchal Bayesian methods, mixed-effects Gaussian process regression models, and the random forest model, which has rarely been addressed in previous studies. A comprehensive exploration of uncertainty is crucial for understanding the reliability and limitations of models' predictions, especially when dealing with complex ecological processes, such as tree growth and crown development [20]. Failing to account for various sources of uncertainty, such as measurement errors, model misspecification, and parameter variability, may lead to overconfident or misleading inferences [21,22]. Uncertainty analysis not only helps to understand the reliability and limitations of models' predictions but also serves as an important evaluation metric for synthesis science, deepening our understanding of ecosystem processes and functions [23].

Bayesian inference is particularly well suited for this purpose, as it is based on probability theory and provides estimates of parameter uncertainty in the form of probability distributions, explicitly considering prior knowledge [24]. This contrasts with the frequentist approach, which treats model parameters as fixed and provides point estimates [25,26]. Bayesian methods have been increasingly applied in ecological and forestry research, providing new perspectives for modeling complex ecological data.

In addition to hierarchical Bayesian models, Gaussian process models have gained increasing attention in ecological and environmental modeling due to their flexibility and ability to capture complex nonlinear relationships [27]. Gaussian process models are nonparametric Bayesian models that define a prior distribution over functions, which can be updated based on observed data to obtain a posterior distribution [28]. The key advantage of Gaussian process models lies in their ability to provide a natural measure of uncertainty in predictions, taking into account both the inherent variability in the data and the uncertainty in the model itself [29]. Moreover, Gaussian process models can seamlessly incorporate various types of input data, such as remote sensing data and spatial information, making them particularly suitable for modeling ecological processes at different scales [30].

In addition to Bayesian and Gaussian process models, random forest models offer a powerful alternative for ecological modeling. Random forest models are an ensemble learning technique that builds multiple decision trees using different subsets of the training data and combines their predictions to improve robustness and reduce overfitting. This method does not rely on strict parametric assumptions, making it a versatile choice for a wide range of ecological applications, including the analysis of tree growth, species distribution, and environmental influences.

Hierarchical models, such as hierarchical Bayesian models and Gaussian process mixed-effects models, offer several advantages in addressing the challenges associated with modeling complex ecological data [31,32]. These advantages include (1) improving estimates for repeated measurements by avoiding underfitting or overfitting, (2) handling imbalanced sampling by automatically accounting for varying uncertainty across groups, (3) explicitly estimating variation at different levels, and (4) retaining variation by avoiding the pitfalls of pre-averaging data [33,34].

The main objective of this research is to develop a suitable LiDAR-based individual *HCB* prediction model for *Picea crassifolia* Kom. with canopy cover around 0.6, specifically in the Dayekou Basin. Therefore, we strive to construct a relatively complex model with strong predictive capability while accounting for plot-level variability. Compared to existing *HCB* models, the innovations of the proposed model in this study include the following: (1) focusing on the natural secondary forests of *Picea crassifolia*, a regionally important forest type, to fill the gap in the current growth model system; (2) introducing novel spatial structure indices to reveal the intrinsic relationship between spatial patterns and tree growth; and (3) comparing the performance of Bayesian models, Gaussian process regression models, and random forest models to provide a reference for model selection in future *HCB* research.

2. Methods

2.1. Data

2.1.1. Study Area

The study area was the Dayekou Observation Station of the Xishui forest farm in Sunan Yuguzu Autonomous County, Gansu Qilian Mountain National Nature Reserve (38°32′ N, 100°15′ E) (Figure 1). This area is a natural secondary forest in the shallow mountain area of the Qilian Mountains, belonging to the water conservation forest of Heihe River Basin, with an altitude of 2700–3000 m. The climate of this area is an alpine, semi-arid, and semi-humid mountain forest steppe climate. The slope exposed to the Sun is mountain grassland, while the northern slope is the forest landscape. The dominant species in the forest is *Picea crassifolia* Kom. The ground is covered with a moss layer ~10 cm thick, and the vegetation is in good condition.

2.1.2. Plot Setting

A 100 m × 100 m grid was laid out along the slope with an azimuth angle of 122 degrees. The grid was divided into sixteen 25 m × 25 m plots, and the stakes were set up at corners and at the approximate center points of each sample plot (shown as Figure 2). During the plot layout, a compass and measuring tape were used, with the compass for determining direction and the tape for measuring distances. During individual tree measurements, all trees with a *DBH* greater than 2.5 cm within the plot were surveyed, including leaning trees, standing dead trees, and snags. *DBH* was measured using a diameter tape, while *HCB* was measured using a laser rangefinder (TruPulse 200, Laser Technology, Inc., Centennial, CO, USA). For shorter trees, a measuring pole was used to determine the tree height and the height to the base of the crown. Crown width was measured using a tape measure along two main directions. Additionally, leaf area index (LAI) measurements were conducted using the LAI2000 (LI-COR Biosciences, Lincoln, NE, USA) and a fisheye camera.

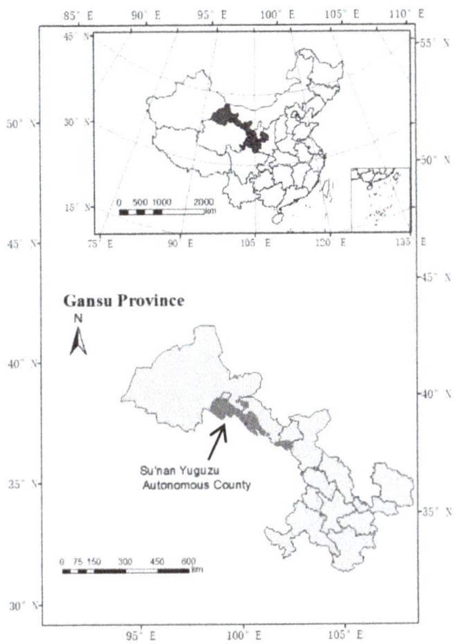

Figure 1. Location of the sample plot.

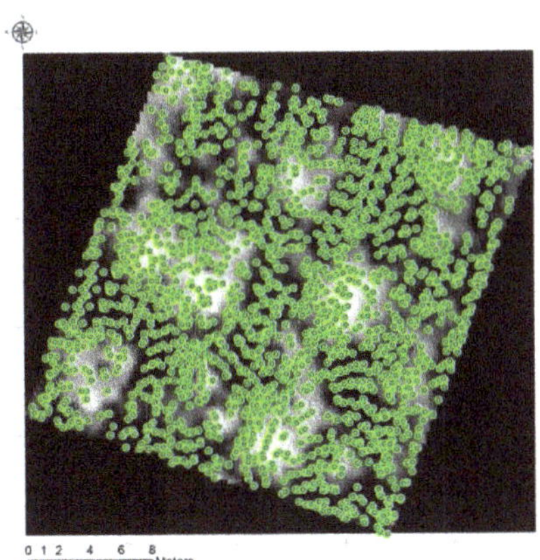

Figure 2. Distribution of sampled trees within the sample plots.

2.1.3. ALS Data

The full-waveform LiDAR data were collected using a Riegel LMS-Q560 laser scanner (RIEGL Laser Measurement Systems GmbH, Horn, Austria) mounted on an airplane, with a 50 kHz pulse repetition frequency, a 49 kHz scanning frequency, and a 30° maximum scanning angle. The ground-measured data were also collected. The wavelength was 1550 nm, and the pulse length and laser beam divergence were 3.5 ns and 0.5 mrad, respectively. The average flight height was 3699 m, and the average flight speed was 230 km/h^{-1}. The mean density of point clouds was 4.34 m^{-2}. The spatial distribution of 510 neighbor-smoothed *Picea crassifolia* Kom. trees is shown in Figure 2. The data are summarized in Table 1.

Table 1. Descriptive statistics of measurements of tree variables (*hL*, LiDAR-derived total tree height; caL, LiDAR-derived crown projection area; *HCB*, height to crown base; *DBH*, diameter at breast height; *SD*, standard deviation).

Variable	Min	Max	Mean	SD
hL (m)	4.62	22.15	13.84	3.20
caL (m^2)	3.19	38.63	12.21	5.47
HCB (m)	0.90	10.20	4.80	3.52
DBH (cm)	3.60	81.10	22.57	8.54
aCI	1.36	3.10	1.93	0.34
Hegyi	0.24	11.86	2.13	1.43
Aggregation index	2.34	3.28	2.71	0.24
SCI	0.00	2.64	1.13	0.60

The point cloud was generated from LiDAR waveforms provided by Pang et al. [35]. Ground points were classified using the Ground algorithm in TerraScan 005 (Terrasolid, Helsinki, Finland), following Axelsson's method (1999) [36]. These points were used to create a digital elevation model (DEM) at a 0.25 m resolution using a triangulation algorithm. Similarly, a digital surface model (DSM) at a 0.25 m resolution was generated using ground and vegetation points with the highest hit z algorithm, assigning the highest z-value within each cell. The canopy height model (CHM) at a 0.25 m resolution was obtained by subtracting the DEM from the DSM [37]. To smooth pits in the CHM, a neighbor smoothing algorithm was applied [38].

Crown tops were identified from the CHM using a local maxima algorithm with a 2.0 m window size, and LiDAR-derived tree height (*hL*) was estimated as the height of these crown tops. Crown edges were delineated using a region-growing algorithm, and the LiDAR-derived crown area (*caL*) was calculated as the average horizontal diameter in the west–east and north–south directions [37].

2.2. Models

2.2.1. *HCB* Base Model

We applied the base model in logistic form, similar to Water and Hanns's research [39], and the logistic form base model was introduced with variables, such as individual *hL* (LiDAR-based tree height), *caL* (LiDAR-based crown projection area), *DBH* (diameter at breast height), as follows:

$$HCB = hL/(1.0+ exp(\beta_0 + \beta_1 DBH + \beta_2 caL)) \quad (1)$$

2.2.2. Included Spatial Covariates

We calculated four spatial indices to characterize the competitive environment and spatial structure of the forest stand, namely Hegyi's competition index (*CI*), the aggregation index (*R*), and the angular competition index (*aCI*), as follows:

$$hcb \ = \ hL/(1.0{+}exp(\beta_0 + \beta_1 DBH{+}\beta_2 caL{+}\beta_3 S)) \quad (2)$$

where S represents the spatial covariate included in the model.

Hegyi's CI [40] quantifies the competitive pressure on a target tree from its neighbors. For each target tree *i*, CI_i was calculated as follows:

$$CI_i = \sum \frac{D_j}{D_i * dist_{ij}} \quad (3)$$

where D_i and D_j are the diameter at breast height of the target tree *i* and competitor tree *j*, respectively, and $dist_{ij}$ is the distance between trees *i* and *j*. Only trees within a 7 m radius were considered competitors.

The aggregation index (R) [41] measures the spatial distribution pattern of trees within a plot as follows:

$$R = \left(\frac{1}{n}\right) * \sum r_i \tag{4}$$

where n is the number of trees in the plot and r_i is the distance from tree i to its nearest neighbor.

The aCI [42] considers both the size and spatial arrangement of neighboring trees Las follows:

$$aCI = \sum \frac{(a_1 + a_2 * c_{ij})}{\pi} \tag{5}$$

where $a_1 = arctan\left(\frac{H_j}{d_{ij}}\right)$, $a_2 = arctan\left(\frac{H_j - H_i}{d_{ij}}\right)$, and C_{ij} is 1 if $H_j > H_i$ and 0 otherwise. H_i and H_j are the heights of the target and competitor trees, respectively, and d_{ij} is the distance between them. The four nearest neighbors were considered for each target tree.

All indices were calculated using R, with tree locations specified by X and Y coordinates. The results provide a comprehensive characterization of the competitive environment and spatial structure of the forest stand.

2.2.3. HCB Hierarchical Bayesian Model

(1) Bayes' rule

Bayes' rule is usually expressed as follows:

$$p(y, \theta) = p(y|\theta)p(\theta) = p(\theta|y)p(y) \tag{6}$$

where $y = (y_1, y_2, y_3, \ldots)$ is a vector of data and $\theta = (\theta_1, \theta_2, \theta_3, \ldots)$ is a vector of the parameters. p represents the density function or probability distribution, $p(y, \theta)$ is the joint probability of y and θ and, $p(\theta)$ is the marginal density of θ.

Unlike a statistical reference, a Bayesian reference uses conditional probability distribution to describe uncertainty in the estimated parameters. In order to estimate parameters, posterior distribution should be obtained as follows:

$$p(\theta|y) = \frac{P(y|\theta)p(\theta)}{p(y)} \tag{7}$$

where $p(\theta|y)$ is the posterior distribution of parameters, $p(\theta)$ is the prior distribution of parameters, $P(y|\theta)$ is the likelihood function, and $p(y)$ is the average likelihood, which is used for normalization to guarantee that the integral value of posterior distribution of the defined domain is 1.

Therefore, the posterior distribution is proportional to the product of likelihood and prior distribution, as follows:

$$p(\theta|y) \propto P(y|\theta)p(\theta) \tag{8}$$

In our study, the relationship between HCB, DBH, and hL was described as follows:

$$HCB = hL/(1.0 + \exp(\beta_0 + \beta_1 DBH + \beta_2 caL)) \tag{9}$$

$$HCB \sim N(f(hL, DBH, caL : \beta_0, \beta_1, \beta_2) \tag{10}$$

where $f(hL, DBH, caL : \beta_0, \beta_1, \beta_2) = hL/(1.0 + exp(\beta_0 + \beta_1 DBH + \beta_2 caL))$ gives the mean distribution of HCB. Therefore, the likelihood function can be expressed as follows:

$$p(y|\beta_0, \beta_1, \beta_2) = \Pi_i \frac{1}{\sqrt{2\pi\sigma}} exp\left(\frac{-(HCB_i - f(hL_i, DBH_i, caL_i))^2}{2\sigma^2}\right) \tag{11}$$

where i is the number of observations.

(2) Hierarchical Bayesian model development

We developed a hierarchical Bayesian model with random effects to predict the height to crown base (*HCB*) of trees. The model was formulated as follows:

$$HCB = LH/(1 + exp(\sum(\beta_i X_i) + \sum(u_j X_j))) \quad (12)$$

where hL is the tree height, X_i represents fixed-effect predictors (*DBH*, *caL*, and spatial competition index), X_j represents predictors with random effects, β_i represents fixed-effect coefficients, and u_j represents random-effects coefficients varying by plot.

We assumed normal priors for fixed effects ($\beta_i \sim N(0, 10^2)$) and random effects ($u_j \sim N(0, \sigma_u^2)$), with a half-normal prior on the random-effects standard deviation ($\sigma_u \sim HalfNormal(1)$). The model error was assumed to be normally distributed, with a half-normal prior on its standard deviation ($\sigma \sim HalfNormal(1)$).

(3) Parameter estimation

We implemented the model using PyMC3 running on Python version 3.9 and estimated the parameters via Markov chain Monte Carlo (MCMC) sampling, with 2000 tuning steps and 2000 sampling steps. We explored all possible combinations of random effects on predictors (excluding tree height) and compared the models using the root-mean-square error (*RMSE*), Akaike information criterion (*AIC*), Bayesian information criterion (*BIC*), and log-likelihood.

(4) Uncertainty analysis of Hierarchical Bayesian Model

In Bayesian modeling using MCMC sampling, an uncertainty analysis was conducted for comprehensive decision-making based on full datasets. The uncertainty in *HCB* estimation was analyzed based on three main aspects, as follows:

a. Posterior predictive intervals

A total of 1000 parameter samples were drawn from the posterior distribution, and 90% posterior predictive intervals were calculated for all observations. The proportion of observed values falling within their respective intervals was assessed to confirm the reliability of the model and the uncertainty estimates.

b. Marginal posterior distributions

The marginal posterior distributions for each model parameter were summarized and visualized. The posterior distributions of regression coefficients β_1 to β_3 were examined for their significance in relation to zero, in order to determine the effects of crown diameter, live crown length, and total height on *HCB*. The posterior distributions of the random-effect variance parameters σ_1 and σ_2 were analyzed to assess how well both inter-plot and intra-plot *HCB* variations were characterized. The pseudocode of hierarchical Bayesian model was provided in Appendix A.

2.3. *Mixed-Effects Gaussian Process Regression Models*

2.3.1. Gaussian Process Regression (GPR)

GPR is a nonparametric Bayesian machine learning method used to model the nonlinear relationships between input features and target variables [27]. Given a training dataset $D = \{(x_i, y_i)|i = 1, 2, \cdots, n\}$, where $x_i \in R^d$ represents the input feature vector (*DBH*, *hL*, *caL*, and one of the four spatial variables) and $y_i \in R$ represents the corresponding target value (*HCB*), the goal of GPR is to learn a mapping function $f: R^d \to R$ from the input space to the target space.

In GPR, the mapping function f is assumed to be a Gaussian process with a prior distribution defined as follows:

$$f(x) \sim GO(m(x), k(x, x')) \quad (13)$$

where $m(x)$ is the mean function and $k(x, x')$ is the covariance function, also known as the kernel function. Commonly used kernel functions include the squared exponential kernel, the Matérn kernel, and the periodic kernel [43]. Given the training dataset D, the posterior distribution of GPR can be obtained using Bayes' theorem, as follows:

$$p(f|D) = \frac{p(D|f)p(f)}{p(D)} \tag{14}$$

where $p(D|f)$ is the likelihood function, $p(f)$ is the prior distribution, and $p(D)$ is the marginal likelihood. By maximizing the log-marginal likelihood, the hyperparameters of the Gaussian process, such as the parameters of the kernel function and the noise variance, can be estimated.

For a new input x_*, GPR can predict the distribution of the corresponding target value as follows:

$$p(y_*|x_*,\ D) = \int p(y_*|x_*,\ f)p(f|D)df \tag{15}$$

The mean and variance of this predictive distribution can be analytically derived, representing the prediction and the uncertainty estimation, respectively.

In this study, GPR was employed to develop a model for predicting *HCB*. The input features included *caL*, *hL*, *DBH*, and one of the four spatial pattern indices, while the target variable was *HCB*. We selected the squared exponential kernel as the covariance function and optimized the kernel hyperparameters using maximum likelihood estimation.

2.3.2. Mixed-Effects Gaussian Process Regression

To account for plot-level random effects, we extended the GPR model to a mixed-effects Gaussian process regression (ME-GPR) [27]. Consider m plots, each containing n_j tree samples, where $j = 1,2,\cdots,m$. *Let* $x_{ij} \in R^d$ denote the input features of the $y_{ij} \in R$ representing the corresponding height to crown base. The ME-GPR model can be expressed as follows:

$$y_{ij} = f(x_{ij}) + b_j + \varepsilon_{ij} \tag{16}$$

where $f(x_{ij})$ is the fixed-effects component from the GPR model, b_j is the random effect for the j-th plot following a Gaussian distribution $b_j \sim N(0,\ \sigma_b^2)$, and ϵ_{ij} is independent and identically distributed Gaussian noise, $\varepsilon_{ij} \sim N(0,\ \sigma_n^2)$.

In ME-GPR, we assume that the fixed-effects component $f(x_i)$ has a Gaussian process prior, as follows:

$$f(x) \sim GP(0,\ k(x,x')) \tag{17}$$

where $k(x,x')$ is the covariance function. The prior distribution of the random effects (b_j) is as follows:

$$b \sim N\left(0,\sigma_b^2 I\right) \tag{18}$$

where $b = [b_1,b_2,\cdots,b_m]^T$ is the vector of all plot random effects, and I is the identity matrix.

Given the training dataset $D = \{(x_{ij},y_{ij})|i = 1,2,\cdots,n_j;j = 1,2,\cdots,m\}$, the posterior distribution of ME-GPR can be obtained through Bayes' theorem, as follows:

$$p(f,b|D) = \frac{p(D|f,b)p(f)p(b)}{p(D)} \tag{19}$$

where $p(D\,|\,f,\,b)$ is the likelihood function, $p(f)$ and $p(b)$ are the priors of fixed and random effects, respectively, and $p(D)$ is the marginal likelihood. Hyperparameters of the Gaussian process and the variance of random effects σ_b^2 can be simultaneously estimated by maximizing the log-marginal likelihood.

For a new input x^*, ME-GPR predicts the distribution of its corresponding target value as follows:

$$p(y_*|x_*,\ D) =\uparrow \iint p(y_*|x_*\ ,\ f,\ b_*)p(f,\ b_*|D)df\ db_* \tag{20}$$

where b_* is the plot random effect associated with the new input x_*. The mean and variance of the predictive distribution can be obtained through numerical integration or approximate inference methods [44].

In this study, we employed ME-GPR to construct a prediction model for height to crown base, incorporating tree-level input features (*hL, DBH, caL*, and spatial pattern indices) and plot-level random effects. We selected the squared exponential kernel as the covariance function and used maximum likelihood estimation to optimize the model's hyperparameters. The pseudocode of mixed-effects Gaussian process regression model was provided in Appendix A.

2.3.3. Uncertainty Analysis of Mixed-Effects Gaussian Process Regression Models

To complement the ME-GPR model, a comprehensive uncertainty analysis was conducted based on the full dataset. This analysis aimed to quantify and characterize the various sources of uncertainty in the model predictions, focusing on the following three main aspects:

a. Posterior predictive distributions

For each new input x^*, the posterior predictive distribution $p(y^* \mid x^*, D)$ was computed using Equation (21). From these distributions, 95% credible intervals were calculated for each prediction. These intervals represent the ranges within which the true *HCB* value is expected to lie with 95% probability, given the model and data. The proportion of observed values falling within their respective credible intervals was calculated in order to assess the reliability of the model's uncertainty estimates.

b. Marginal posterior distributions

The posterior distributions of the model parameters were analyzed to assess parameter uncertainty. This included examining the posterior distributions of the fixed-effects coefficients (associated with *caL, hL, DBH*, and spatial pattern indices) to determine their significance and variability, analyzing the posterior distributions of the random-effects variance components to characterize the inter-plot variations in *HCB*, investigating the posterior distributions of the kernel function hyperparameters to understand the learned structure of the Gaussian process, and computing credible intervals and summary statistics for each parameter to quantify their uncertainty.

2.4. Random Forest Model

2.4.1. Random Forest *HCB* Model

The random forest technique is an ensemble learning technique that constructs multiple decision trees, each trained on a random subset of the data and features, to perform regression or classification tasks. The final prediction is obtained by averaging the results from individual trees, thereby reducing the risk of overfitting—a common issue with single decision trees [45]. This approach is particularly suitable for predicting height to crown base (*HCB*), a key parameter in forestry that influences light penetration, understory dynamics, and timber quality.

In this study, the random forest model was applied to predict *HCB* using key predictor variables, including LiDAR-derived tree height (*hL*), diameter at breast height (*DBH*), crown projection area (*caL*), and the spatial competition index (*SCI*). Model parameters, such as the number of trees (n_estimators = 500) and the number of features considered at each split (mtry), were optimized through 10-fold cross-validation to achieve a balance between accuracy and computational efficiency.

2.4.2. Bootstrap Method for Uncertainty Estimation

To quantify the uncertainty of the random forest model's predictions, a bootstrap method was applied to each cross-validation training set. For each training set in a cross-validation fold, 100 bootstrap samples were generated by resampling the training data with replacement. For each bootstrap sample, a random forest model was trained, and the trained model was used to predict the response variable for the corresponding test data (the held-out fold).

For each test data point, the mean and standard deviation (SD) of predictions across all 100 bootstrap iterations were calculated. The mean prediction represents the central tendency of the RF predictions for each test sample. The standard error (SE) was calculated as the standard deviation of the predictions, which provides an estimate of the uncertainty associated with each prediction.

2.5. Model Validation

For model validation, new data could be applied; however, as our data were limited in number, it was not practical to perform such a validation. In our study, we applied a cross-validation approach to validate the Bayes models and Gaussian models, similar to the research of Fu [46]. Each time, we removed one plot and used the remaining plots to fit the *HCB* model and predict the removed one (leave-one-plot-out cross-validation (LOPCV)). This process was repeated $N-1$ times, where N is the total number of plots. This method was chosen because it allowed us to assess the model's ability to generalize to new, unseen plots, which is critical given the inclusion of random effects related to plot variability. For the machine learning algorithm (random forest model), we used a 10-fold cross-validation method. This approach is particularly suited for machine learning models that do not explicitly account for hierarchical structure, allowing for a robust assessment of the model's overall predictive accuracy. The predictive performance of the model was assessed using *RMSE* and *MAE*. The model with the smallest *RMSE* and *MAE* was chosen as the final model to predict *HCB*. We performed all computations using R software, version 4.3.2 [47]. Equations (21) and (22) are as follows:

$$RMSE = \sqrt{\bar{e}^2 + \sigma_e^2} \tag{21}$$

$$MAE = |\sum_{i=1}^{N} e_i|/N = |\sum_{i=1}^{N} (HCB_i - \hat{HCB_i})|/N \tag{22}$$

where N is the number of observations, *RMSE* is the root-mean-square error, and *MAE* is the mean absolute error.

3. Results

3.1. Bayes Model Development

3.1.1. Hierarchical Bayes Base Model

The base model yielded parameter estimates (Table 2) for the two main coefficients (b1 and b2) and the error term (sigma). The first coefficient, b1, had a mean of 0.0113 (SD = 0.0032), with a 95% confidence interval of [0.0050, 0.0175]. The second coefficient, b2, showed a mean of 0.0262 (SD = 0.0058), with a 95% confidence interval of [0.0149, 0.0377]. Both coefficients were statistically significant, as their confidence intervals did not include zero. These results suggest that both predictors have a significant effect on the outcome variable, with the model accounting for some of the variability in the data.

Table 2. Parameter estimates from the Bayesian base model.

Parameter	Mean	Std. Dev	95% CI
b_1	0.0113	0.0032	(0.0050, 0.0175)
b_2	0.0262	0.0058	(0.0149, 0.0377)
$log(\sigma)$	0.5784	0.0313	(0.5195, 0.6410)
σ	1.7840	0.0559	(1.6813, 1.8984)

The Bayesian base model yielded the following equation:

$$HCB = \frac{X_0}{1 + exp(0.0113X_1 + 0.0262X_2)} + \varepsilon \quad (23)$$

where:

HCB is the dependent variable;

X_0, X_1, and X_2 are the independent variables;

ε is the error term, normally distributed with a mean of 0 and a standard deviation of 1.7840.

Both coefficients were statistically significant (β_1: 95% CI [0.0050, 0.0175]; β_2: 95% CI [0.0149, 0.0377]). The model demonstrated modest predictive power (*MAE* = 3.17, *RMSE* = 1.78), suggesting a nonlinear relationship between the predictors and *HCB*. This structure may reflect complex underlying processes, although the moderate fit indicates that additional factors could improve the model's performance.

3.1.2. Spatial Variables Introduced

The results show a performance ranking of five indices based on LOPCV (Table 3):

Table 3. Comparisons of five spatial indices.

Rank	LOO	P Loo	Dse	Loo Scale
SCI	−1135.88	4.61	0.00	log
aCI	−1138.261	4.70	3.80	log
Aggregation index	−1352.79	145.21	8.31	log
Hegyi	−1356.54	142.81	7.55	log

Note: LOO represents the LOPCV information criterion (LOOIC), p loo represents the effective number of parameters, and Dse represents the difference in standard error.

SCI performed best, with the lowest LOO value (−1135.88) and p loo (effective number of parameters) (4.61), along with a very low Dse (difference in standard error), because lower LOO values represent better performance, low p loo values represent a simple model, and lower Dse indicates proximity to the best model. The *aCI* was a close second, with slightly higher LOO (−1138.26) and p loo (4.70). The Hegyi indices performed significantly worse, with much higher LOO values (both above −1356). *SCI* and *aCI* demonstrated superior predictive accuracy and model parsimony. The aggregation index received a warning flag, suggesting potential issues. All of these models use a logarithmic LOO scale.

In summary, *SCI* and *aCI* outperformed the other indices across all metrics, indicating that they are likely the most suitable for accurate and reliable predictions in this context.

3.1.3. Hierarchical Bayes Model

(1) Random effect selection

To investigate the influence of plot-level random effects on the prediction of *HCB*, we fitted a series of mixed-effects logistic regression models with different combinations of random effects. The model performance metrics, including the Akaike information criterion (*AIC*), Bayesian information criterion (*BIC*), and *log-likelihood*, were calculated for each model (Table 4).

Although the results showed that the model with random effects on all three predictor variables had the lowest *AIC* and *BIC* values, the difference in *AIC* between random effects on all three variables (1676.60) and *DBH* and *SCI* (1796.77) was not very big. Considering the convergence cost, the model with random effects on *DBH* and *SCI* was screened out as the most effective *HCB* hierarchical Bayes model, indicating the best balance between model fit and complexity.

Table 4. Performance of different random effects.

Random Effects	AIC	BIC	Log-Likelihood
(*DBH*)	1924.37	1949.78	−956.19
(*caL*)	1934.86	1960.26	−961.43
(*SCI*)	2009.87	2035.27	−998.93
(*DBH*, *caL*)	1810.43	1840.07	−898.21
(*DBH*, *SCI*)	1796.77	1826.41	−891.38
(*caL*, *SCI*)	1853.03	1882.67	−919.51
(*DBH*, *caL*, *SCI*)	1676.60	1710.47	−830.30

(2) Hierarchical Bayes model

We implemented a hierarchical Bayes model on the dataset using LOPCV. The model achieved an average *MSE* of 2.81 and an *RMSE* of 1.68, indicating moderate predictive accuracy.

The hierarchical Bayesian model yielded the following equation:

$$HCB = \frac{hL}{1 + exp((0.0166 + u_1)DBH + 0.0272caL - (0.0902 + u_2)SCI)} + \varepsilon \quad (24)$$

where:

u_1 and u_2 are the random-effect variables;

ε is the error term, normally distributed, with a mean of 0 and a standard deviation of 1.7840.

(3) Uncertainty analysis of hierarchical Bayes model

All parameters' posterior distributions exhibit distinct bell-shaped curves, indicating some uncertainty in the model's parameter estimation but relatively stable estimates. The estimated values of b_1 and b_2 are positive, with 95% HDI not including zero, suggesting potential statistical significance for these parameters. The estimated value of b_3 is negative, but its 95% HDI includes zero, indicating that b_3's effect may not be statistically significant. The large estimated value of sigma indicates potentially high residual variance in the model. Most of the parameters have relatively narrow HDI intervals, suggesting high precision in estimation. However, b_3's wider HDI interval indicates greater uncertainty in its estimation (Figure 3).

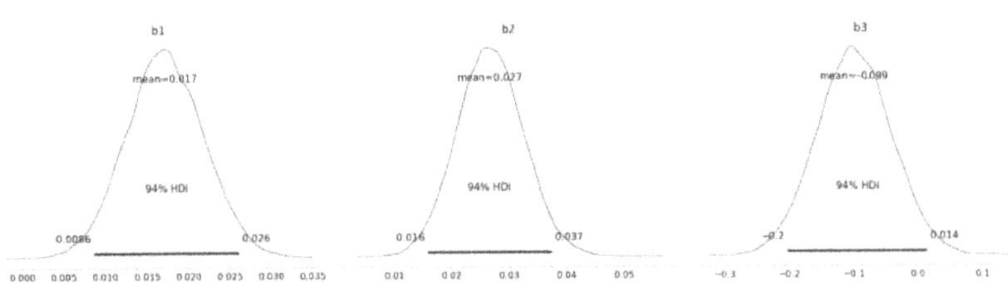

Figure 3. Figures of hierarchical Bayesian uncertainty quantification.

The distribution of the model's residuals shows an approximately normal shape, centered near zero (Figure 4). Most residuals fall within the range of −2 to 2, indicating a reasonable model fit. There is a slight positive skew, suggesting that the model may underestimate some higher values. The peak of the distribution occurs around zero, with the majority of residuals clustered within ±1 units. While outliers exist beyond ±3 units, they are infrequent, pointing to generally good predictive accuracy, with some exceptions.

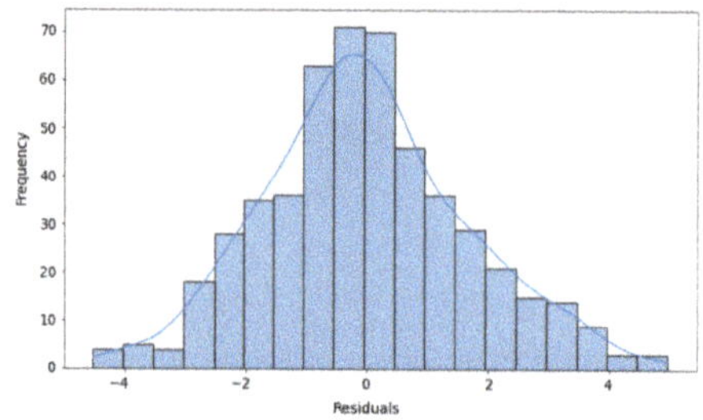

Figure 4. Distribution of residuals of the *HCB* hierarchical Bayes model.

3.2. Gaussian Process Model

3.2.1. Gaussian Process Base Model

For the Gaussian process model, the kernel is a combination of a radial basis function (RBF) and a white kernel, as follows:

$$k(x,\ x') = 0.716^2 RBF(length\ scale = 1.85) + WhiteKernel(noise\ level = 0.649) \quad (25)$$

The Gaussian process base model demonstrated moderate predictive performance. The model's accuracy is reflected in its *MAE* of 2.70 and *RMSE* of 1.64. The model's complexity and fit are captured by an AIC of 415.15 and BIC of 449.27, with a log-likelihood of −194.57.

The kernel function, which defines the covariance structure, combines an RBF term with a length scale of 1.85 and a white noise term. This structure suggests that the model both captures smooth trends in the data and accounts for noise. The mean prediction interval of 6.66 and mean confidence interval of 3.33 provide insights into the model's uncertainty estimates. The mean total variance of 2.90 indicates the overall variability in the model's predictions.

3.2.2. Random Effect Selection

The comparison of GPR with different random effects revealed that the model incorporating random effects on *SCI* outperformed the other models (Table 5). This model demonstrated the lowest Akaike information criterion (AIC) of 401.03, indicating the best balance between model fit and complexity. The kernel for this model, a rational quadratic function with alpha = 0.0332 and length scale = 0.0736, captures both large-scale and small-scale variations in the data. These results indicate that accounting for random effects in SCI provides the most effective model for describing the underlying patterns in the dataset, potentially reflecting important spatial competition dynamics in the system under study.

Table 5. Results of random selection of the mixed-effects Gaussian process model.

Random Effects	AIC	BIC	Log-Likelihood
SCI	401.03	429.91	−189.52
caL	405.53	434.40	−191.76
caL, SCI	406.41	435.29	−1928.21
DBH, LCA	406.74	435.62	−192.37
DBH, caL, SCI	407.16	436.03	−192.58
DBH, SCI	407.29	436.16	−192.64
DBH	408.40	437.27	−193.20

3.2.3. Gaussian Mixed-Effects Process Model

We implemented a GPR with random effects on the dataset using LOPCV. Figure 5 illustrates the true versus predicted values with uncertainty, demonstrating the model's performance across the range of observed values. The model achieved an average MAE of 1.25 and an RMSE of 1.59, indicating moderate predictive accuracy.

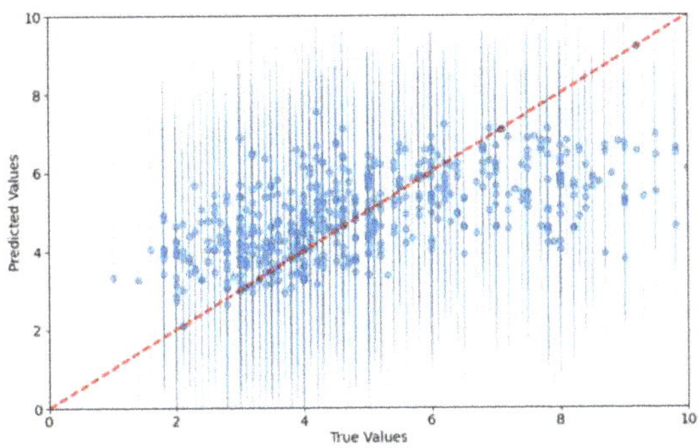

Figure 5. True vs. predicted values of the Gaussian mixed-effects process model, with uncertainty.

3.2.4. Uncertainty Analysis of Gaussian Mixed-Effects Process Model

The model's uncertainty estimation was robust, with a 95% prediction interval coverage of 0.95, closely aligning with the expected 95% coverage. Figure 6 depicts the predictions with uncertainty (a) and the relative uncertainty versus true values (b), revealing a consistent pattern of uncertainty across the range of true values. The average relative uncertainty was 0.3389, indicating that the model's predictions had an average uncertainty of ~33.89% relative to the predicted values. The average error variance was 2.61, reflecting the model's estimate of the inherent variability in the predictions.

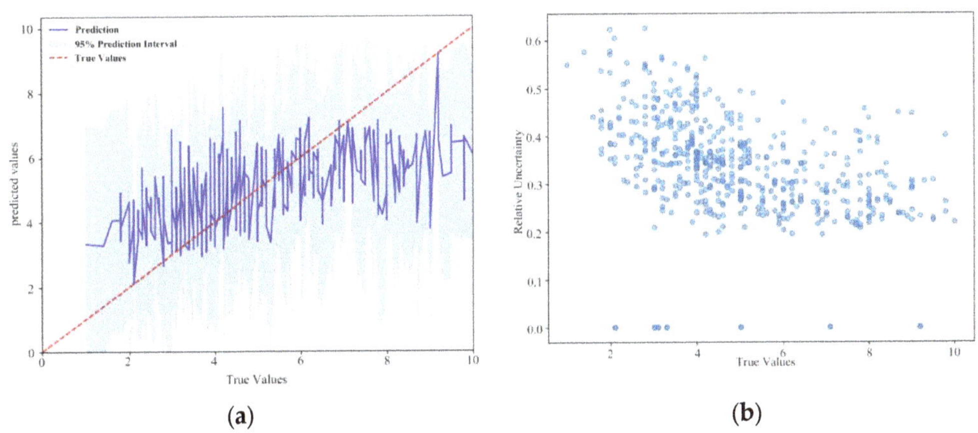

Figure 6. The predictions with uncertainty (**a**) and the relative uncertainty versus true values (**b**).

3.3. *Random Forest Model*

3.3.1. Random Forest *HCB* Model Hyperparameter Turning and Model Validation

In the process of tuning the random forest model, various values of the hyperparameters mtry and min.node.size were evaluated using a 10-fold cross-validation approach. Given the relatively small sample size, the number of trees (ntree) was set to 500. The results (Table 6) indicated that the optimal parameter combination was *mtry* = 3 and *min.node.size* = 5, which yielded the lowest *RMSE* of 1.639, suggesting the best predictive performance among the tested configurations. Additionally, the mean absolute error (MAE) for the optimal configuration was 1.300, indicating a relatively low average prediction error. Therefore, the best-performing *HCB* prediction model based on the random forest approach was the one configured with *mtry* = 3, *min.node.size* = 5, and *ntree* = 500.

Table 6. Results of hyperparameter tuning of the random forest *HCB* model.

mtry	Min Node Size	RMSE	MAE
2	1	1.651	1.309
2	3	1.650	1.308
2	5	1.653	1.312
3	1	1.652	1.309
3	3	1.642	1.301
3	5	1.639	1.300
4	1	1.643	1.301
4	3	1.642	1.301
4	5	1.643	1.308

3.3.2. Uncertainty Analysis of Random Forest Model

The random forest model's predictions were analyzed for uncertainty using a bootstrap-based approach combined with 10-fold cross-validation. The resulting plot (Figure 7) displays the predicted values against the true values for the test data, including 95% confidence intervals for each prediction. The results showed the following: (1) predictions for lower true values (below 3.0) generally show wider intervals, indicating higher uncertainty in this region; (2) for intermediate true values (between 3.0 and 6.0), the intervals are narrower, suggesting more confident predictions in this range; (3) predictions for higher true values (above 6.0) tend to have slightly increased uncertainty, as shown by the wider error bars, which may reflect the model's reduced ability to generalize in this range due to fewer training samples or increased data variability.

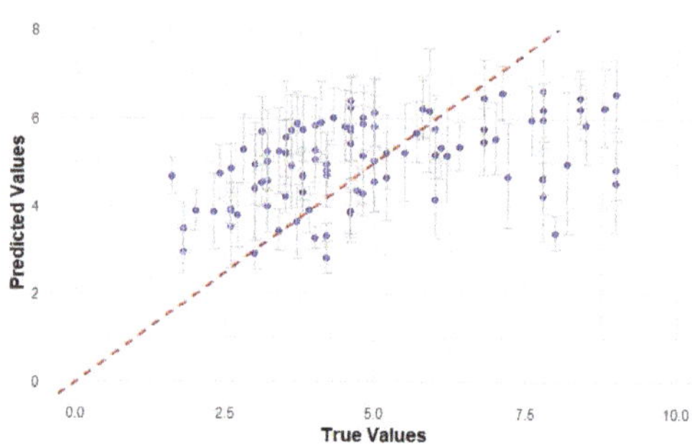

Figure 7. Predicted vs. true values, with 95% confidence intervals.

4. Discussion

This study employed hierarchical Bayesian, Gaussian process regression, and random forest models to predict *HCB* in *Picea crassifolia* forests. All models fitted the data well, explaining the majority of variation in *HCB*. The hierarchical Bayesian model achieved an *RMSE* of 1.6773 m, the Gaussian mixed-effects process model achieved an *RMSE* of 1.5902 m, and the random forest model achieved an *RMSE* of 1.6394 m. These results suggest that the Gaussian mixed-effects process model performs best, offering a robust alternative for *HCB* prediction despite the differences in methodology.

Regarding the influence of variables, the hierarchical Bayesian, Gaussian process, and random forest models consistently identified LiDAR-derived tree height, *DBH*, and LiDAR-derived crown projection area as the primary factors affecting *HCB*. Tree height exhibited a significant positive correlation with *HCB*, indicating that taller trees tend to have higher *HCB*. This relationship is consistent with the findings of Sharma and Parton [48] in their study on boreal forest species, suggesting a consistent pattern across different forest ecosystems. *DBH* also showed a positive correlation with *HCB*, which may be attributable to larger-diameter trees generally being older and having undergone a longer natural pruning process. Rijal et al. [49] observed a similar relationship in their study on Scots pine, further corroborating our findings. Interestingly, the live crown projection area was negatively correlated with *HCB*, implying that trees with larger crowns have relatively lower *HCB*. This inverse relationship might be explained by the allocation balance of carbohydrates between stem and crown organs during tree growth, as discussed by Mäkelä and Valentine [50]. The consistency of these relationships across models strengthens the reliability of our findings.

From an ecological perspective, factors influencing *HCB* can be categorized into two types: (1) tree growth and development characteristics, such as genetic traits, age, and growth rate; and (2) environmental conditions, including stand density and site index. These factors interact, creating spatial heterogeneity in *HCB*. Our study used plot-level random effects to account for some of this spatial variation. Notably, the superiority of the *SCI* in the Gaussian process model suggests that local competition dynamics play a crucial role in shaping *HCB* patterns. This highlights the importance of considering spatial relationships in the structural modeling of forests. Inter-plot differences in *HCB* may be related to the microenvironment, stand structure, soil fertility, and disturbance history. These factors, while not explicitly modeled in our study, likely contributed to the unexplained variance and represent an important avenue for future research. Understanding these spatial dynamics is crucial for developing more accurate predictive models and informing forest management practices aimed at optimizing stand structure and timber quality.

The findings of this study are generally consistent with those of previous research on other tree species and regions. For instance, Hasenauer and Monserud [51] found that *HCB* in Austrian spruce plantations was primarily governed by tree height and *DBH*. Similarly, Rijal [49] identified tree height as the most important variable for predicting HCB in Finnish pine forests. This suggests that the factors influencing *HCB* identified in this study may have some universality.

Both hierarchical Bayesian and Gaussian process models demonstrated strong performance in *HCB* prediction, each with distinct strengths. The hierarchical Bayesian model excels in its flexibility to incorporate prior knowledge, comprehensive quantification of uncertainty sources, and applicability to complex data structures, including spatial data and hyperparameters [32]. This makes it particularly suitable for capturing the multi-level nature of forest ecosystems. On the other hand, the Gaussian process model showed superior ability in fitting nonlinear relationships, providing probabilistic prediction intervals, and exhibiting strong adaptability [27]; its performance in our study, particularly with the *SCI* as a random effect, highlights its capacity to capture complex spatial competition dynamics.

The random forest model, while not outperforming the GPR model, still provided a competitive alternative for *HCB* prediction. The random forest approach is particularly advantageous due to its ability to handle large datasets, identify variable importance, and generate robust predictions without extensive assumptions about the underlying data distribution.

The *HCB* models developed in this study provide important references for the precision management of *Picea crassifolia* forests. Firstly, *HCB* is a crucial indicator for assessing usable timber volume, directly affecting the wood's grade and value. Accurate *HCB* prediction aids in formulating rational harvesting plans to maximize volume utilization. Secondly, *HCB* significantly influences stand canopy closure, understory light conditions, and canopy structure, which are critical for stand regeneration and mixed-species regulation. Optimizing *HCB* can create favorable light and spatial conditions for seedling regeneration. Furthermore, changes in *HCB* are closely related to the spread of forest fires and pest outbreak processes. Accurate *HCB* estimation can inform forest disaster early warning and control measures [16].

Future research could consider the following directions: (1) expanding the scale of plot surveys to include more tree species and forest types, enhancing the models' universality; (2) integrating remote sensing techniques to obtain high-spatiotemporal-resolution image data for monitoring and predicting dynamic HCB changes [14]; (3) developing model-based *HCB* management decision systems to directly serve production practices; and

(4) incorporating more physiological and ecological processes to enhance models' mechanistic understanding of *HCB* formation processes. Through multidisciplinary integration, the breadth and depth of *HCB* research can be continuously expanded.

The uncertainty analysis provides a more comprehensive understanding of the reliability of *HCB* predictions. Based on the current training data, the hierarchical Bayesian model and the Gaussian process regression model can provide reasonable and credible interval estimates and uncertainty measures, which serve as important references for the application of models' results in practice. However, when using these models for estimation, it is crucial to fully consider the interval nature of the predictions and not to overly rely on point estimates [52]. The posterior distribution of the parameters in the hierarchical Bayesian model suggests that tree size has a stable promoting effect on *HCB*, while the influence of crown structure is weaker and potentially negative. The variation in *HCB* within and between plots is generally controllable. These findings are consistent with the results reported by Smith [53], who found that tree diameter and height were the primary drivers of *HCB* variation in a mixed coniferous forest.

While this study provides valuable insights into the factors influencing *HCB* in Qinghai spruce forests, its focus on natural spruce stands within a specific region may limit the generalizability of the findings to other forest types or geographic areas. Variations in ecological conditions, species composition, and forest management practices across different regions could influence the relationships between tree height, DBH, crown area, and *HCB*. For instance, the environmental conditions in boreal forests, temperate hardwood forests, or mixed-species stands might result in different dynamics of tree growth and competition [54]. Therefore, caution should be exercised when applying these findings beyond the specific context of Qinghai spruce forests. Expanding plot surveys to encompass a broader range of forest types could enhance the models' general applicability [55]. Additionally, investigating how the identified relationships manifest under varying climatic and ecological conditions could provide a more comprehensive understanding of *HCB* dynamics.

In summary, due to the complexity of the *HCB* formation process, the existing models still struggle to exhaustively consider all influencing factors. Genetic, environmental, and disturbance variables not included in these models may be important sources of prediction errors [56]. This limitation presents an opportunity for future research to explore more comprehensive influencing factors through mechanistic studies, optimize the model structure and parameter space, and further improve the accuracy and robustness of the predictions [57]. Additionally, future studies could benefit from incorporating the spatial indices (such as *SCI* and *aCI*) that showed promise in our analysis, potentially leading to a more nuanced understanding of spatial competition dynamics in HCB formation.

5. Conclusions

This study applied hierarchical Bayesian, Gaussian process regression, and random forest models to predict LiDAR-based individual tree height to crown base (*HCB*) in *Picea crassifolia* Kom. forests. The Gaussian process model showed the best performance (*RMSE* = 1.5902 m), followed closely by the random forest and hierarchical Bayesian models. LiDAR-derived tree height and crown projection area and *DBH* were consistently identified as key factors influencing *HCB* across all models. The integration of spatial competition indices, such as the *SCI*, highlighted the role of local competition dynamics in shaping *HCB* patterns. While the models provide insights for precision forest management, future studies should aim to include other factors, such as genetic variability and environmental factors, to further enhance the models' accuracy and applicability.

Author Contributions: Formal analysis, Z.Y., H.Y. and Z.Z.; methodology, G.D.; software, G.D., X.W. and H.Y.; supervision, H.Z. and Z.Y.; validation, X.W.; visualization, Z.Z.; writing—original draft, Z.Y., H.Y., X.W., G.D. and H.Z.; writing—review and editing, Z.Y. and H.Z. All authors have read and agreed to the published version of the manuscript.

Funding: This research was funded by the following projects: Shanxi Province Basic Research Program, Youth Science Research Project: No. 202203021212425; Shanxi Agricultural University Doctoral Research Startup Project: No. 2021BQ110; the National Program on Key Basic Research Project (973 Program): No. 2007CB714400.

Data Availability Statement: Data are available from the authors upon reasonable request and with permission from the Research Institute of Forest Resource Information Techniques, Chinese Academy of Forestry.

Acknowledgments: We thank the Research Institute of Forest Resource Information Techniques, the Chinese Academy of Forestry for data support. We sincerely thank the four anonymous reviewers for their valuable comments and suggestions.

Conflicts of Interest: The authors declare that they have no conflicts of interest.

Appendix A

1. Hierarchical Bayesian model

(1) Set up environment:
(2) Define Functions:

calculate_metrics(y_true, y_pred): Calculate MAE and RMSE between true and predicted values.
mixed_effects_model(X_scaled, y, plot_index, random_effects):

Define the Bayesian model:
Return the fitted model and posterior samples.

cross_validation_by_plot(X_scaled, y, plot_index, random_effects):

Loop over each unique plot:
Store predictions for each plot.
Return all predictions.

(3) Main Analysis:

Load Data:
Define Random Effects Options: Create combinations of random effects terms for model testing.
Initialize an empty list to store results.
Loop through each random effects combination:

Train and Evaluate Model:
Handle any errors encountered during model fitting.
Save intermediate results to an external file (optional for tracking progress).

Sort and Display Results:

2. Gaussian mixed-effects model

(1) Set up environment:
(2) Define Functions:

calculate_metrics(y_true, y_pred): Calculate MAE and RMSE between true and predicted values.
create_features(X, y):

Generate polynomial features.
Select relevant features using SelectKBest.

process_fold(plot, X_selected, y_scaled, plot_effects, affected_params, scaler_y):

Remove one plot: Exclude all samples from the current plot as the test set.
Train the model: Use data from the remaining plots to fit the model.
Apply random effects to the selected parameters in the training data.
Define a set of Gaussian Process kernels.
Perform grid search to find the best kernel.

Predict outcomes for the excluded plot using the best model.
Return evaluation metrics and predictions.

gaussian_process_model(X, y, plot_effects, affected_params):

Standardize features and response variable.
Generate polynomial features and select the best.
Perform cross-validation by removing one plot at a time
Calculate mean evaluation metrics (MAE, RMSE) across all plots.
Return the model performance metrics and predictions.

(3) Main Analysis:

Load Data
Define Parameters: Specify parameter names and initialize a list for results.
Base Model: Fit the base Gaussian Process model without random effects and store the results.
Random Effects Models:

Loop through all combinations of random effects:

Apply selected random effects to affected parameters.
Fit the Gaussian Process model using gaussian_process_model with leave-one-plot-out cross-validation.
Store model name and results.

Sort Results: Sort models based on mean RMSE and MAE.

References

1. Maltamo, M.; Bollandsås, O.M.; Vauhkonen, J.; Breidenbach, J.; Gobakken, T.; Næsset, E. Comparing different methods for prediction of mean crown height in Norway spruce stands using airborne laser scanner data. *Forestry* **2010**, *83*, 257–268. [CrossRef]
2. Rais, A.; Jacobs, M.; Van de Kuilen, J.W.; Pretzsch, H. Crown structure of European beech (*Fagus sylvatica*): A noncausal proxy for mechanical–physical wood properties. *Can. J. For. Res* **2021**, *51*, 834–841. [CrossRef]
3. Krajnc, L.; Farrelly, N.; Harte, A.M. The influence of crown and stem characteristics on timber quality in softwoods. *For. Ecol. Manag.* **2019**, *435*, 8–17. [CrossRef]
4. Vanninen, P.; Mäkelä, A. Needle and stem wood production in Scots pine (*Pinus sylvestris*) trees of different age, size and competitive status. *Tree Physiol.* **2000**, *20*, 527–533. [CrossRef] [PubMed]
5. Valentine, H.T.; Gregoire, T.G.; Burkhart, H.E.; Hollinger, D.Y. A stand-level model of carbon allocation and growth, calibrated for loblolly pine. *Can. J. For. Res.* **1994**, *24*, 1575–1585. [CrossRef]
6. Mäkelä, A. Derivation of stem taper from the pipe theory in a carbon balance framework. *Tree Physiol.* **2002**, *22*, 891–905. [CrossRef]
7. Hyytiäinen, K.; Hari, P.; Kokkila, T.; Mäkelä, A.; Tahvonen, O.; Taipale, J. Connecting a process-based forest growth model to stand-level economic optimization. *Can. J. For. Res.* **2004**, *34*, 2060–2073. [CrossRef]
8. Maguire, D.A.; Kershaw, J.A.; Hann, D.W. Predicting the effects of silvicultural regime on branch size and crown wood core in Douglas-fir. *For. Sci.* **2005**, *51*, 503–514. [CrossRef]
9. Kershaw, J.A.; Maguire, D.A.; Hann, D.W. Longevity and duration of radial growth in Douglas-fir branches. *Can. J. For. Res.* **2008**, *20*, 1690–1695. [CrossRef]
10. Jiang, L.; Liu, R. Segmented taper equations with crown ratio and stand density for Dahurian Larch (*Larix gmelinii*) in northeastern China. *J. For. Res.* **2011**, *22*, 347–352. [CrossRef]
11. Nyland, R.D. *Silviculture: Concepts and Applications*; Waveland Press: Long Grove, IL, USA, 2016.
12. Mäkinen, H.; Isomäki, A. Thinning intensity and growth of Scots pine stands in Finland. *For. Ecol. Manag.* **2004**, *201*, 311–325. [CrossRef]
13. Fernández-Alonso, J.M.; Alberdi, I.; Álvarez-González, J.G.; Vega, J.A.; Cañellas, I.; Ruiz-González, A.D. Canopy fuel characteristics in relation to crown fire potential in pine stands: Analysis, modelling and classification. *Eur. J. For. Res.* **2013**, *132*, 363–377. [CrossRef]
14. Andersen, H.E.; McGaughey, R.J.; Reutebuch, S.E. Estimating forest canopy fuel parameters using LIDAR data. *Remote Sens. Environ.* **2005**, *94*, 441–449. [CrossRef]
15. Cruz, M.G.; Alexander, M.E.; Wakimoto, R.H. Assessing canopy fuel stratum characteristics in crown fire prone fuel types of western North America. *Int. J. Wildland Fire* **2003**, *12*, 39–50. [CrossRef]
16. Agee, J.K.; Skinner, C.N. Basic principles of forest fuel reduction treatments. *For. Ecol. Manag.* **2005**, *211*, 83–96. [CrossRef]
17. Yang, Z.; Liu, Q.; Luo, P.; Ye, Q.; Duan, G.; Sharma, R.P.; Fu, L. Prediction of individual tree diameter and height to crown base using nonlinear simultaneous regression and airborne LiDAR data. *Remote Sens.* **2020**, *12*, 2238. [CrossRef]

18. Maltamo, M.; Karjalainen, T.; Repola, J.; Vauhkonen, J. Incorporating tree- and stand-level information on crown base height into multivariate forest management inventories based on airborne laser scanning. *Silva Fenn.* **2018**, *52*, 10006. [CrossRef]
19. Yan, Y.; Wang, J.; Liu, S.; Gaire, D.; Jiang, L. Modeling the influence of competition, climate, soil, and their interaction on height to crown base for Korean pine plantations in Northeast China. *Eur. J. For. Res.* **2024**, *9*, 1–4. [CrossRef]
20. Clark, J.S. Uncertainty and variability in demography and population growth: A hierarchical approach. *Ecology* **2003**, *84*, 1370–1381. [CrossRef]
21. Regan, H.M.; Colyvan, M.; Burgman, M.A. A taxonomy and treatment of uncertainty for ecology and conservation biology. *Ecol. Appl.* **2002**, *12*, 618–628. [CrossRef]
22. Xu, C.; Gertner, G.Z. Uncertainty and sensitivity analysis for models of complex systems. *Ecol. Model.* **2008**, *213*, 385–397. [CrossRef]
23. Harmon, M.E.; Fasth, B.; Halpern, C.B.; Lutz, J.A. Uncertainty analysis: An evaluation metric for synthesis science. *Ecosphere* **2015**, *6*, 1–12. [CrossRef]
24. Reichert, P.; Omlin, M. On the usefulness of overparameterized ecological models. *Ecol. Model* **1997**, *95*, 289–299. [CrossRef]
25. Ellison, A.M. Bayesian inference in ecology. *Ecol. Lett.* **2004**, *7*, 509–520. [CrossRef]
26. McCarthy, M.A. *Bayesian Methods for Ecology*; Cambridge University Press: Cambridge, UK, 2007. [CrossRef]
27. Rasmussen, C.E.; Williams, C.K.I. *Gaussian Processes for Machine Learning*; MIT Press: Cambridge, MA, USA, 2006; pp. 79–104.
28. Chu, W.; Ghahramani, Z. Gaussian processes for ordinal regression. *J. Mach. Learn. Res.* **2005**, *6*, 1019–1041.
29. Quiñonero-Candela, J.; Rasmussen, C.E.; Williams, C.K. Approximation methods for gaussian process regression. In *Large-Scale Kernel Machines*; MIT Press: Cambridge, MA, USA, 2007; pp. 203–224. [CrossRef]
30. Camps-Valls, G.; Verrelst, J.; Munoz-Mari, J.; Laparra, V.; Mateo-Jimenez, F.; Gomez-Dans, J. A survey on Gaussian processes for earth-observation data analysis: A comprehensive investigation. *IEEE Geosci. Remote Sens. Mag.* **2016**, *4*, 58–78. [CrossRef]
31. Cressie, N.; Calder, C.A.; Clark, J.S.; Hoef, J.M.V.; Wikle, C.K. Accounting for uncertainty in ecological analysis: The strengths and limitations of hierarchical statistical modeling. *Ecol. Appl.* **2009**, *19*, 553–570. [CrossRef]
32. Gelman, A.; Carlin, J.B.; Stern, H.S.; Dunson, D.B.; Vehtari, A.; Rubin, D.B. *Bayesian Data Analysis*; Chapman and Hall/CRC: London, UK, 2013. [CrossRef]
33. Gelman, A. *Data Analysis Using Regression and Multilevel/Hierarchical Models*; Cambridge University Press: Cambridge, UK, 2007; Available online: https://www.jstor.org/stable/20461880 (accessed on 12 October 2024).
34. Qian, S.S.; Cuffney, T.F.; Alameddine, I.; McMahon, G.; Reckhow, K.H. On the application of multilevel modeling in environmental and ecological studies. *Ecology* **2010**, *91*, 355–361. [CrossRef]
35. Pang, Y.; Chen, E.; Liu, Q.; Xiao, Q.; Zhong, K.; Li, X.; Ma, M. WATER: Dataset of airborne LiDAR mission at the super site in the Dayekou watershed flight zone on Jun. 23, 2008. In *Chinese Academy of Forestry; Institute of Remote Sensing Applications, Chinese Academy of Sciences*; Cold and Arid Regions Environmental and Engineering Research Institute, Chinese Academy of Sciences; Heihe Plan Science Data Center: Lanzhou, China, 2008.
36. Axelsson, P. Processing of laser scanner data—Algorithms and applications. *ISPRS J. Photogramm.* **1999**, *54*, 138–147. [CrossRef]
37. Liu, Q.; Fu, L.; Wang, G.; Li, S.; Li, Z.; Chen, E.; Pang, Y.; Hu, K. Improving estimation of forest canopy cover by introducing loss ratio of laser pulses using airborne LiDAR. *IEEE Trans. Geosci. Remote* **2020**, *58*, 567–584. [CrossRef]
38. Liu, Q. Study on the Estimation Method of Forest Parameters Using Airborne Lidar. Ph.D. Thesis, Chinese Academy of Forestry, Beijing, China, 2009. (In Chinese).
39. Walters, D.K.; Hann, D.W. Taper equations for six conifer species in southwest Oregon. *For. Res. Lab. Or. State Univ. Corvallis Res. Bull.* **1986**, *56*, 41.
40. Hegyi, F. A simulation model for managing jack-pine stands Growth Model Tree Stand Simul. *R. Coll. Res. Notes* **1974**, *30*, 74–90.
41. Clark, P.h.J.; Evans, F.C. Distance to nearest neighbor as a measure of spatial relationships in populations. *Ecology* **1954**, *35*, 445–453. [CrossRef]
42. Hui, G.; Hu, Y.; Zhao, Z.; Yuan, S.; Liu, W. A Forest Competition Index Based on Intersection Angle. *Ann. For. Sci.* **2013**, *49*, 68–73. (In Chinese)
43. Duvenaud, D. Automatic Model Construction with Gaussian Processes. Ph.D. Dissertation, Cambridge University Press, Cambridge, UK, 2014. [CrossRef]
44. Hensman, J.; Matthews, A.; Ghahramani, Z. Scalable variational Gaussian process classification. In Proceedings of the Eighteenth International Conference on Artificial Intelligence and Statistics, San Diego, CA, USA, 9–12 May 2015; Volume 38, pp. 351–360.
45. Breiman, L. Random Forests. *Mach. Learn.* **2001**, *45*, 5–32. [CrossRef]
46. Fu, L.; Liu, Q.; Sun, H.; Wang, Q.; Li, Z.; Chen, E.; Pang, Y.; Song, X.; Wang, G. Development of a system of compatible Individual tree diameter and aboveground biomass prediction models using error-in-variable regression and airborne LiDAR Dat-a. *Remote Sens.* **2018**, *10*, 325. [CrossRef]
47. R Core Team. *R: A Language and Environment for Statistical Computing*; R Foundation for Statistical Computing: Vienna, Austria, 2018; Available online: https://www.R-project.org/ (accessed on 12 October 2024).
48. Sharma, M.; Parton, J. Modeling stand density effects on taper for jack pine and black spruce plantations using dimensional analysis. *For. Sci.* **2009**, *55*, 268–282. [CrossRef]
49. Rijal, B.; Weiskittel, A.R.; Kershaw, J.A. Development of height to crown base models for thirteen tree species of the North American Acadian Region. *For. Chron.* **2012**, *88*, 60–73. [CrossRef]

50. Mäkelä, A.; Valentine, H.T. Crown ratio influences allometric scaling in trees. *Ecology* **2006**, *87*, 2967–2972. [CrossRef]
51. Hasenauer, H.; Monserud, R.A. A crown ratio model for Austrian forests. *For. Ecol. Manag.* **1996**, *84*, 49–60. [CrossRef]
52. Hooten, M.B.; Hobbs, N.T. A guide to Bayesian model selection for ecologists. *Ecol. Monogr.* **2015**, *85*, 3–28. [CrossRef]
53. Smith, J.P.; Hoffman, C.M.; Landhäusser, S.M. Modelling crown base height and crown ratio in mixed-species stands of Douglas-fir and western hemlock. *For. Ecol. Manag.* **2019**, *447*, 115–124.
54. Kneeshaw, D.; Kobe, R.K.; Coates, K.D.; Messier, C. Sapling size influences shade tolerance ranking among southern boreal tree species. *J. Ecol.* **2002**, *90*, 17–24. [CrossRef]
55. Temesgen, H.; LeMay, V.M.; Mitchell, S.J. Tree crown ratio models for multi-species and multi-layered stands of southeastern British Columbia. *For. Chron.* **2005**, *81*, 133–141. [CrossRef]
56. Lines, E.R.; Zavala, M.A.; Purves, D.W.; Coomes, D.A. Predictable changes in aboveground allometry of trees along gradients of temperature, aridity and competition. *Glob. Ecol. Biogeogr.* **2012**, *21*, 1017–1028. [CrossRef]
57. Xing, Z.; Bourque, C.P.A.; Swift, D.E.; Clowater, C.W.; Krasowski, M.; Meng, F.R. Carbon and biomass partitioning in balsam fir (*Abies balsamea*). *Tree Physiol.* **2005**, *25*, 1207–1217. [CrossRef]

forests

MDPI

Article

A Study on the Growth Model of Natural Forests in Southern China Under Climate Change: Application of Transition Matrix Model

Xiangjiang Meng [1,2], Zhengrui Ma [1,2], Ying Xia [3], Jinghui Meng [4], Yuhan Bai [5] and Yuan Gao [4,*]

1 Chongqing Academy of Forestry, Chongqing 401147, China; mxj517@sina.com (X.M.); ahnu_mzr1988@163.com (Z.M.)
2 Chongqing Wuling Mountain Forest Ecosystem National Positioning Observatory, Chongqing 401147, China
3 Forestry Bureau of Chongqing Wulong District, Chongqing 408500, China; 13996785558@163.com
4 College of Forestry, Beijing Forestry University, Beijing 100083, China; jmeng@bjfu.edu.cn
5 Academy of Forestry Inventory and Planning, National Forestry and Grassland Administration, Beijing 100714, China; 18810791382@163.com
* Correspondence: yuangao9313@163.com

Abstract: This study establishes a climate-sensitive transition matrix growth model and predicts forest growth under different carbon emission scenarios (representative concentration pathways RCP2.6, RCP4.5, and RCP8.5) over the next 40 years. Data from the Eighth (2013) and Ninth (2019) National Forest Resource Inventories in Chongqing and climate data from Climate AP are utilized. The model is used to predict forest growth and compare the number of trees, basal area, and stock volume under different climate scenarios. The results show that the climate-sensitive transition matrix growth model has high accuracy. The relationships between the variables and forest growth, mortality, and recruitment correspond to natural succession and growth. Although the number of trees, basal area, and stock volume do not differ significantly for different climate scenarios, the forest has sufficient seedling regeneration and large-diameter trees. The growth process aligns with succession, with pioneer species being replaced by climax species. The proposed climate-sensitive transition matrix growth model fills the gap in growth models for natural secondary forests in Chongqing and is an accurate method for predicting forest growth. The model can be used for long-term prediction of forest stands to understand future forest growth trends and provide reliable references for forest management. Forest growth can be predicted for different harvesting intensities to determine the optimal intensity to guide natural forest management in Chongqing City. The results of this study can help formulate targeted forest management policies to deal more effectively with climate change and promote sustainable forest health.

Keywords: climate sensitivity; forest management; natural forests; transition matrix growth model

1. Introduction

Chongqing City is located in the upper reaches of the Yangtze River at the heart of the Three Gorges Reservoir area [1,2]. It is the last ecological barrier in the upper reaches of the Yangtze River, ensuring the ecological security of the river's middle and lower reaches [3–5]. The region has favorable hydrothermal conditions and significant vertical climate variability. Thus, Chongqing City ranks third in terms of species richness in China, following Yunnan and Sichuan provinces [6].

The forest area in Chongqing City is 4.53 million hectares, accounting for 55.4% of the total land area of Chongqing. Natural forests cover 2.59 million hectares, representing 72.98% and 74.43% of the total area and stand volume, respectively [7]. However, natural forests in Chongqing City have historically suffered from anthropogenic and natural disturbances and are highly degraded [8,9]. For example, the average stocking rate of natural forests in Chongqing City is 59.41 m^3/ha, far below the average value of 100.79 m^3/ha in China [10]. Additionally, most trees in natural forests originated from vegetative propagation and are short and have a small diameter, curved trunks, and low utilization values [11]. The degraded forest structure and species composition have decreased the forest functions. For example, although the total forest area in Chongqing has increased annually due to plantation establishment, carbon neutrality has not been achieved. Li et al. attributed this to the lower carbon sequestration rate of natural forests than of plantations resulting from the degradation of natural forests [12]. Therefore, improving the quality of natural forests is critical to local ecological security and carbon neutrality in Chongqing City.

The Chinese central government has focused extensively on restoring degraded natural forests. For instance, China banned all commercial logging in natural forests in 2017 [13,14] to ensure natural regeneration. In 2019, China implemented the 'Program for Natural Forest Conservation and Restoration System', which encourages management strategies to restore degraded natural forests. Formulating a sound forest management strategy is crucial to increasing the quality of natural forests because it improves forest structure, species composition, and forest functions, e.g., carbon sequestration, soil, and water conservation, and the protection of rare species habitat [15–20].

Understanding forest dynamics under different management scenarios is required for prescribing sound forest management strategies [21,22]. Forest growth models facilitate forest management decision-making by predicting forest dynamics under different forest management or climate change scenarios [23–27]. Therefore, they are widely used to support management decision-making in plantations and natural forests. For instance, López et al. proposed a transition matrix model to determine the harvesting cycle for natural European beech forests in the Spanish province of Navarre [28]. Burkhart and Margarida developed a stand model to predict plantation stand density for different management objectives [29]. Many forest growth simulators have been developed to facilitate forest management, e.g., Forest Vegetation Simulator (FVS), SILVA (version 2.2), and VisForest [30–32].

Forest growth and yield models include stand, diameter class, and individual tree models [33–37]. Transition matrix growth models have been extensively used in natural forests due to their complex structure and species composition. These models represent a trade-off between individual growth models and stand models in terms of model performance and cost [38]. For example, Sheng et al. used a transition matrix growth model to simulate the growth of mixed needle-broadleaf forests in northeast China [39]. Liang (2010) uses a multi-age, multi-species matrix growth model and the net present value to evaluate different forest management regimes in the boreal forest in Alaska [37].

Many authors have found that climate change significantly influences forest structure and species composition, altering forest functions and dynamics [40–43]. For instance, Lasch et al. predicted that climate warming would shift the natural species composition toward more drought-tolerant species and reduce forest diversity and groundwater recharge [44]. Therefore, it is essential to consider climate change in formulating forest management regimes and include climate variables, such as precipitation, temperature, and sunlight, as independent variables in the growth model.

Although many forest growth models have been developed to facilitate forest management in Chongqing, most were created for plantations, e.g., *Pinus massoniana* Lamb. [45], and *Cinnamomum camphora* Endl. [46]. A lack of research exists on growth and harvest modeling of natural forests in Chongqing, particularly forest growth models that consider climate change. The lack of modeling hinders the development of forest management strategies for natural forests in Chongqing. Therefore, the objectives of this study are to (1) develop climate-sensitive natural forest growth models and (2) predict forest growth trends under three different carbon emission scenarios (representative concentration pathways RCP 2.6, RCP 4.5, and RCP 8.5).

2. Materials and Methods

2.1. Study Area

The study area is located in Chongqing City (105°17′–110°11′ E, 28°10′–32°13′ N), which covers 82,400 km^2, with an altitude range of 170–2600 m [47] (Figure 1). Chongqing City is characterized by hilly terrain and a subtropical humid monsoon climate, with abundant precipitation and sufficient heat [48]. The average annual precipitation is 1000–1350 mm, and the average annual temperature is 16–18 °C [12]. The dominant tree species are *Quercus*, *Pinus*, *Tsuga*, *Acer*, *Betula*, *Populus*, and *Salix* [46,49,50].

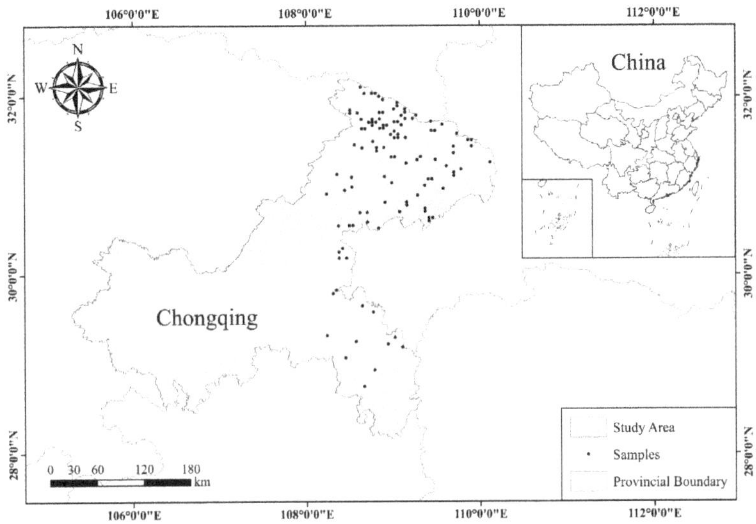

Figure 1. Distribution of sample plots in Chongqing City.

2.2. Data

We utilized data from the Eighth (2013) and Ninth (2018) Chinese National Forest Resource Inventories (CNFIs) of Chongqing City. A systematic sample distribution was used with a 4 × 4 km grid and a plot size of 0.067 ha (Figure 1). Tree-level and plot-level data were collected in the plots. Tree-level information included species identification and diameter at breast height (DBH), and plot-level data encompassed elevation, slope, aspect, soil depth, tree count, dominant species, and forest type. The 131 plots were located in natural forests that had experienced no human intervention (no logging or planting). We randomly selected 90 plots for model development and the remaining 41 for validation.

Climate factors, such as average growing season temperature (March–September) and mean annual precipitation (MAP), significantly impact forest structure and species composition [44,51–53]. We incorporated these variables to evaluate the effects of climate change on forest dynamics. Climate data were obtained from Climate AP, which projects long-term changes in forest dynamics and predicts monthly temperature and precipitation. Future climate scenarios were modeled using the Canadian Earth System Model 2 (CanESM2) general circulation model (GCM) developed by the Canadian Centre for Climate Modelling and Analysis. It provides historical climate data for the Asia-Pacific region and predicts changes under different emission scenarios (RCP 2.6, RCP 4.5, and RCP 8.5), representing low, moderate, and high radiative forcing and CO_2 emission levels [54].

2.3. Methods

2.3.1. Establishment of the Transition Matrix Model

The tree species were categorized into six groups based on the CNFI results, considering species frequency within sample plots and locally dominant species. The groups included oak, *Cupressus funebris*, *Pinus massoniana*, softwoods, hardwoods, and other species (those with less than 5% frequency). The model incorporated several variables: species diversity (H_1), size diversity (H_2), DBH, slope (SLcos), basal area (B), number of trees per hectare (N), mean annual temperature (MAT), and MAP. Stepwise regression was employed to identify the optimal variables. The variance inflation factor (VIF) was used to assess multicollinearity in the growth, recruitment, and mortality models. Values below 5 indicated minimal multicollinearity [55]. All selected variables had VIF values below this threshold, ensuring robust model predictions.

The transfer matrix model predicts the number of trees in a specific diameter class in future forests. It can be used to monitor the dynamic growth and development of forests by predicting growth, mortality, and recruitment [37].

The general expression of the transition matrix growth model is

$$y_{t+1} = G_t(y_t - h_t) + R_t + \epsilon_t \tag{1}$$

where $y_t = [y_{ijt}]$ represents the number of surviving trees of species i (i = 1, 2, 3, ... sp) and diameter class j (j = 1, 2, 3, ... dc) at time t. $h_t = [h_{ijt}]$ represents the number of trees of species i and diameter class j harvested at time t. A value of 0 is used if no harvesting occurs. R_t represents the number of trees in the minimum diameter class during a specific interval (5 years), and ϵ_t represents the random error.

G and R are expressed as

$$G = \begin{bmatrix} G_1 & & & \\ & G_2 & & \\ & & \ddots & \\ & & & G_m \end{bmatrix}, Gi = \begin{bmatrix} \alpha_{i1} & & & & \\ b_{i1} & \alpha_{i2} & & & \\ & \ddots & \ddots & & \\ & & b_{i,n-2} & \alpha_{i,n-1} & \\ & & & b_{i,n-1} & \alpha_{in} \end{bmatrix}$$
$$R = \begin{bmatrix} R_1 \\ R_2 \\ \vdots \\ R_m \end{bmatrix}, R_i = \begin{bmatrix} R_i \\ 0 \\ \vdots \\ 0 \end{bmatrix} \tag{2}$$

where α_{ij} denotes the likelihood that trees of species i in diameter class j remain in the same class during the growth period from time t to t + 1. b_{ij} represents the probability that trees of the same species and diameter class fall in the next diameter class in the same period. R_i indicates the number of individuals of species i entering the minimum diameter class within a specified interval (5 years). The value of b_{ij} is calculated by dividing the annual growth rate g by the diameter class width (5 cm). The relationship between α_{ij} and b_{ij} can be expressed as follows:

$$\alpha_{ij} = 1 - b_{ij} - m_{ij} \tag{3}$$

where m_{ij} represents the probability of tree mortality within a given time interval. Early studies on transition matrix growth models assumed that the growth, mortality, and recruitment probabilities (α_{ij}, b_{ij}, and m_{ij}) were constant across stands. However, subsequent research demonstrated that tree species responded differently to varying environmental conditions, prompting the development of more refined regression models for improved accuracy [38,56].

The dependent variable $g_{ij} + 1$ was selected to improve model fitting when growth rates are zero or close to zero. Since tree growth is typically nonlinear and follows exponential growth or power functions, the following expression was used to capture these dynamics more accurately:

$$\log\left(g_{ij}+1\right) = \gamma_{i1} + \gamma_{i2}DBH + \gamma_{i3}H_1 + \gamma_{i4}H_2 + \gamma_{i5}B + \gamma_{i6}SLcos + \gamma_{i7}MAT + \gamma_{i8}MAP + \mu_{ij} \tag{4}$$

where DBH represents the diameter at breast height of trees, H_1 indicates species diversity, H_2 denotes size diversity, B stands for basal area, MAT represents the mean annual temperature, MAP stands for the mean annual precipitation, and SLcos represents slope × cos(aspect) [57]. γ represents the coefficients, and μ_{ij} represents the error term.

H_1 and H_2 are calculated as follows:

$$H_1 = -\sum_{i=1}^{m} \frac{B_i}{B} \ln\left(\frac{B_i}{B}\right) \quad (5)$$

$$H_2 = -\sum_{j=1}^{n} \frac{B_j}{B} \ln\left(\frac{B_j}{B}\right) \quad (6)$$

where B_i and B_j represent the basal area at DBH for species i and diameter class j, respectively, and B denotes the total basal area at DBH for all trees in the sample plot. m indicates the number of species, and n refers to the number of diameter classes.

Since transitions within a stand are always non-negative, the distribution is truncated. As a result, the Tobit model was employed [58,59].

$$R = \Omega\left(\frac{\beta_i X_i}{\sigma_i}\right)\beta_i X_i + \sigma_i \omega\left(\frac{\beta_i X_i}{\sigma_i}\right) \quad (7)$$

$$\beta_i X_i = \beta_{i1} + \beta_{i2} N_i + \beta_{i3} B + \beta_{i4} H_1 + \beta_{i5} H_2 + \beta_{i6} SLcos + \beta_{i7} MAT + \beta_{i8} MAP + v_i \quad (8)$$

where N_i represents the number of trees per hectare for species i; Ω and ω refer to the cumulative distribution function and probability density function of the standard normal distribution, respectively. σ_i denotes the standard deviation of the residual v_i obtained from the parameter β estimation.

Since forest tree mortality follows a binary [0, 1] distribution—survival or death—the Probit model is used to model tree mortality [60].

$$m_{ij} = \frac{M_{ij}}{T} = \frac{1}{T}\Omega(\delta_{i1} + \delta_{i2} DBH + \delta_{i3} B + \delta_{i4} H_1 + \delta_{i5} H_2 + \delta_{i6} SLcos + \delta_{i7} MAT + \delta_{i8} MAP + \xi_{ij}) \quad (9)$$

where M_{ij} denotes the mortality probability for trees of species i and diameter class j in a given time interval, δ_i represents the parameter, and ξ_{ij} is the error term. The descriptive statistics of the variables are listed in Table 1.

Table 1. Descriptive statistics of the variables for the sample plots.

	DBH (cm)	B (m^2/ha)	H_1	H_2	N (Trees/ha)	MAT (°C)	MAP (mm)
Max	76.2	45.07	1.32	2.31	2985	18.36	1870.20
Min	5.00	1.74	0.00	0.00	45	7.20	439.8
Mean	11.05	12.81	0.64	1.31	992	12.86	1422.10
SD	5.65	8.76	0.38	0.41	496	3.02	239.12

2.3.2. Model Validation

We used data from the Eighth National Forest Resource Inventory (41 validation plots) and the transition matrix growth model to predict forest growth over a five-year period. The results were compared with field data from the Ninth Chinese National Forest Resource Inventory of the same permanent plots. The model accuracy was assessed by calculating the root mean square error (RMSE) and mean absolute error (MAE).

$$RMSE = \sqrt{\frac{1}{m}\sum_{n=1}^{m}(y_n - \hat{y}_n)^2} \quad (10)$$

$$MAE = \frac{1}{m}\sum_{n=1}^{m} |y_n - \hat{y}_n| \quad (11)$$

where $\hat{y}_n$ represents the predicted value, y_n represents the measured value, and m represents the number of sample plots.

2.3.3. Prediction of Natural Forest Dynamics Under Different Climate Change Scenarios

RCP2.6, RCP4.5, and RCP8.5 have been used in climate modeling to describe various potential trajectories of future greenhouse gas concentrations. These scenarios enable researchers to analyze possible climate outcomes under different levels of greenhouse gas emissions [54].

The RCP 2.6 involves significant global initiatives to curb greenhouse gas emissions, with emissions expected to peak by 2020, followed by a sharp decline. This pathway aims to limit global warming to 1.5 °C to 2 °C above pre-industrial levels by 2100, necessitating rapid transformations in energy systems, land use, and technology. In contrast, RCP 4.5 represents a moderate scenario, where emissions peak around 2040, followed by a decrease to stabilize atmospheric concentrations by the latter half of the century, resulting in a temperature increase of approximately 2.5 °C by 2100. RCP 8.5, known as the business-as-usual scenario, assumes that emissions will continue to rise throughout the century, leading to significant climate impacts and potential temperature increases exceeding 4 °C by 2100. This scenario reflects high population growth, sluggish economic development, and ongoing reliance on fossil fuels [61,62].

We selected a representative sample plot to predict future forest growth. This plot received average annual precipitation ranging from 1314 mm to 1791.8 mm, with average annual temperatures of 12.58–16.2 °C. The number of trees and diameter class distribution for various species were consistent with the average conditions of all sample plots in Chongqing City.

We used the climate-sensitive transition matrix growth model to predict the growth and development of natural forests in Chongqing City under different climate scenarios over the next 40 years. The model included key indicators, such as the number of trees, basal area, and stock volume. We compared the predicted and measured results under different climate scenarios to assess the impact of climate change on the natural forests of Chongqing and provide a scientific basis for future forestry management and protection.

3. Results

3.1. Model Parameters and Validation Results

The results of the growth model (Equation (4) and Table 2) indicate that the DBH is significantly positively related, and B is significantly negatively related to the growth of all tree species, except for *Cupressus funebris*, with which positive relationships are observed. H_1 is significantly positively related to the growth of oak, *Cupressus funebris*, and softwoods, whereas H_2 is significantly negatively related to the growth of most tree species. SLcos is significantly negatively related to the growth of *Pinus massoniana* and other tree species and positively related to the growth of the remaining species. MAT shows a negative relationship with the growth of *Pinus massoniana* and *Cupressus funebris* and a positive one with the other species. MAP is positively related to the growth of *Pinus massoniana* and other tree species groups and negatively related to the growth of the remaining species.

According to the recruitment model (Equation (7) and Table 3), N is significantly positively related to the recruitment of all tree species. H_1 is significantly negatively related to the recruitment of *Cupressus funebris* and other tree species and positively related to the recruitment of the rest. H_2 is significantly positively related to the recruitment of hardwoods and negatively related to that of the rest. MAT has a negative relationship with the recruitment of *Cupressus funebris* and a positive one with that of the rest. MAP is significantly positively related to the recruitment of oaks, *Pinus massoniana*, and other tree species and positively related to that of the rest. B has a positive relationship with the recruitment of *Cupressus funebris* and a negative one with that of the rest. SLcos is negatively related to the recruitment of oak species, *Cupressus funebris*, and softwoods and positively related to that of the remaining species.

Table 2. Estimates of parameters and the attendant lack-of-fit statistics for the tree diameter growth model.

	Oak	Others	*Pinus massoniana*	Hardwoods	Softwoods	*Cupressus funebris*
Intercept	6.34×10^{-1} ***	2.36×10^{-1} **	9.96×10^{-1} ***	8.23×10^{-1} ***	1.24×10^{0} ***	1.77×10^{0} ***
DBH	1.82×10^{-2} ***	2.22×10^{-2} ***	3.44×10^{-2} ***	1.53×10^{-2} ***	1.58×10^{-2} ***	4.53×10^{-2} ***
SLcos [1]	4.18×10^{-4}	-1.61×10^{-3} **	-1.27×10^{-3} **	1.11×10^{-3} **	4.79×10^{-4}	2.84×10^{-3} ***
H_1	1.31×10^{-1} ***	-6.79×10^{-2} *	-2.54×10^{-2}	5.18×10^{-2}	1.07×10^{-1} **	6.03×10^{-1} ***
H_2	-2.67×10^{-1} ***	-4.07×10^{-2}	-1.00×10^{-1} **	2.17×10^{-2}	-1.68×10^{-1} ***	-4.81×10^{-1} ***
MAT	3.27×10^{-2} ***	3.10×10^{-2} ***	-2.77×10^{-2} ***	6.53×10^{-3}	1.40×10^{-2} ***	-7.34×10^{-2} ***
MAP	-1.60×10^{-5}	2.29×10^{-4} ***	2.82×10^{-4} ***	-1.27×10^{-4} **	-3.09×10^{-4} ***	-4.43×10^{-5}
B	-5.12×10^{-3} ***	-8.97×10^{-3} ***	-1.76×10^{-2} ***	-1.16×10^{-2} ***	-9.27×10^{-3} ***	8.48×10^{-3} **
AIC [2]	147.11	1749.17	2480.55	168.30	1186.71	150.28
BIC [3]	189.07	1797.03	2530.82	210.25	1231.19	188.93
R^2 [4]	0.26	0.15	0.17	0.27	0.15	0.42
LogLink [5]	−64.56	−865.59	−1231.28	−75.15	−584.35	−66.14

Note: Significance levels: * $p < 0.10$; ** $p < 0.05$; *** $p < 0.01$; [1] SLcos = Slope × cos(Aspect); [2] AIC: Akaike information criterion; [3] BIC: Bayesian information criterion; [4] R^2: Nagelkerke's pseudo R-squared; [5] LogLink: log-likelihood value.

Table 3. Estimates of parameters and the lack-of-fit statistics for the tree recruitment model.

	Oak	Others	*Pinus massoniana*	Hardwoods	Softwoods	*Cupressus funebris*
Intercept	-6.39×10^{2} ***	-4.12×10^{2} ***	-1.04×10^{3} ***	3.33×10^{2} **	4.04×10^{2} ***	1.21×10^{3} ***
N [2]	4.09×10^{-1} ***	1.42×10^{-1} ***	2.88×10^{-1} ***	9.45×10^{-1} ***	9.28×10^{-2} ***	3.32×10^{-1} ***
SLcos	-4.86×10^{0} ***	1.66×10^{0} ***	1.36×10^{0} *	4.79×10^{0} ***	-3.18×10^{-1}	-5.40×10^{-1}
H_1	-2.27×10^{2} ***	3.36×10^{1} **	-3.23×10^{2} ***	-3.46×10^{2} ***	-1.53×10^{2} ***	4.44×10^{2} ***
H_2	-3.40×10^{2} ***	-7.29×10^{1} ***	-8.56×10^{2} ***	1.48×10^{2} ***	-1.06×10^{2} ***	-4.15×10^{2} ***
MAT	6.25×10^{1} ***	3.11×10^{1} ***	1.33×10^{2} ***	4.52×10^{0}	1.44×10^{1} ***	-7.63×10^{1} ***
MAP	3.34×10^{-1} ***	1.11×10^{-1} ***	2.45×10^{-1} ***	-2.11×10^{-2}	-2.96×10^{-2}	-1.37×10^{-2}
B	-1.73×10^{0}	-3.58×10^{0} ***	-2.24×10^{1} ***	-3.32×10^{1} ***	-1.43×10^{1} ***	7.82×10^{0} ***
logSigma [1]	5.94×10^{0} ***	5.00×10^{0} ***	6.35×10^{0} ***	5.77×10^{0} ***	5.11×10^{0} ***	4.13×10^{0} ***
AIC	13,386.16	15,124.19	27,852.04	13,499.68	11,999.50	5665.27
BIC	13,431.39	15,173.15	27,903.57	13,544.88	12,045.40	5704.19
R^2	0.31	0.37	0.35	0.50	0.51	0.93
LogLink	−6684.08	−7553.10	−13,917.02	−6740.84	−5990.75	−2823.63

Note: Significance levels: * $p < 0.10$; ** $p < 0.05$; *** $p < 0.01$; [1] logSigma: log of the standard deviation of residuals; [2] N: number of plots with recruitment, the total number of plots.

The mortality model results (Equation (9) and Table 4) indicate that DBH is negatively related to the mortality of all tree species. H_1 is positively related to the mortality of oak species, *Pinus massoniana*, and *Cupressus funebris* and negatively related to that of the rest. H_2 has a positive relationship with the mortality of all tree species, except for oaks. B is positively related to the mortality of all tree species except for hardwoods. MAT has positive or negative relationships with different tree species. MAP is significantly negatively correlated with the mortality of other tree species and not significantly with that of the rest.

The transition matrix growth models for natural secondary forests in Chongqing have high R^2 values. The RMSE is 1.45 m^2/ha, and the MAE is 1.43 m^2/ha. Therefore, the proposed transition matrix growth model accurately predicts forest growth in this region.

Table 4. Estimates of parameters and the lack-of-fit statistics for the tree mortality model.

	Oak	Others	*Pinus massoniana*	Hardwoods	Softwoods	*Cupressus funebris*
Intercept	-2.32×10^{0} **	1.20×10^{0} *	-7.57×10^{0} ***	-1.55×10^{0} **	-2.77×10^{0} ***	-7.90×10^{-1}
DBH	-1.20×10^{-1} ***	-1.03×10^{-2}	-1.06×10^{-1} ***	-7.68×10^{-2} ***	-1.72×10^{-2} *	-2.52×10^{-1} *
SLcos	-3.06×10^{-3}	1.50×10^{-2} ***	-4.79×10^{-3} *	-6.31×10^{-3} **	7.09×10^{-4}	6.11×10^{-1}
H_1	3.84×10^{-1}	-6.82×10^{-1} ***	2.87×10^{-1} *	-4.62×10^{-1} **	-4.45×10^{-3}	1.88×10^{1}
H_2	-9.19×10^{-2}	1.58×10^{-1}	1.94×10^{0} ***	9.89×10^{-1} ***	5.87×10^{-1} ***	2.49×10^{0}
MAT	6.34×10^{-2}	-1.77×10^{-1} ***	2.67×10^{-1} ***	3.42×10^{-2}	5.87×10^{-2} ***	-1.06×10^{0}
MAP	1.06×10^{-4} *	-6.47×10^{-4} ***	2.57×10^{-4}	-2.11×10^{-4}	-1.41×10^{-4}	-3.87×10^{-2}
B	3.93×10^{-2} ***	2.88×10^{-2} ***	2.17×10^{-2} ***	-2.37×10^{-2} ***	1.41×10^{-2} *	2.57×10^{0}
AIC	427.05	754.76	1074.69	522.18	736.70	38.87
BIC	467.26	798.27	1120.49	562.36	777.50	73.47
R^2	0.12	0.19	0.34	0.12	0.05	0.67
LogLink	−205.52	−369.38	−529.34	−253.09	−360.35	−11.44

Note: Significance levels: * $p < 0.10$; ** $p < 0.05$; *** $p < 0.01$.

3.2. Prediction of Forest Dynamics Under Different Climate Scenarios

We selected a sample plot representative of natural forests in Chongqing City from the CNFI. The plot contained oak, hardwoods, softwoods, *Pinus massoniana*, *Cupressus funebris*, and other species. Oak was the dominant species, with a volume of 98.84 m^3/ha. We used the transition matrix growth model to predict the forest dynamics under different climate change scenarios over the next 40 years.

At the beginning of the forest dynamic prediction, the forest stand is dominated by small-diameter trees, primarily oak (Figure 2). As the prediction progresses, the proportion of large-diameter trees increases, regardless of the climate change scenario. Hardwoods and oak are the dominant species. The number of small-diameter trees, most of which are *Cupressus funebris*, increases significantly.

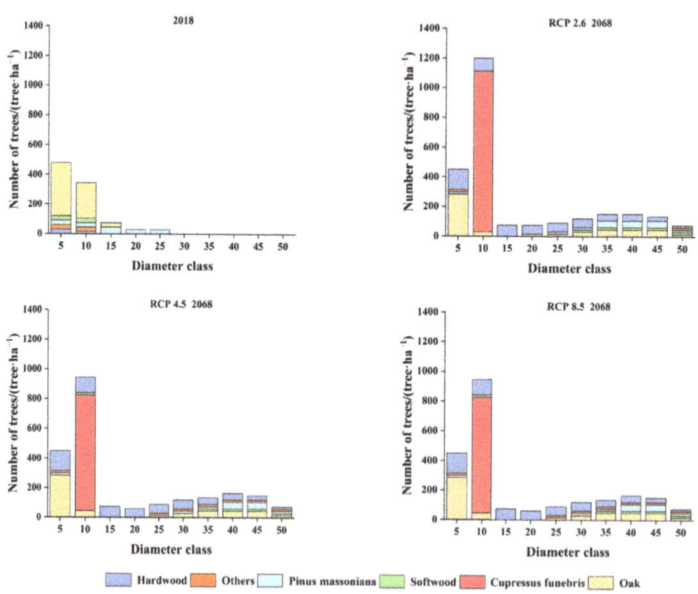

Figure 2. The initial diameter distribution (2018) and the simulated diameter distributions after 40 years under different climate scenarios (RCP 2.6, RCP 4.5, and RCP 8.5).

Regardless of the climate change scenario, the number of trees per hectare increases for the next 15 years, followed by a decrease (Figure 3). As the prediction continues, the number of hardwoods and *Cupressus funebris* increases significantly, whereas oak increases for the next 15 years, followed by a decrease. The changes in other tree species are more subtle.

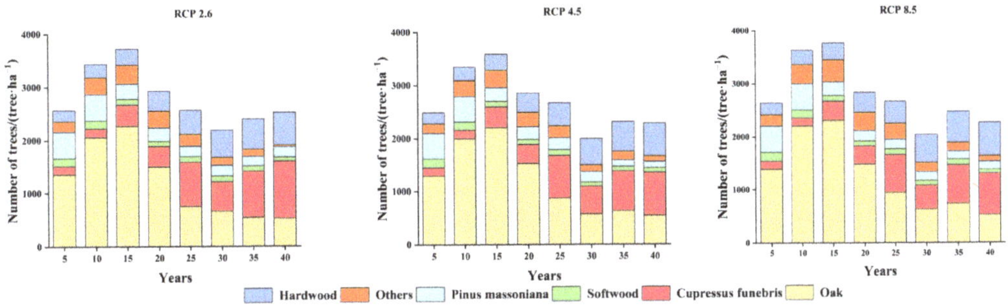

Figure 3. The diameter distribution over the next 40 years.

The volume shows an increasing trend over time (Figure 4), but different tree species exhibit different trends. For instance, the hardwoods increase, and the oaks increase for the first 25 years, followed by stabilization. Although the number of *Cupressus funebris* increases (Figure 2), the stand volume does not change significantly and shows a decreasing trend (Figure 4), suggesting that most *Cupressus funebris* are small individuals. In contrast, the number of oaks decreases significantly, and the stand volume remains relatively stable, suggesting the presence of large-diameter trees.

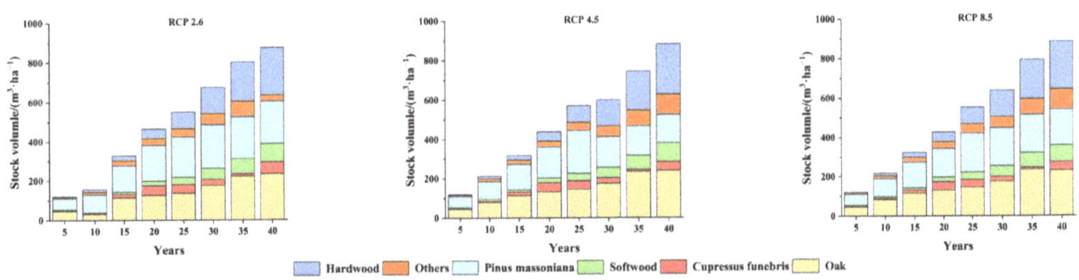

Figure 4. The change in forest stand volume under different climate scenarios (RCP 2.6, RCP 4.5, RCP 8.5) over the next 40 years.

The difference in the diameter distribution and the number of *Cupressus funebris* decreases in the small-diameter class, and the number of large-diameter trees increases from RCP 2.6 to RCP 4.5 to RCP 8.5 (Figure 2).

However, under different climatic scenarios, there is no significant difference in the change in the number of trees in the forest, which showed a tendency to increase at the beginning and then decrease (Figure 3). A comparison of the changes in forest stock volume under different climate scenarios shows that forest stock volume increases gradually with increasing emission scenarios, i.e., the RCP8.5 scenario has the largest forest stock (Figure 4).

4. Discussion

4.1. Climate-Sensitive Transition Matrix Growth Model

Our results showed that the DBH was positively correlated with tree growth (Table 2) and negatively correlated with tree mortality (Table 4). Yang et al. observed the same relationship in an individual-tree growth model for *Casuarina equisetifolia* L. in Hainan Province, China [63]. Zhou et al. also reported a negative correlation between DBH and mortality [64]. The reason is that larger trees tend to dominate the stand and exhibit greater vitality and survivability, resulting in lower mortality and higher growth rates.

The B was significantly negatively correlated with tree growth and tree recruitment in most tree groups, and a significant positive correlation occurred with mortality (Table 2). The same findings were reported by many authors [65–69]. For instance, Rozendaal et al. reported that the growth rate of trees decreased as the basal area increased in mature tropical forests in Amazonia and tropical Africa [65]. Bradford and David observed a positive correlation between basal area and tree mortality in *Pinus ponderosa* Douglas ex C. Lawson forests in northern Arizona [66]. Klopcic et al. investigated European beech and found that a larger basal area increased competition among trees, reducing tree recruitment in Slovenian forests [67]. Chen et al. attributed this relationship to competition. As the basal area increased, competition among species and individual trees increased, leading to higher mortality rates and lower growth rates [68].

A higher H_1 was associated with a higher tree growth rate in most tree groups (Table 2) and lower mortality (Table 4). Many authors observed similar results [70–73]. For instance, Zeller et al. reported that higher species diversity improved stand productivity based on the national forest inventory data from Germany and the United States [70]. Hisano et al. observed that forest mortality increased as species diversity increased in natural forests in western Canada [74]. The positive correlation between biodiversity and productivity and the negative correlation between biodiversity and mortality can be explained by the "niche difference" hypothesis. This theory suggests that tree species that are positively correlated utilize resources differently, resulting in more efficient resource use within the community and demonstrating niche complementarity [75,76]. Therefore, this effect reduced tree mortality and increased tree growth. The complementarity effects of different tree species increase interspecies competition and niche differentiation, enabling better resource utilization and promoting stand growth [71]. Species diversity (H_1) was generally negatively correlated with recruitment. Searle et al. found that higher diversity resulted in the alteration of stand density and size proportions among species, leading to higher mortality rates [77].

H_2 was negatively correlated with tree growth (Table 2), suggesting that a higher diversity resulted in reduced tree growth. In addition, we observed that H_2 was positively correlated with tree mortality but negatively correlated with recruitment. The same result was observed by Du et al. [36] and Liang et al. [78], who analyzed oak-dominated natural forests in Shanxi Province, China, and Douglas fir and western hemlock forests in Oregon and Washington, USA, respectively [36,78]. The reason for the positive correlation between H_2 and mortality and the negative relationship between H_2 and recruitment is that a larger size diversity reduces the ecological niches for tree recruitment and their competition for light, water, and other resources [79–82].

MAT was negatively correlated with the growth of the conifer tree species, i.e., *Pinus massoniana* and *Cupressus funebris* and positively correlated with the growth of other tree groups (broad-leaved tree species). Many authors have reported that high temperatures promote tree growth [83–86]. However, high summer temperatures can cause drought conditions, and high winter temperatures can increase evaporation, reducing soil moisture and slowing tree growth [87–89]. The reason for the negative correlation between MAT and the growth of *Pinus massoniana* and *Cupressus funebris* is that conifer tree species are more vulnerable to drought in summer and evaporation in winter than broad-leaved tree species [90]. Many authors have documented a negative correlation between MAT and tree growth of conifer tree species, e.g., Alaskan white spruce [91], *Pinus tabulaeformis* Carr. [33], Norway spruce, and Douglas-fir [92]. Our study also indicated that MAT was positively correlated with tree mortality, which can be attributed to drought caused by high temperatures. McDowell et al. found that rising temperatures led to significant forest mortality [93]. We observed that MAT was significantly positively correlated with tree recruitment, suggesting that increased temperatures can trigger and promote natural regeneration. The same positive relationship was observed by Du et al., who reported that higher temperatures increased tree regeneration [94]. However, many authors stated that natural regeneration is complex and influenced by many factors [95–97]. For instance, Maleki et al.

(2024) conducted a study in boreal mixed forests in western Quebec, Canada, and found that aspen (*Populus tremuloides*), which regenerates by root suckers, was largely unaffected by most climate scenarios. Recruitment is negatively impacted by the temperature in the current year but positively influenced by warm weather two years prior, aligning with seed production [97].

MAP was positively correlated with the growth of *Pinus massoniana* and other tree species but negatively correlated with the growth of oak, hardwoods, and softwoods. The reason for the positive relationship between MAP and the growth of *Pinus massoniana* is that water is a limiting factor for *Pinus massoniana* [86]. In contrast, MAP was negatively correlated with the growth of the other tree species, e.g., oak, hardwoods, and softwoods group. This finding might be attributed to the fact that water is not the limiting factor for these tree species, especially for broad-leaved tree species. Moreover, many authors have found that MAP was negatively correlated with tree growth in humid environments because of sufficient water for tree growth [98,99]. Additionally, we observed that MAP was positively correlated with tree recruitment, indicating that a higher MAP facilitated tree regeneration. Tingstad et al. also found that sufficient precipitation improved the regeneration of most tree species [100]. Our study showed no significant correlation between MAP and mortality, possibly because local precipitation levels were sufficient for the growth of tree species. Clifford et al. observed that tree mortality due to drought was negligible when annual precipitation exceeded 600 mm in the southwestern United States [101].

4.2. Forest Dynamics Prediction Under Different Climate Change Scenarios

In the early prediction stages, the forest stand was dominated by small-diameter trees, mostly oaks (Figure 2). In the later prediction stage (2068), large-diameter trees occurred (e.g., 50 cm DBH class), and climax tree species (*Cupressus funebris* and hardwoods) dominated the forest stand. This result was consistent with forest succession, which posits that succession moves toward shade-tolerant or climax tree species [102].

In the initial stage of forest growth prediction, the number of trees increased, followed by a decrease (Figure 3). The reason is that the forest composition and structure are simple in the early stage, with less competition among trees, providing ample growth space and resulting in lower mortality, especially for pioneer species with higher growth and recruitment rates [103]. Van der Sande et al. observed similar results; the recruitment increased with higher resource availability and ample growth space [104]. As the forest succession continues, competition for light and nutrients among trees increases, leading to higher mortality rates and fewer trees [105,106]. Notably, the reduction in the number of trees was confined to pioneer tree species (softwoods and *Pinus massoniana*), whereas the number of climax tree species (hardwoods and *Cupressus funebris* increased. This trend confirms forest succession, i.e., pioneer species (*Pinus massoniana* and softwoods) are replaced by shade-tolerant species or climax tree species (*Cupressus funebris* and hardwoods) [107].

The stand volume showed an increasing trend (Figure 4); however, different species exhibited different growth trends. For instance, *Pinus massoniana* showed a slightly decreasing trend, whereas hardwoods and *Cupressus funebris* exhibited an increasing trend. The latter was attributed to the climax species' growth characteristics, i.e., slow growth under low light conditions and rapid growth when canopy gaps appeared and sufficient light was available [108–110]. Many researchers observed that although shade-intolerant or pioneer species survived in large numbers, shade-tolerant hardwood species had higher recruitment rates under the canopy and dominated the canopy as the forest grew [111,112].

In contrast to other tree species groups in our study, oak initially showed an increasing trend for the first 15 years, followed by a decreasing trend (Figure 3). In the initial stage, ample light existed in the understory, facilitating the regeneration and growth of oak. However, as succession progressed, the light level in the understory decreased, hindering the natural regeneration of oak [113]. The stand volume of oak increased over time (Figure 4) because the existing oaks were highly competitive, resulting in a rapid growth rate [114].

The increasing trend of the stand volume and the decrease in the number of trees suggest large-diameter oak trees in this stand. Oaks are important economic tree species. Due to their hard, attractive wood, they are widely used for high-quality furniture [115–117]. The large-diameter oaks in this stand suggest an increase in the stand value. Furthermore, hardwood species were also found in the large-diameter classes.

A comparison of the diameter distributions under the three different climate scenarios indicates that the number of *Cupressus funebris* decreased with increasing emissions (Figure 2). Lindenmayer et al. predicted that some conifer species would experience widespread mortality due to significant climate change [118]. Furthermore, similar results have been reported in recent studies. Many dominant conifer species, particularly those with narrow temperature tolerances, are at a high risk of extinction under future climate scenarios [119,120]. Many studies suggest climate change impacts forest growth, especially in drier regions [90,121,122]. Babst et al. conducted a study of 1000 sites in Europe and parts of North Africa and found that tree growth in high-latitude and high-altitude regions was more sensitive to the temperature, with warm summers promoting growth. In contrast, growth in low-altitude and southern European areas was more limited by water availability, with rainfall improving growth and drought inhibiting it [121]. Our results showed no significant differences in the tree number between the three climate scenarios (Figure 3). The likely reason is that differences in temperature and precipitation in Chongqing City under the different emission scenarios do not influence forest structure and dynamics because of sufficient precipitation in Chongqing City [49]. As the emissions increased, the stock volume proportion of hardwoods and oaks increased for all species under different climate scenarios (Figure 4). This observation aligns with existing research findings that showed climate change and rising temperatures reduced the number of pioneer species (softwoods, conifers, etc.) and increased the number of hardwood species [123,124]. Additionally, many authors observed that rising temperatures improved forest growth and tree productivity. For instance, Pretzsch et al. found that the growth rates of European beech and Norway spruce increased since 1870 as temperatures have increased due to climate change [125]. Feng et al. reported that climate change increased the carbon storage capacity in the eastern part of Sichuan and Chongqing [126]. Eggers et al. examined forest growth in 15 European countries and noted that climate change has led to accelerated forest growth and increased carbon storage capacity [127].

5. Conclusions

This study established a climate-sensitive transition matrix growth model for the natural forests of Chongqing. The model was used to predict forest growth over the next 40 years. We compared the number of trees, basal area at breast height, and stock volume under different climate scenarios. The simulation results indicated that succession occurred without human intervention, increasing the hardwood stock volume and decreasing the proportion of softwoods and conifers. The forests contained many small-diameter trees, ensuring sufficient seedling regeneration. The lack of significant differences in the parameters between different climate scenarios was attributed to the minimal changes in temperature and precipitation in the study area as emissions increased, resulting in a negligible impact on forest growth.

The proposed climate-sensitive transition matrix growth model fills a gap in growth models for Chongqing natural forests, providing a more accurate tool for predicting future forest growth. This model enables long-term forecasts of stand growth trends, offering valuable insights for future forest management. The model can be combined with forest management measures, such as harvesting and replanting, to guide natural forest management and determine the optimal management strategy in Chongqing City. We plan to optimize forest harvesting intensity and determine the optimal trees to harvest to support the management of natural forests in Chongqing City.

Author Contributions: All authors made significant contributions to the manuscript. Conceptualization, Y.G. and X.M.; methodology, J.M. and Z.M.; validation, Y.X. and Y.B.; database organization,

X.M. and Z.M.; writing—original draft preparation, X.M.; X.M. and Y.G. are the main authors who developed and revised the manuscript. All authors have read and agreed to the published version of the manuscript.

Funding: This research was funded by the National Natural Science Foundation of China (grant No. 32271871) and the Research and Demonstration of Key Technologies for Precision Management of Oak Natural Secondary Forests in the Three Gorges Reservoir Area (Yulin Scientific Research 2022-13).

Data Availability Statement: The data presented in this study are available upon reasonable request from the corresponding author. The data are not publicly available due to privacy restrictions.

Acknowledgments: The authors would like to thank their colleagues at the Academy of Forest Inventory and Planning, National Forestry, Chongqing Academy of Forestry and Grassland Administration for their kind support during data collection.

Conflicts of Interest: The authors declare no conflict of interest.

References

1. Zhou, L.; Zhang, H.; Bi, G.; Su, K.; Wang, L.; Chen, H.; Yang, Q. Multiscale perspective research on the evolution characteristics of the ecosystem services supply-demand relationship in the Chongqing section of the three gorges reservoir area. *Ecol. Indic.* **2022**, *142*, 109227. [CrossRef]
2. Yu, Y.; Zhou, L.; Zhou, W.; Ren, H.; Kharrazi, A.; Ma, T.; Zhu, B. Decoupling environmental pressure from economic growth on city level: The Case Study of Chongqing in China. *Ecol. Indic.* **2017**, *75*, 27–35. [CrossRef]
3. Liu, Y.; Yue, W.; Fan, P.; Zhang, Z.; Huang, J. Assessing the urban environmental quality of mountainous cities: A case study in Chongqing, China. *Ecol. Indic.* **2017**, *81*, 132–145. [CrossRef]
4. Yang, C.; Zeng, W.; Yang, X. Coupling coordination evaluation and sustainable development pattern of geo-ecological environment and urbanization in Chongqing municipality, China. *Sustain. Cities Soc.* **2020**, *61*, 102271. [CrossRef]
5. Xiang, S.J.; Wang, Y.; Deng, H.; Yang, C.M.; Wang, Z.F.; Gao, M. Response and Multi-Scenario Prediction of Carbon Storage to Land Use/Cover Change in the Main Urban Area of Chongqing, China. *Ecol. Indic.* **2023**, *142*, 109205. [CrossRef]
6. Wang, F.; Yuan, X.; Sun, Y.; Liu, Y. Species distribution modeling based on MaxEnt to inform biodiversity conservation in the Central Urban Area of Chongqing Municipality. *Ecol. Indic.* **2024**, *158*, 111491. [CrossRef]
7. National Forestry and Grassland Administration. *China Forest Resources Report*; National Forestry and Grassland Administration: Beijing, China, 2014.
8. Fang, X.; Tan, W.; Gao, X.; Chai, Z. Close-to-nature management positively improves the spatial structure of Masson pine forest stands. *Web Ecol.* **2021**, *21*, 45–54. [CrossRef]
9. Zhang, H.; Lei, X. Forest management and accurate improvement of forest quality. *Land Green.* **2017**, *33*, 13–15.
10. Bulletin of China's Land Greening Status 2023. *Land Green.* **2024**, *3*, 6–9.
11. Meng, X.; He, B.; Wang, L.; Chen, B.; Pan, C. Research on the Countermeasures for Accurately Lifting Forest Quality in Chongqing. *For. Grass Res. Res.* **2017**, *s1*, 59–63+69.
12. Li, L.; Qi, Z.; Zhong, T. Forest Transition and Its Dynamics in Subtropical Chongqing, China since 1990s. *Land* **2021**, *10*, 777. [CrossRef]
13. Viña, A.; McConnell, W.; Yang, H.; Xu, Z.; Liu, J. Effects of conservation policy on China's forest recovery. *Sci. Adv.* **2016**, *2*, e1500965. [CrossRef]
14. Hua, F.; Xu, J.; Wilcove, D.S. A new opportunity to recover native forests in China. *Conserv. Lett.* **2018**, *11*, e12396. [CrossRef]
15. Hou, J.; Yin, R.; Wu, W. Intensifying Forest Management in China: What does it mean, why, and how? *For. Policy Econ.* **2019**, *98*, 82–89. [CrossRef]
16. Schelhaas, M.; Nabuurs, G.; Hengeveld, G.; Reyer, C.; Hanewinkel, M.; Zimmermann, N.; Cullmann, D. Alternative forest management strategies to account for climate change-induced productivity and species suitability changes in Europe. *Reg. Environ. Change* **2015**, *15*, 1581–1594. [CrossRef]
17. Amato, A.; Bradford, J.; Fraver, S.; Palik, B. Forest management for mitigation and adaptation to climate change: Insights from long-term silviculture experiments. *For. Ecol. Manag.* **2011**, *262*, 803–816.
18. Chen, W.; Chen, J.M.; Price, D.T.; Cihlar, J.; Liu, J. Carbon Offset Potentials of Four Alternative Forest Management Strategies in Canada: A Simulation Study. *Mitig. Adapt. Strateg. Glob. Change* **2000**, *5*, 143–169. [CrossRef]
19. Bengston, D.N. Changing forest values and ecosystem management. *Soc. Nat. Resour.* **1994**, *7*, 515–533. [CrossRef]
20. Jandl, R.; Lindner, M.; Vesterdal, L.; Bauwens, B.; Baritz, R.; Hagedorn, F.; Johnson, D.W.; Minkkinen, K.; Byrne, K. How strongly can forest management influence soil carbon sequestration? *Geoderma* **2007**, *137*, 253–268. [CrossRef]
21. Kuuluvainen, T.; Grenfell, R. Natural disturbance emulation in boreal forest ecosystem management—Theories, strategies, and a comparison with conventional even-aged management. *Can. J. For. Res.* **2012**, *42*, 1185–1203. [CrossRef]
22. Rauscher, H.M. Ecosystem management decision support for federal forests in the United States: A review. *For. Ecol. Manag.* **1999**, *114*, 173–197. [CrossRef]

23. Vanclay, J.K. Growth models for tropical forests: A synthesis of models and methods. *For. Sci.* **1995**, *41*, 7–42. [CrossRef]
24. Vanclay, J.K. A growth model for north Queensland rainforests. *For. Ecol. Manag.* **1989**, *27*, 245–271. [CrossRef]
25. Hamidi, S.K.; Weiskittel, A.; Bayat, M.; Fallah, A. Development of individual tree growth and yield model across multiple contrasting species using nonparametric and parametric methods in the Hyrcanian forests of northern Iran. *Eur. J. For. Res.* **2021**, *140*, 421–434. [CrossRef]
26. Porté, A.; Bartelink, H.H. Modelling mixed forest growth: A review of models for forest management. *Ecol. Modell.* **2002**, *150*, 141–188. [CrossRef]
27. Peng, C. Growth and yield models for uneven-aged stands: Past, present and future. *For. Ecol. Manag.* **2000**, *132*, 259–279. [CrossRef]
28. López, I.; Ortuño, S.F.; Martín, Á.J.; Fullana, C. Estimating the sustainable harvesting and the stable diameter distribution of European beech with projection matrix models. *Ann. For. Sci.* **2007**, *64*, 593–599. [CrossRef]
29. Burkhart, H.E.; Margarida, T. *Modeling Forest Trees and Stands*; Burkhart, H.E., Tomé, M., Eds.; Springer: Dordrecht, The Netherlands, 2012.
30. Pretzsch, H.; Biber, H.; Ďurský, J. The single tree-based stand simulator SILVA: Construction, application and evaluation. *For. Ecol. Manag.* **2002**, *162*, 3–21. [CrossRef]
31. Crookston, N.; Dixon, G. The forest vegetation simulator: A review of its structure, content, and applications. *Comput. Electron. Agric.* **2005**, *49*, 60–80. [CrossRef]
32. Tang, L.; Peng, X.; Chen, C.; Huang, H.; Lin, D. Three-dimensional Forest growth simulation in virtual geographic environments. *Earth Sci. Inform.* **2019**, *12*, 31–41. [CrossRef]
33. Weiskittel, A.; Hann, D.W.; Kershaw, J.A.; Vanclay, J. *Forest Growth and Yield Modeling*; Wiley: Hoboken, NJ, USA, 2011.
34. Zhang, Y.; Zhang, Q.; Kang, X.; Li, Y.; Zhang, M.; Xu, G.; Yang, Y. Methods for determining equilibrium curve of uneven-aged selection theropencedrymion—Taking sample plot data in Jingouling Forest Farm as a case. *J. Beijing For. Univ.* **2015**, *37*, 53–60.
35. Brzeziecki, B.; Pommerening, A.; Mi'scicki, S.; Drozdowski, S.; Zybura, H.; Collins, B. A common lack of demographic equilibrium among tree species in Białowieza National Park (NE Poland): Evidence from long-term plots. *J. Veg. Sci.* **2016**, *27*, 460–469. [CrossRef]
36. Du, X.; Chen, X.; Zeng, W.; Meng, J. A climate-sensitive transition matrix growth model for uneven-aged mixed-species oak forests in North China. *Forestry* **2021**, *94*, 258–277. [CrossRef]
37. Liang, J. Dynamics and management of Alaska boreal forest: An all-aged multi-species matrix growth model. *For. Ecol. Manag.* **2010**, *260*, 491–501. [CrossRef]
38. Liang, J.; Picard, N. Matrix Model of Forest Dynamics: An Overview and Outlook. *For. Sci.* **2013**, *59*, 359–378. [CrossRef]
39. Sheng, Q.; Liu, Z.; Dong, L. A climate-spatial matrix growth model for major tree species in Lesser Khingan Mountains and responses of forest dynamics change to different representative concentration path scenarios. *Front. For. Glob. Change* **2023**, *6*, 1309189. [CrossRef]
40. Yin, Y.; Ma, D.; Wu, S. Climate change risk to forests in China associated with warming. *Sci. Rep.* **2018**, *8*, 493. [CrossRef]
41. Vlam, M.; Baker, P.J.; Bunyavejchewin, S.; Zuidema, P. Temperature and rainfall strongly drive temporal growth variation in Asian tropical forest trees. *Oecologia* **2014**, *174*, 1449–1461. [CrossRef]
42. Shestakova, T.A.; Mutke, S.; Gordo, J.; Camarero, J.; Sin, E.; Pemán, J.; Voltas, J. Weather as the main driver for masting and stem growth variation in stone pine supports compatible timber and nut co-production. *Agric. For. Meteorol.* **2021**, *298–299*, 108287. [CrossRef]
43. Liang, R.; Sun, Y.; Qiu, S.; Wang, B.; Xie, Y. Relative effects of climate, stand environment and tree characteristics on annual tree growth in subtropical Cunninghamia lanceolata forests. *Agric. For. Meteorol.* **2023**, *342*, 109711. [CrossRef]
44. Lasch, P.; Lindner, M.; Erhard, M.; Suckow, F.; Wenzel, A. Regional impact assessment on forest structure and functions under climate change—The Brandenburg case study. *For. Ecol. Manag.* **2002**, *162*, 73–86. [CrossRef]
45. He, Y.; Chen, Y.; Lin, Y. A Study of Individual Tree Growth Model of Pinus massoniana in Shibao Town, Chongqing City. *J. Sichuan For. Sci. Technol.* **2019**, *34*, 56–60.
46. Yang, Y.; Fujihara, M.; Li, B.; Yuan, X.; Hara, K.; Da, L.; Tomita, M.; Zhao, Y. Structure and diversity of remnant natural evergreen broad-leaved forests at three sites affected by urbanization in Chongqing metropolis, Southwest China. *Landsc. Ecol. Eng.* **2014**, *10*, 137–149. [CrossRef]
47. Bao, H.; Li, L.; Lizieri, C. City profile: Chongqing (1997–2017). *Cities* **2019**, *94*, 161–171. [CrossRef]
48. Yao, R.; Qing, L.; Luo, Z.; Jiang, L.; Yang, Y. An integrated study of urban microclimates in Chongqing, China: Historical weather data, transverse measurement and numerical simulation. *Sustain. Cities Soc.* **2015**, *14*, 187–199. [CrossRef]
49. Dodson, J.; Shi, G.; Lu, F.; Yang, H. A 40,000 year record of vegetation, environment and climate change from Chongqing, Central China. *Palaeogeogr. Palaeoclimatol. Palaeoecol.* **2021**, *573*, 110441. [CrossRef]
50. Han, G.; Yang, Y.; Yan, S. Vegetation Activity Trend and Its Relationship with Climate Change in the Three Gorges Area, China. *Adv. Meteorol.* **2013**, *2013*, 235378. [CrossRef]
51. Mérian, P.; Lebourgeois, F. Size-mediated climate–growth relationships in temperate forests: A multi-species analysis. *For. Ecol. Manag.* **2011**, *261*, 1382–1391. [CrossRef]
52. Bugmann, H.; Pfister, C. Impacts of interannual climate variability on past and future forest composition. *Reg. Environ. Change* **2000**, *1*, 112–125.

53. Muscarella, R.; Kolyaie, S.; Morton, D.; Zimmerman, J.; Uriarte, M. Effects of topography on tropical forest structure depend on climate context. *J. Ecol.* **2020**, *108*, 145–159. [CrossRef]
54. Taylor, K.; Stouffer, R.; Meehl, G. An overview of CMIP5 and the experiment design. *Bull. Am. Meteorol. Soc.* **2012**, *93*, 485–498. [CrossRef]
55. Yu, H.; Jiang, S.; Land, K. Multicollinearity in hierarchical linear models. *Soc. Sci. Res.* **2015**, *53*, 118–136. [CrossRef] [PubMed]
56. Liang, J.; Buongiorno, J.; Monserud, R.A.; Kruger, E.L.; Zhou, M. Effects of diversity of tree species and size on forest basal area growth, recruitment, and mortality. *For. Ecol. Manag.* **2007**, *243*, 116–127. [CrossRef]
57. Stage, A.R. An Expression for the Effect of Aspect, Slope, and Habitat Type on Tree Growth. *For. Sci.* **1976**, *22*, 457–460.
58. Tobin, J. Estimation of Relationships for Limited Dependent Variables. *Econometrica* **1958**, *26*, 24–36. [CrossRef]
59. Li, R.; Weiskittel, A.R.; Kershaw, J.A. Modeling annualized occurrence, frequency, and composition of ingrowth using mixed-effects zero-inflated models and permanent plots in the Acadian Forest Region of North America. *Can. J. For. Res.* **2011**, *41*, 2077–2089. [CrossRef]
60. Ai, C.; Norton, E. Interaction terms in logit and probit models. *Econ. Lett.* **2003**, *80*, 123–129. [CrossRef]
61. Moss, R.H.; Edmonds, J.A.; Hibbard, K.A.; Manning, M.R.; Rose, S.K.; Van Vuuren, D.P.; Carter, T.R.; Emori, S.; Kainuma, M.; Kram, T. The next generation of scenarios for climate change research and assessment. *Nature* **2010**, *463*, 747. [CrossRef]
62. Chong-Hai, X.; Ying, X. The projection of temperature and precipitation over China under RCP scenarios using a CMIP5 multi-model ensemble. *Atmos. Ocean. Sci. Lett.* **2012**, *5*, 527–533. [CrossRef]
63. Yang, X.; Du, Z.; Zeng, W.; Meng, J. Development of Mixed-Effects Individual-Tree Diameter Increment Model for *Casuarina equisetifolia* Considering the Effects of Tree-Size Diversity, Tree Density Reduction, and Climate. *J. Sustain. For.* **2023**, *4*, 553–572. [CrossRef]
64. Zhou, X.; Chen, Q.; Ram, P.S.; Wang, Y.; He, P.; Guo, J.; Lei, Y.; Fu, L. A climate sensitive mixed-effects diameter class mortality model for Prince Rupprecht larch (*Larix gmelinii* var. principis-rupprechtii) in northern China. *For. Ecol. Manag.* **2021**, *491*, 119091. [CrossRef]
65. Rozendaal, D.M.A.; Phillips, O.L.; Lewis, S.L.; Affum-Baffoe, K.; Alvarez-Davila, E. Competition influences tree growth, but not mortality, across environmental gradients in Amazonia and tropical Africa. *Ecology* **2020**, *101*, e03052. [CrossRef] [PubMed]
66. Bradford, J.; David, M.B. A window of opportunity for climate-change adaptation: Easing tree mortality by reducing forest basal area. *Front. Ecol. Environ.* **2017**, *15*, 11–17. [CrossRef]
67. Klopcic, M.; Poljanec, A.; Boncina, A. Modelling natural recruitment of European beech (*Fagus sylvatica* L.). *For. Ecol. Manag.* **2012**, *284*, 142–151. [CrossRef]
68. Chen, Q.; Zhang, L.; Fu, S. Effects of tree age and basal area on boreal forest tree mortality in Canada. *Chin. J. Ecol.* **2011**, *22*, 2477–2481.
69. Klos, R.; Wang, G.; Bauerle, W.; Rieck, J. Drought impact on forest growth and mortality in the southeast USA: An analysis using Forest Health and Monitoring data. *Ecol. Appl.* **2009**, *19*, 699–708. [CrossRef] [PubMed]
70. Zeller, L.; Liang, J.; Pretzsch, H. Tree species richness enhances stand productivity while stand structure can have opposite effects, based on forest inventory data from Germany and the United States of America. *For. Ecosyst.* **2018**, *5*, 4. [CrossRef]
71. Chamagne, J.; Tanadini, M.; Frank, D.; Matula, R.; Paine, C.; Philipson, C.; Svátek, M.; Turnbull, L.; Volařík, D.; Hector, A. Forest diversity promotes individual tree growth in central European forest stands. *J. Appl. Ecol.* **2017**, *54*, 71–79. [CrossRef]
72. Aussenac, R.; Bergeron, Y.; Gravel, D.; Drobyshev, I. Interactions among trees: A key element in the stabilising effect of species diversity on forest growth. *Funct. Ecol.* **2019**, *33*, 360–367. [CrossRef]
73. Morin, X.; Fahse, L.; Scherer-Lorenzen, M.; Bugmann, H. Tree species richness promotes productivity in temperate forests through strong complementarity between species. *Ecol. Lett.* **2011**, *14*, 1211–1219. [CrossRef]
74. Hisano, M.; Chen, H.; Searle, E.; Reich, P. Species-rich boreal forests grew more and suffered less mortality than species-poor forests under the environmental change of the past half-century. *Ecol. Lett.* **2019**, *22*, 999–1008. [CrossRef]
75. Fridley, J. Diversity effects on production in different light and fertility environments: An experiment with communities of annual plants. *J. Ecol.* **2003**, *91*, 396–406. [CrossRef]
76. Kahmen, A.; Renker, C.; Unsicker, S.; Buchmann, N. Niche complementarity for nitrogen: An explanation for the biodiversity and ecosystem functioning relationship? *J. Ecol.* **2006**, *87*, 1244–1255. [CrossRef]
77. Searle, E.B.; Chen, H.Y.; Paquette, A. Higher tree diversity is linked to higher tree mortality. *Proc. Natl. Acad. Sci. USA* **2022**, *119*, e2013171119. [CrossRef] [PubMed]
78. Liang, J.; Buongiorno, J.; Monserud, R.A. Growth and yield of all-aged Douglas-fir western hemlock forest stands: A matrix model with stand diversity effects. *Can. J. For. Res.* **2005**, *35*, 2368–2381. [CrossRef]
79. Young, B.; Liang, J.; Stuart Chapin, F. Effects of species and tree size diversity on recruitment in the Alaskan boreal forest: A geospatial approach. *For. Ecol. Manag.* **2011**, *262*, 1608–1617. [CrossRef]
80. Grubb, P.J. The maintenance of species-richness in plant communities: The importance of the regeneration niche. *Biol. Rev.* **1977**, *52*, 107–145. [CrossRef]
81. Denslow, J.S. Tropical rainforest gaps and tree species diversity. *Annu. Rev. Ecol. Syst.* **1987**, *18*, 431–451. [CrossRef]
82. Man, C.D.; Comeau, P.G.; Pitt, D.G. Competitive effects of woody and herbaceous vegetation in a young boreal mixedwood stand. *Can. J. For. Res.* **2008**, *38*, 1817–1828. [CrossRef]
83. Way, D.; Montgomery, R. Photoperiod limits on tree climate responses. *Plant Cell Environ.* **2015**, *38*, 1725–1736. [CrossRef]

84. Saxe, H.; Cannell, M.G.R.; Johnsen, Ø.; Ryan, M.G.; Vourlitis, G. Tree and forest functioning in response to global warming. *New Phytol.* **2001**, *149*, 369–399. [CrossRef] [PubMed]
85. Soja, A.J.; Tchebakova, N.M.; French, N.H.; Flannigan, M.D.; Shugart, H.H.; Stocks, B.J.; Sukhinin, A.I.; Parfenova, E.I.; Chapin, F.S., III; Stackhouse, P.W., Jr. Climate-induced boreal forest change: Predictions versus current observations. *Glob. Planet. Change* **2007**, *56*, 274–296. [CrossRef]
86. Huang, X.; Dai, D.; Xiang, Y.; Yan, Z.; Teng, M.; Wang, P.; Zhou, Z.; Zeng, L.; Xiao, W. Radial growth of *Pinus massoniana* is influenced by temperature, precipitation, and site conditions on the regional scale: A meta-analysis based on tree-ring width index. *Ecol. Indic.* **2021**, *126*, 107659. [CrossRef]
87. Adams, H.; Collins, A.; Briggs, S.; Vennetier, M.; Dickman, L.; Sevanto, S.; Garcia-Forner, N.; Powers, H.; McDowell, N. Experimental drought and heat can delay phenological development and reduce foliar and shoot growth in semiarid trees. *Glob. Change Biol.* **2015**, *21*, 4210–4220. [CrossRef]
88. Restaino, C.; Peterson, D.; Littell, J. Increased water deficit decreases Douglas fir growth throughout western US forests. *Proc. Natl. Acad. Sci. USA* **2016**, *113*, 9557–9562. [CrossRef]
89. Li, J.; Huang, J.; Tardif, J.C.; Liang, H.; Jiang, S.; Zhu, H.; Zhou, P. Spatially heterogeneous responses of tree radial growth to recent El Niño southern-oscillation variability across East Asia subtropical forests. *Agric. For. Meteorol.* **2020**, *287*, 107939. [CrossRef]
90. Berkelhammer, M.; Still, C.J.; Ritter, F.; Winnick, M.; Anderson, L.; Carroll, R.; Carbone, M.; Williams, K.H. Persistence and plasticity in conifer water-use strategies. *J. Geophys. Res. Biogeosci.* **2020**, *125*, e2018JG004845. [CrossRef]
91. Barber, V.A.; Juday, G.P.; Finney, B.P. Reduced growth of Alaskan white spruce in the twentieth century from temperature-induced drought stress. *Nature* **2000**, *405*, 668. [CrossRef]
92. Ols, C.; Hervé, J.; Bontemps, J. Recent growth trends of conifers across Western Europe are controlled by thermal and water constraints and favored by forest heterogeneity. *Sci. Total Environ.* **2020**, *742*, 140453. [CrossRef]
93. McDowell, N.G.; Williams, A.P.; Xu, C.; Pockman, W.T.; Dickman, L.T.; Sevanto, S.; Pangle, R.E.; Limousin, J.M.; Plaut, J.J.; Mackay, D.S.; et al. Multi-scale predictions of massive conifer mortality due to chronic temperature rise. *Nat. Clim. Change* **2016**, *6*, 295–300. [CrossRef]
94. Du, H.; Li, M.-H.; Rixen, C.; Zong, S.; Stambaugh, M.; Huang, L.; He, H.S.; Wu, Z. Sensitivity of recruitment and growth of alpine treeline birch to elevated temperature. *Agric. For. Meteorol.* **2021**, *304–305*, 108403. [CrossRef]
95. Yamamoto, S. Forest gap dynamics and tree regeneration. *J. For. Res.* **2000**, *5*, 223–229. [CrossRef]
96. Swaine, M.; Hall, J. The mosaic theory of forest regeneration and the determination of forest composition in Ghana. *J. Trop. Ecol.* **1988**, *4*, 253–269. [CrossRef]
97. Maleki, K.; Marchand, P.; Charron, D.; Bergeron, Y. Seedling recruitment in response to stand composition, interannual climate variability, and soil disturbance in the boreal mixed woods of Canada. *J. Veg. Sci.* **2024**, *35*, e13237. [CrossRef]
98. Giardina, F.; Konings, A.G.; Kennedy, D.; Alemohammad, S.; Oliveira, S.; Uriarte, M.; Gentine, P. Tall Amazonian forests are less sensitive to precipitation variability. *Nature Geosci.* **2018**, *11*, 405–409. [CrossRef]
99. Schuur, E.A.G. Productivity and global climate revisited: The sensitivity of tropical forest growth to precipitation. *Ecology* **2003**, *84*, 1165–1170. [CrossRef]
100. Tingstad, L.; Olsen, S.L.; Klanderud, K.; Vandvik, V.; Ohlson, M. Temperature, precipitation and biotic interactions as determinants of tree seedling recruitment across the tree line ecotone. *Oecologia* **2015**, *179*, 599–608. [CrossRef]
101. Clifford, M.J.; Royer, P.D.; Cobb, N.S.; Breshears, D.D.; Ford, P.L. Precipitation thresholds and drought-induced tree die-off: Insights from patterns of Pinus edulis mortality along an environmental stress gradient. *New Phytol.* **2013**, *200*, 413–421. [CrossRef]
102. Lienard, J.; Florescu, I.; Strigul, N. An Appraisal of the Classic Forest Succession Paradigm with the Shade Tolerance Index. *PLoS ONE* **2015**, *10*, e0117138. [CrossRef]
103. Chi, X.; Tang, Z.; Xie, Z.; Guo, Q.; Zhang, M.; Ge, J.; Xiong, G.; Fang, J. Effects of size, neighbors, and site condition on tree growth in a subtropical evergreen and deciduous broad-leaved mixed forest, China. *Ecol. Evol.* **2015**, *5*, 5149–5161. [CrossRef]
104. Van der Sande, M.T.; Peña-Claros, M.; Ascarrunz, N.; Arets, E.J.M.M.; Licona, J.C.; Toledo, M.; Poorter, L. Abiotic and biotic drivers of biomass change in a Neotropical forest. *J. Ecol.* **2017**, *105*, 1223–1234. [CrossRef]
105. Kozlowski, T. Growth Characteristics of Forest Trees. *J. For.* **1963**, *61*, 655–662.
106. Swanson, M.E.; Franklin, J.F.; Beschta, R.L.; Crisafulli, C.M.; DellaSala, D.A.; Hutto, R.L.; Lindenmayer, D.B.; Swanson, F.J. The forgotten stage of forest succession: Early-successional ecosystems on forest sites. *Front. Ecol. Environ.* **2011**, *9*, 117–125. [CrossRef]
107. Aldrich, P.R.; Parker, G.R.; Ward, J.S.; Michler, C.H. Spatial dispersion of trees in an old-growth temperate hardwood forest over 60 years of succession. *For. Ecol. Manag.* **2003**, *180*, 475–491. [CrossRef]
108. Dillaway, D.N.; Stringer, J.W.; Rieske, L.K. Light availability influences root carbohydrates, and potentially vigor, in white oak advance regeneration. *For. Ecol. Manag.* **2007**, *250*, 227–233. [CrossRef]
109. D'Amato, A.W.; Catanzaro, P.F.; Fletcher, L.S. Early Regeneration and Structural Responses to Patch Selection and Structural Retention in Second-Growth Northern Hardwoods. *For. Sci.* **2015**, *61*, 183–189. [CrossRef]
110. Harmer, R.; Morgan, G. Development of *Quercus robur* advance regeneration following canopy reduction in an oak woodland. *Int. J. For. Res.* **2007**, *80*, 37–49. [CrossRef]
111. Boggess, W.R.; Bailey, L.W. Brownfield Woods, Illinois: Woody vegetation and changes since 1925. *Am. Midl. Nat.* **1964**, *71*, 392–401. [CrossRef]

112. Shotola, S.J.; Weaver, G.T.; Robertson, P.A.; Ashby, W.C. Sugar maple invasion of an old-growth oak-hickory forest in southwestern Illinois. *Am. Midl. Nat.* **1992**, *127*, 125–138. [CrossRef]
113. Dey, D.; Jacobs, D.; McNabb, K.; Miller, G.; Baldwin, V.; Foster, G. Artificial Regeneration of Major Oak (Quercus) Species in the Eastern United States—A Review of the Literature. *For. Sci.* **2008**, *54*, 77–106. [CrossRef]
114. Paquette, A.; Bouchard, A.; Cogliastro, A. Morphological plasticity in seedlings of three deciduous species under shelterwood under-planting management does not correspond to shade tolerance ranks. *For. Ecol. Manag.* **2007**, *214*, 278–287. [CrossRef]
115. Smith, R.L.; McDaniel, P.W.; Fell, D. Opportunities for the utilization of alternative species in secondary wood manufacturing. *For. Prod. J.* **2005**, *55*, 71.
116. Dumitriu, G.-D.; Cotea, V.V.; López, D.L.; Peinado, R.A.; Teodosiu, C.; Niculaua, M. Characterization of 'Fetească neagră' red wines aged with oak staves: Major volatile compounds, aromatic series and sensory analysis. *Mitt. Klosterneubg.* **2018**, *68*, 97–106.
117. Soriano, A.; Alañón, M.; Alarcón, M.; García-Ruíz, A.; Díaz-Maroto, M.; Pérez-Coello, M. Oak wood extracts as natural antioxidants to increase shelf life of raw pork patties in modified atmosphere packaging. *Food Res. Int.* **2018**, *111*, 524–533. [CrossRef]
118. Lindenmayer, D.B.; Laurance, W.F.; Franklin, J.F. Global decline in large old trees. *Science* **2012**, *338*, 1305–1306. [CrossRef]
119. Dakhil, M.A.; Xiong, Q.; Farahat, E.A.; Zhang, L.; Pan, K.; Pandey, B.; Huang, D. Past and future climatic indicators for distribution patterns and conservation planning of temperate coniferous forests in southwestern China. *Ecol. Indic.* **2019**, *107*, 105559. [CrossRef]
120. Dakhil, M.A.; Halmy, M.W.A.; Liao, Z.; Pandey, B.; Zhang, L.; Pan, K.; El-Barougy, R.F. Potential risks to endemic conifer montane forests under climate change: Integrative approach for conservation prioritization in southwestern China. *Landsc. Ecol.* **2021**, *36*, 3137–3151. [CrossRef]
121. Babst, F.; Poulter, B.; Trouet, V.; Tan, K.; Neuwirth, B.; Wilson, R.; Carrer, M.; Grabner, M.; Tegel, W.; Levanic, T.; et al. Site-and species-specific responses of forest growth to climate across the European continent. *Glob. Ecol. Biogeogr.* **2013**, *22*, 706–717. [CrossRef]
122. Zhao, M.; Running, S.W. Drought-induced reduction in global terrestrial net primary production from 2000 through 2009. *Science* **2010**, *329*, 940–943. [CrossRef]
123. Eyre, F.H. *Forest Cover Types of the United States and Canada*; SAF: Washington, DC, USA, 1980; ISBN 978-0686306979.
124. Rossi, S. Local adaptations and climate change: Converging sensitivity of bud break in black spruce provenances. *Int. J. Biometeorol.* **2015**, *59*, 827–835. [CrossRef]
125. Pretzsch, H.; Biber, P.; Schütze, G.; Uhl, E.; Rötzer, T. Forest stand growth dynamics in Central Europe have accelerated since 1870. *Nat. Commun.* **2014**, *5*, 496. [CrossRef] [PubMed]
126. Feng, H.; Lang, P.; Deng, Z.; Zhao, W.; Hua, M.; Zhu, X.; Wang, Z. The impact of climate change and human activities to vegetation carbon sequestration variation in Sichuan and Chongqing. *Environ. Res.* **2023**, *238*, 117138. [CrossRef] [PubMed]
127. Eggers, J.; Lindner, M.; Zudin, S.; Zaehle, S.; Liski, J. Impact of changing wood demand, climate and land use on European forest resources and carbon stocks during the 21st century. *Glob. Change Biol.* **2008**, *14*, 2288–2303. [CrossRef]

forests

MDPI

Article

An Algorithm for Determining Pith Position Based on Crown Width Size

Jianfeng Yao [1,2,3], Xiaowei Shang [1], Xuefan Hu [4], Yingshan Jin [4,*], Liming Cai [2,3,5], Zhuofan Li [2,3,6], Fang Li [4] and Fang Liang [4]

1 College of Computer and Information Technology, Xinyang Normal University, Xinyang 464000, China; yaojf@xynu.edu.cn (J.Y.); 2024331109@xynu.edu.cn (X.S.)
2 Henan Dabieshan National Field Observation and Research Station of Forest Ecosystem, Zhengzhou 450046, China; limingcai@xynu.edu.cn (L.C.); lizhuofan@mails.ccnu.edu.cn (Z.L.)
3 Xinyang Academy of Ecological Research, Xinyang 464000, China
4 Beijing Key Laboratory of Greening Plants Breeding, Beijing Academy of Forestry and Landscape Architecture, Beijing 100102, China; hufanzi@163.com (X.H.); dlfl2004@163.com (Fang Li)
5 College of Mathematics and Statistics, Xinyang Normal University, Xinyang 464000, China
6 College of Tourism, Xinyang Normal University, Xinyang 464000, China
* Correspondence: j_ysh@163.com

Abstract: To accurately estimate the pith position, a method was proposed for estimating the pith position by the crown width. The crown widths of 120 trees and radiuses of each disc extracted at the height of 1.3 m from these trees were measured in four directions. The crown and radius ratios of the length of each direction to the total length in that direction and the opposite direction were calculated. Using the crown ratio as an independent variable, as well as the radius ratio as a dependent variable, the linear, logarithmic, exponential, and polynomial models were built. The model with the highest R^2 was selected as the radius ratio model. The geometric center method and the crown width method were applied to estimate the pith position, and the estimation errors were calculated, respectively. The R^2 of the linear, logarithmic, exponential, and polynomial models were 0.405, 0.379, 0.403, 0.404, respectively, and the linear model was chosen as the radius ratio model. The prediction error based on the crown width was 7.6%, and that of the geometric center method was 10.1%. The findings indicate that the crown width method can improve the accuracy of estimating the pith position.

Keywords: pith position; crown width; radius ratio; linear model

1. Introduction

According to the growth law of a tree trunk, the growth of all the annual rings is centered on the pith [1]. Consequently, to accurately measure tree rings, the measurement path must intersect the pith. For instance, when using an increment corer to measure the annual rings of standing trees, the initial step is to identify the pith's location [2–4]. However, estimating the pith position in standing trees is challenging. Typically, the geometric center method is employed to approximate the pith's location, which takes the geometric center of the tree trunk as the location of the pith. However, this method is inaccurate because the radial growth rate of trees varies in different directions, causing the pith to deviate from the geometric center of the trunk [5]. To solve this problem, several scholars have proposed alternative methods for estimating the pith position. Zhang Qiaosong developed a method based on the shapes of tree trunks, concluding that when the trunk's cross-section is nearly circular, the pith is located at the geometric center, whereas, for an elliptical cross-section, the pith is near the shortest axis of the ellipse [6]. Although this method does not require specialized instruments, it still results in some deviation between the estimated and actual pith positions. Other researchers, such as Bhandarkar, Wu and Liew, Andreu and Rinnhofer, and Longuetau, have utilized CT imaging to estimate the pith position [7–10]. Bhandarkar et al. used an edge detection Sobel operator to process a trunk CT image to change the Hough accumulation array, and its maximum value was the location of the pith. Longuetaud et al. applied medical CT technology, performed interpolation enhancement based on Hough transform, and completed the automatic positioning of the pith. Marichal et al. proposed a fully automated technique for wood pith detection (APD) based on the concentric shape of the structure of wood ring slices. The method estimated the ring's local orientations using the 2D structure tensor and found the pith position [11]. Perlin et al. introduced a method for localizing the tree pith based on the principle that the radial propagation speed of ultrasonic pulses is higher than the tangential speed [12]. A total of 20 fixed ultrasonic pulse receivers were placed at equal intervals outside the tree disc, and a grid was established on the cross-section. The sum of all ultrasonic pulse velocities related to each grid point was calculated, and the assumption that the pith was located at the convergence point of the maximum sum was proposed. The final result was only 0.5 cm different from the actual pith distance. Additionally, Wang Danyu employed two-dimensional LiDAR point cloud data to determine the outer contour of the tree trunk and subsequently estimate the pith position [13]. The results show that a straight line through the pith can be found within a 1.5% error range. Although the CT, ultrasound, and LiDAR methods offer high accuracy, they require expensive equipment and involve complex procedures. Therefore, these methods have not been widely used in the field.

The pith of a tree trunk is composed of the xylem that develops during the first year of growth, and its position is influenced by the varying growth rates of the trunk in different directions [14–17]. Research indicates that factors affecting the radial growth rate of tree trunks include the root system, crown width, and so on. Tree roots play a crucial role not only in absorbing and transporting water and nutrients but also in synthesizing and storing nutrients essential for growth [18,19]. The distribution of tree roots varies according to soil fertility and moisture content, which differ in various directions [20–23]. Consequently, the direction with more favorable soil conditions tends to support more vigorous root systems and exhibit a faster radial growth rate, causing the pith to deviate from that direction [24]. Crown width directly affects its efficiency in obtaining resources such as light, water, and nutrients [25,26]. Variations in the intensity and duration of sunlight in different directions can lead to differences in crown size. A larger crown width in a particular direction correlates with a faster radial growth rate in that direction, resulting in the pith deviating from it. Compared to the root system, crown width is more straightforward to measure. Therefore, this study attempts to estimate the position of the tree pith using crown width.

2. Materials and Methods

2.1. Overview of the Study Area

The research area is located in Shuizhang Village, Mujiayu Town, Miyun District, Beijing, within the Miyun Reservoir Water Conservation Forest Demonstration Zone. Geographically, this area is situated at 116°53′ E and 40°25′ N. The region has a sub-humid, warm temperate, continental monsoon climate, characterized by hot, rainy summers and cold, dry winters. The average annual precipitation is 600 mm, and the average annual temperature is 11.8 °C. There are low mountains at an altitude of approximately 120 m a.s.l., with slopes ranging from 10° to 35°. The south-facing slopes are characterized by thin soil and exposed rocks, whereas the north-facing slopes feature shaded, deep, and loose soil. In 1980, a *Pinus tabuliformis* plantation was established on a north slope. The age of *Pinus tabuliformis* was 46 years. This plantation exhibits an average tree height of 7.56 m and an average trunk diameter of 14.8 cm, with a stand density of approximately 915 trees per hectare and a canopy density of about 0.8. Conversely, a *Platycladus orientalis* plantation was established on a south slope in 1990. The age of *Platycladus orientalis* was 37 years. The *Platycladus orientalis* trees have an average height of 6.87 m and an average trunk diameter of 9.1 cm. This plantation has a stand density of about 2550 trees per hectare and a canopy density of approximately 0.95.

2.2. Data Collection

In July 2024, a total of 120 trees were selected as analytical wood materials at the Shuizhang plot site in Miyun District, Beijing. This selection comprised 60 *Pinus tabulaeformis* and 60 *Platycladus orientalis*. Initially, the diameter at breast height (DBH) of all the trees within each plot was measured and ranked from largest to smallest. The DBH was divided into three diameter grades according to the arrangement order, namely dominant wood, medium wood, and inferior wood. From each diameter class, 20 standard trees were selected, resulting in a total of 120 standard trees designated as analytical tree samples. The crown widths in the four directions of east, west, south, and north were measured with a handheld laser rangefinder (UT398A). Subsequently, the location on the trunk at the height of 1.3 m, as well as the north–south orientation, was marked on each tree. A disc approximately 5 cm thick was cut from the marked position on the trunk, and the relevant wood information was recorded. The discs were polished until the annual ring lines were clear. The width of the annual rings in four directions were measured with the annual ring analysis instrument, Lintab 6.0. Detailed statistical data regarding the analytical wood and crown width are presented in Table 1.

Table 1. Statistical data of single tree experiment.

Tree Type	Variable		Mean	Variance	Minimum	Maximum
Pinus tabulaeformis	DBH/cm		13.68	4.99	19.2	9.6
	CW/m	E	1.95	0.51	3.67	0.46
		S	1.99	0.52	3.62	0.74
		W	1.90	0.42	4.12	0.8
		N	1.98	0.75	3.92	0.32
Platycladus orientalis	DBH/cm		9.89	4.97	16.6	6
	CW/m	E	1.31	0.55	3.6	0.11
		S	1.37	0.35	3.13	0.31
		W	1.16	0.38	2.62	0.23
		N	1.17	0.31	2.62	0.25

In the table, DBH stands for diameter at breast height. CW stands for crown width. E, S, W, and N, respectively, represent east, south, west, and north.

2.3. Data Processing

Firstly, based on the width of the annual rings in each direction of the disc, the radiuses of each direction were calculated. Then, the total crown width and diameter of each tree were calculated in both the east–west direction and north–south direction, respectively. Lastly, the crown ratio NCR (the northern crown width to the north–south crown width), SCR (the southern crown width to the north–south crown width), ECR (the eastern crown width to the east–west crown width), and WCR (the western crown width to the east–west crown width), as well as the radius ratio NRR (the northern radius to the north–south diameter), SRR (the southern radius to the north–south diameter), ERR (the eastern radius to the east–west diameter), and WRR (the western radius to the east–west diameter) were calculated. The calculation formulas for the crown width and radius ratios are shown in Equations (1)–(8). The diagrammatic sketch for calculating the crown ratio NCR and radius ratio NRR are shown in Figure 1.

$$NCR = \frac{C_n}{C_{ns}} \tag{1}$$

$$SCR = 1 - NCR \tag{2}$$

$$ECR = \frac{C_e}{C_{ew}} \tag{3}$$

$$WCR = 1 - ECR \tag{4}$$

$$NRR = \frac{R_n}{D_{ns}} \tag{5}$$

$$SRR = 1 - NRR \tag{6}$$

$$ERR = \frac{R_e}{D_{ew}} \tag{7}$$

$$WRR = 1 - ERR \tag{8}$$

where C_n is the crown width length in the north direction; C_{ns} is the total length of the north–south crown width; C_e is the crown width length in the east direction; C_{ew} is the total length of the east–west crown width; R_n is the radius length in the north direction; D_{ns} is the length of the north–south diameter; R_e is the radius length in the east direction; and D_{ew} is the length of the east–west diameter.

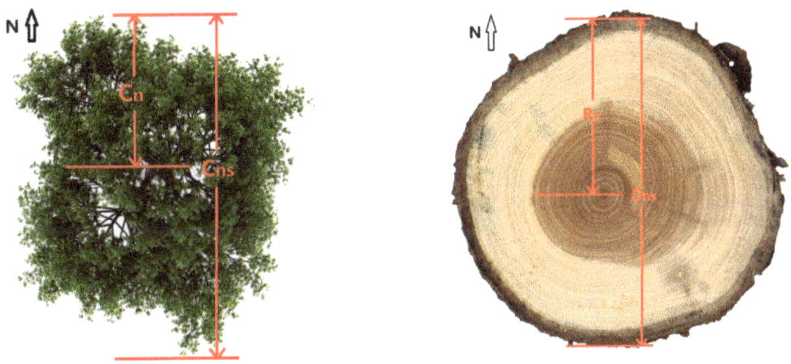

(**a**) (**b**)

Figure 1. Calculation method of the crown ratio *NCR* (**a**) and radius ratio *NRR* (**b**).

2.4. Experimental Methods

2.4.1. Model Construction and Selection

For each tree species, 2/3 of the data were randomly selected as the modeling dataset, while the remaining 1/3 served as the testing dataset. Using the crown ratio as an independent variable, and the radius ratio as a dependent variable, the linear, logarithmic, exponential, and polynomial models were built. These four candidate models are shown in Table 2.

Table 2. Candidate model for pith localization.

Number	Model Type	Expression Formula
M1	Linear model	$y = a \times x + b$
M2	Logarithmic model	$y = a \times \log(x)$
M3	index model	$y = a \times e^{b \times x}$
M4	Polynomial	$y = a \times x^2 + b \times x + c$

The model fitting results were evaluated and analyzed using the adjusted coefficient of determination (R^2), root mean square error (*RMSE*), mean absolute error (*MAE*), and Akaike information criterion (*AIC*). The evaluation indicators are as follows:

$$R^2 = 1 - \frac{\sum_{i=1}^{n}(y_i - \hat{y}_i)^2}{\sum_{i=1}^{n}(y_i - \overline{y})^2}\left(\frac{n-1}{n-p-1}\right) \tag{9}$$

$$RMSE = \sqrt{\frac{\sum_{i=1}^{n}(y_i - \hat{y}_i)^2}{n}} \tag{10}$$

$$MAE = \frac{1}{n}\sum_{i=1}^{n}|y_i - \hat{y}_i| \tag{11}$$

$$AIC = 2p - 2ln(l) \tag{12}$$

where y_i is the true value of the east radius ratio or the true value of the north radius ratio of the *i*th tree; $\hat{y}_i$ is the predicted value of the east radius ratio or the predicted value of the north radius ratio of the *i*th tree; *n* is the number of samples; *p* is the number of model estimated parameters; $\overline{y}$ is the average value of the predicted values of the east radius ratio or the average value of the predicted values of the north radius ratio; *ln(l)* is the log-likelihood of the model.

Based on these four evaluation indicators, the model with higher R^2, lower *RMSE, MAE*, and *AIC* values was selected as the model for estimating radius ratio from the four candidate models.

2.4.2. Method for Determining the Pith Position and Error Analysis

(1) Prediction of Pith Position with Crown Width Method

Firstly, the crown width ratios *NCR* and *ECR* of each test tree were calculated. Then, the radius ratios *NRR* and *ERR* were estimated with the linear model. Finally, the pith position of the test tree was estimated based on *NRR* and *ERR*. The specific process of determining the pith position based on *NRR* and *ERR* was as follows:

(1) The chord was drawn on the disc in the north–south direction;
(2) Point G1 was drawn on the north–south chord, such that the ratio between the distance from the G1 point to the north point of the outer xylem and the length of the north–south chord was *NRR*;
(3) The east–west chord passing through the point G1 was drawn on the disc;
(4) Point P1 was drawn on the east–west chord, such that the ratio between the distance from the P1 point to the east point of the outer xylem and the length of the east–west chord was *ERR*;
(5) Point P1 was the estimated pith position, and the distance between P1 and the real pith was measured using a ruler.

The prediction of the pith position with the crown width method is depicted in Figure 2.

(**a**) North–south direction (**b**) East–west direction

Figure 2. Prediction of pith position with crown width method. L1 in (**a**) is a straight line in the north–south direction, while G1 is the dividing point of the north–south radius obtained according to the crown width method; L2 in (**b**) is the dividing line of the north–south radius passing through point G1, and P1 is the pith position predicted by the crown width method.

(2) Prediction of Pith Position with Geometric Center Method

In the geometric center method, the geometric center of the cross-section of the trunk is taken as the pith position. Therefore, both *NCR* and *ECR* were 0.5. The specific process of determining the pith position with the geometric center method was as follows:

(1) The chord was drawn on the disc in the north–south direction;
(2) The midpoint of the north–south chord was taken as point G2;
(3) The east–west chord passing through point G2 was drawn on the disc;
(4) The midpoint of the east–west chord was taken as point P2;
(5) Point P2 was the estimated pith position, and the distance between P2 and the real pith was measured with a ruler.

The prediction of the pith position with the geometric center method is depicted in Figure 3.

(**a**) North–south direction (**b**) East–west direction

Figure 3. Prediction of pith position with geometric center method. L1 in (**a**) is a straight line in the north–south direction, and G2 is the dividing point of the north–south radius obtained based on the geometric center method; L2 in (**b**) is the dividing line of the north–south radius passing through point G2, and P2 is the predicted pith position of the geometric center method.

(3) Method of Error Analysis

In the experiment, the accuracy of pith prediction was assessed by measuring the distance between the predicted and actual positions of the pith using two different methods. This distance is an absolute value, and it cannot reflect the relative error of the predicted pith position to the real pith positions in the pith rays. Therefore, an error in the predicted pith position was calculated as the ratio of the distance between the predicted pith position and actual pith position to the disc radius. The error calculation method is as follows:

$$\theta = \frac{1}{n}\sum_{i=1}^{n} \frac{2D_i}{DBH} \tag{13}$$

where θ is the average error; n is the number of samples; D_i is the distance between the predicted pith position and the true pith position of the *i*th tree in cm; and DBH is the chest meridian of the *i*th tree.

3. Results

3.1. Modeling Results

Using the crown ratio as the independent variable, and the radius ratio as the dependent variable, the linear, logarithmic, exponential, and polynomial models were built. The fitting curves of these four models are shown in Figure 4, and the model equations and determination coefficients are shown in Table 3.

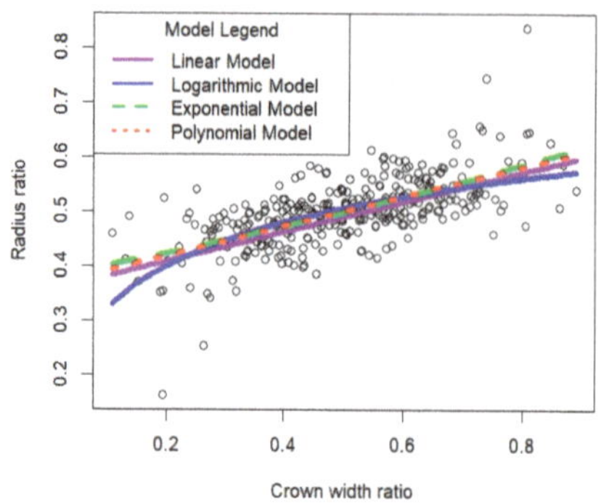

Figure 4. Fitting curves.

Table 3. Model equations and determination coefficients.

Number	Model Type	Equation	R^2
M1	Linear model	$y = 0.364 + 0.272 \times x$	0.405
M2	Logarithmic model	$y = 0.586 + 0.157 \times \log(x)$	0.379
M3	index model	$y = 0.38 \times e^{0.541 \times x}$	0.403
M4	Polynomial	$y = 0.5 + 0.735 \times x + 6.939 \times 10^{-18} \times x^2$	0.404

Using the test dataset, the *RMSE*, *MAE*, and *AIC* of each model were calculated. The results of these statistical indicators are shown in Table 4.

Table 4. Test results of each model.

Model	RMSE	MAE	AIC
M1	0.053	0.035	−406.30
M2	0.054	0.036	−401.60
M3	0.053	0.035	−406.07
M4	0.053	0.035	−404.30

From Table 3, it can be seen that the R^2 of the linear model is the highest. From Table 4, however, the *RMSE*, *MAE*, and *AIC* of the linear model are the lowest. Therefore, the linear model was used to estimate the radius ratio of the test trees.

3.2. Results of the Estimated Pith Position

We measured the distance between the estimated pith position using the crown width method and the geometric center method and calculated the errors with Formula (13). The statistical results are shown in Table 5. From Table 5, the prediction error of the crown width method is 7.6%, while that of the geometric center method is 10.1%. A *t*-test was conducted on the estimated pith position error using the crown width method and geometric center method. The p-value of the *t*-test was 0.065. Therefore, there was a significant difference in the estimated results of these two methods at a significance level of 0.1.

Table 5. Error statistics of two positioning methods.

Method	Variable	Mean	Minimum	Maximum	SD
crown width method	Distance/cm	0.556	0.05	1.80	0.407
	Error/%	7.6	1.3	29.2	6.3
Geometric center method	Distance/cm	0.617	0.10	1.85	0.419
	Error/%	10.1	1.8	24.6	5.8

Kernel density estimation is a nonparametric statistical method for estimating an unknown probability density function [27]. Its basic idea is to calculate the average probability density within a certain range around each observation value. Its weighted function is called a kernel function, which is usually a symmetric function centered on a fixed point, such as a Gaussian function, a trigonometric function, or a rectangular function [27]. This study chose a Gaussian kernel as the kernel function. The formula for a Gaussian kernel function is as follows:

$$K(u) = \frac{1}{\sqrt{2\pi}} e^{-\frac{1}{2}x^2} \quad (14)$$

In Formula (14), x is the independent variable and $K(u)$ is the value of the Gaussian kernel function. The Gaussian kernel formula for kernel density estimation can be expressed as follows:

$$y = \frac{1}{nh} \sum_{i=1}^{n} K\left(\frac{x - x_i}{h}\right) \quad (15)$$

In Formula (15), y is the probability density function estimated based on the sample variable x, n is the number of samples, x_i is the value of variable x at position i, K is the kernel function of the kernel density estimate, and h is the bandwidth, which is a smoothing parameter of the kernel density estimate. The choice of bandwidth affects the final smoothness of the curve. This uses the Scott method to define the bandwidth h, as follows:

$$h = 3.5 \frac{\sigma}{n^{\frac{1}{3}}} \quad (16)$$

In Formula (16), n is the sample size and σ is the standard deviation of the sample data.

The density distribution of the difference and error in the predicted distance of the two methods is shown in Figure 5. In Figure 5a, the horizontal axis represents the distance between the predicted pith and the true pith; the vertical axis represents the density distribution, that is, the frequency of data occurrence in a unit interval. Specifically, the value on the Y axis represents the data distribution density near the horizontal axis. In Figure 5b, the horizontal axis represents the error between the predicted pith and the true pith; the vertical axis represents the density of the corresponding error. The density curve of distance and error obtained by the crown width method reaches its peak earlier. This comparison indicates that the crown width method has a higher accuracy in predicting the true pith position.

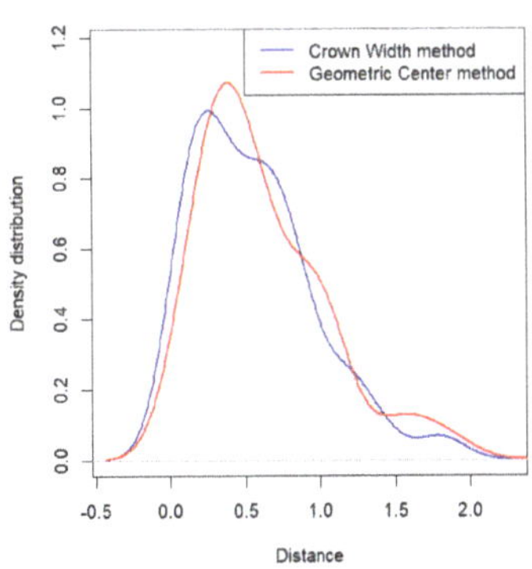

(a)

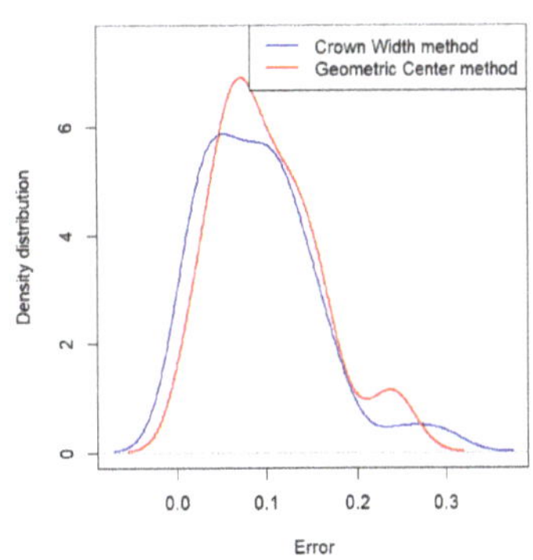

(b)

Figure 5. Curves of density distribution for distance difference and error. (**a**) Distance density curves. (**b**) Error density curve.

4. Discussion

Accurate pith position is a prerequisite for measuring the annual rings and ages of living trees. At present, the geometric center method is commonly used to estimate the position of the pith. However, this method is inaccurate. To solve this problem, several scholars have proposed the CT method, ultrasound method, LiDAR method, and so on [28–31]. The error of the CT scanning method is within 7.5 pixels. The absolute accuracy of the estimated pith position is related to the resolution of the CT scanner. The relative error of the ultrasound method is only 4%. The relative error of LiDAR method is as low as 1.5%. However, the LiDAR method requires the trunk to be a regular circle or ellipse. Although these methods offer high accuracy to the geometric center method, they require expensive equipment and involve complex procedures. Therefore, these methods have not been widely used in the field. So, further research is necessary to determine how to accurately, simply, and quickly estimate the pith position of standing trees.

The main reason for the deviation of the pith from the geometric center of the trunk is that the growth rates of various directions are different [32,33]. There are many reasons that affect the growth rate of tree trunks, such as root system, crown width, lighting time, soil fertility, and soil moisture content [34–37]. For example, Harris, J.E. & Barton, L. found that the growth direction of trees is closely related to light conditions [38,39]. Especially in areas affected by strong light, the growth direction of trees is obviously biased towards the light source, causing the pith to slightly deviate from the center. In the northern hemisphere, trees may lean towards the south because the light intensity in the south is higher. The roots of trees usually grow towards places with abundant water. In an environment with uneven water distribution, trees may grow towards the water source. This bias may affect the position of the pith. Because crown width is a comprehensive result of factors such as lighting time, root system, soil fertility, and soil moisture content, and it is easy to measure, this paper used crown width to estimate the pith position. The research results indicate that the relative error of the pith position estimated using the crown width was 7.6%. Although the error of crown width method was lower than that of the geometric center method, it was still higher than the requirement for the accurate measurement of tree rings. The reasons for the large error are as follows:

(1) The direction for measuring the crown width was inaccurate. When measuring crown width, the measurement direction is generally subjectively judged by the measuring personnel, so the direction to measure crown width may not be accurate.
(2) At different stages of tree growth, the shape of the crown may change. At the juvenile stage, especially in artificial forests, there is not much difference in the crown width of trees in each direction [40], and the pith is located at the center of the trunk. As the tree grows, the difference in crown width in different directions gradually increases, and the pith gradually deviates from the center of the trunk. Therefore, there is a certain degree of error in estimating the pith position solely based on the current crown width.
(3) The nutrient allocation mechanism in various directions of the tree trunk is not yet clear. Due to the unclear nutrient allocation mechanism in various directions of the tree trunk, this study is based on the radial growth rate in which direction is only affected by the crown width in that direction, without considering the influence of crown width in other directions. Based on experience, even if the crown width in a certain direction is 0, the radial growth rate in that direction is not 0, indicating that the trunk in that direction has obtained some nutrients from other directions. Therefore, the influence of crown width in other directions should also be considered.

In future research, the following methods can be adopted to further improve the estimation accuracy of the pith position:

(1) It is necessary to accurately locate the direction of the crown.
(2) If historical crown width data were measured, historical crown width data can be added to the model for estimating the radius ratio.

(3) The nutrient allocation mechanism of the trunk can be studied, and the impact of crown width in other directions on the radial growth rate of trees in this direction should be clarified. If the crown width in other directions has a significant impact on the radial growth rate of trees in this direction, it is necessary to add the crown width in other directions when establishing a crown width model.
(4) In order to eliminate the differences caused by tree species, experiments should be conducted on more species to increase the generality of this method.

5. Conclusions

The findings have indicated that the radial growth rate of trees in each direction is influenced by the relative size of the crown width in that direction. The larger the crown width in each direction, the faster the radial growth rate, and the longer the radius in that direction. Therefore, the pith position can be estimated by the crown width of the tree in four directions. This method has the advantages of simple measurement, low computational complexity, and high estimation accuracy, and provides a feasible approach to estimating the pith position of living standing trees in the field.

Author Contributions: Conceptualization, J.Y. and X.S.; methodology, X.H.; software, X.S.; validation, L.C. and Z.L.; formal analysis, Y.J.; investigation, F.L. (Fang Li); resources, F.L. (Fang Liang); data curation, X.S.; writing—original draft preparation, X.S.; writing—review and editing, J.Y.; visualization, Z.L.; supervision, X.H.; project administration, J.Y.; funding acquisition, X.H. All authors have read and agreed to the published version of the manuscript.

Funding: This research was funded by Standardization Demonstration of Multifunctional Management Technology for Miyun Reservoir Water Conservation Forest in Beijing (JING [2023]TG 06); the Natural Science Foundation of Henan Province (232300421167); Xinyang Academy of Ecological Research Open Foundation (2023XYQN04); Xinyang Academy of Ecological Research Open Foundation (2023XYZD02); Research on the multifunctionality and driving factors of oak forest ecosystems in Beijing (YZQN202405); Research on Key Techniques for Quercus Planting (YZZD202407); and Postgraduate Education Reform and Quality Improvement Project of Henan Province (YJS2023SZ23).

Data Availability Statement: The raw data supporting the conclusions of this article will be made available by the authors on request.

Conflicts of Interest: The authors declare that they have no conflicts of interest.

References

1. Xue, Y.; He, M.; Xu, C.; Li, R. Tree ring pith location based on gradient Hough transform and K-medoids clustering. *Electron. Technol. Softw. Eng.* **2019**, 70–72.
2. Xia, Q. Study on the Response of Non-Destructive Determination of Tree Ring Characteristics and Meteorological Changes. Ph.D. Thesis, Nanjing Forestry University, Nanjing, China, 2023; p. 90.
3. Tannert, T.; Anthony, R.W.; Kasal, B.; Kloiber, M.; Piazza, M.; Riggio, M.; Rinn, F.; Widmann, B.; Yamaguchi, N. In situ assessment of structural timber using semi-destructive techniques. *Mater. Struct.* **2014**, *47*, 767–785. [CrossRef]
4. Nowak, T.P.; Jasienko, J.; Hamrol-Bielecka, K. In situ assessment of structural timber using the resistance drilling method—Evaluation of usefulness. *Constr. Build. Mater.* **2016**, *102*, 403–415. [CrossRef]
5. González-Cásares, M.; Pompa-García, M.; Padilla-Martínez, J. Pith Eccentricity, Basal Area Increments and Disturbances Inferred from Tree-Ring Growth. *Tree-Ring Res.* **2022**, *78*, 25–35. [CrossRef]
6. Zhang, Q.; Yang, W.; Wu, H.; Wu, C. Preliminary study on the age identification of ancient trees in Guangzhou. *Chin. Gard.* **1985**, 43–46.
7. Bhandarkar, S.M.; Faust, T.D.; Tang, M. CATALOG: A system for detection and rendering of internal log defects using computer tomography. *Mach. Vis. Appl.* **1999**, *11*, 171–190. [CrossRef]
8. Longuetaud, F.; Leban, J.M.; Frédéric, M.; Kerrien, E.; Berger, M.O. Automatic detection of pith on CT images of spruce logs. *Comput. Electron. Agric.* **2004**, *44*, 107–119. [CrossRef]
9. Boukadida, H.; Longuetaud, F.; Colin, F.; Freyburger, C.; Constant, T.; Leban, J.M.; Mothe, F. PithExtract: A robust algorithm for pith detection in computer tomography images of wood—Application to 125 logs from 17 tree species. *Comput. Electron. Agric.* **2012**, *85*, 90–98. [CrossRef]
10. Entacher, K.; Hegenbart, S.; Kerschbaumer, J.; Lenz, C.; Weiglmaier, R. Pith detection on CT-Cross-Section images of logs: An experimental comparison. In Proceedings of the 2008 3rd International Symposium on Communications, Control and Signal Processing, Saint Julian's, Malta, 12–14 March 2008; pp. 478–483.

11. Marichal, H.; Passarella, D.; Randall, G. Automatic Wood Pith Detector: Local Orientation Estimation and Robust Accumulation. *arXiv* **2024**, arXiv:2404.01952.
12. Perlin, P.L.; Valle, D.Â. Pinto A D C R. New method to locate the pith position in a wood cross-section based on ultrasonic measurements. *Constr. Build. Mater.* **2018**, *169*, 733–739. [CrossRef]
13. Wang, D. Radial Positioning of Tree Pith Based on Laser Radar. Ph.D. Thesis, Beijing Forestry University, Beijing, China, 2020; p. 75.
14. Neto, W.; Paes, J.; Vidaurre, G.; Arantes, M.; Leite, F. Dendrometric parameters and eccentricity of the pith in leaning eucalypt trees. *Sci. For.* **2015**, *43*, 781–789.
15. Feng, Z.; Wang, J.; Liu, L.; Roessler, R. A Novel Coniferous Tree Trunk with Septate Pith from the Guadalupian (Permian) of China: Ecological and Evolutionary Significance. *Int. J. Plant Sci.* **2012**, *173*, 835–848. [CrossRef]
16. Falcon-Lang, H.; Kurzawe, F.; Lucas, S. Coniferopsid tree trunks preserved in sabkha facies in the Permian (Sakmarian) Community Pit Formation in south-central New Mexico, U.S.A.: Systematics and palaeoecology. *Rev. Palaeobot. Palynol.* **2014**, *200*, 138–160. [CrossRef]
17. Shunn, K.; Gee, C. Cross-sectioning to the core of conifers: Pith anatomy of living Araucariaceae and Podocarpaceae, with comparisons to fossil pith. *IAWA J.* **2023**, *45*, 1–26. [CrossRef]
18. Li, Y. Analysis of tree root growth and development. *Sci. Technol. Inf.* **2006**, 200.
19. Zhao, T. Effects of Ziziphus Spina-Christi Jujube Scion on Root Growth and Physiological Characteristics of Sour Jujube Rootstock. Ph.D. Thesis, Central South University of Forestry and Technology, Changsha, China, 2022; p. 83.
20. Laclau, J.; Arnaud, M.; Bouillet, J.; Ranger, J. Spatial distribution of Eucalyptus roots in a deep sandy soil in the Congo: Relationships with the ability of the stand to take up water and nutrients. *Tree Physiol.* **2001**, *21*, 129–136. [CrossRef] [PubMed]
21. Ruiz-Sánchez, M.; Plana, V.; Ortuño, M.; Tapia, L.; Abrisqueta, J. Spatial root distribution of apricot trees in different soil tillage practices. *Plant Soil* **2005**, *272*, 211–221. [CrossRef]
22. Schwinning, S.; Litvak, M.; Pockman, W.; Pangle, R.; Fox, A.; Huang, C.; McIntire, C. A 3-dimensional model of Pinus edulis and Juniperus monosperma root distributions in New Mexico: Implications for soil water dynamics. *Plant Soil* **2020**, *450*, 337–355. [CrossRef]
23. Burgess, S.; Adams, M.; Turner, N.; Ong, C. The redistribution of soil water by tree root systems. *Oecologia* **1998**, *115*, 306–311. [CrossRef]
24. Kašpar, J.; Šamonil, P.; Krůček, M.; Daněk, P. Changes in the radial growth of trees in relation to biogeomorphic processes in an old-growth forest on flysch, Czechia. *Earth Surf. Process. Landf.* **2020**, *45*, 2761–2772. [CrossRef]
25. Yang, H. Single tree crown width model of natural oak forest in Hunan based on competition effect. *Cent. South Univ. For. Technol.* **2024**, *44*, 92–101.
26. Liu, Q. Research on the Correlation Between Crown Structure and Tree Radial Growth of Korean Pine Plantation. Ph.D. Thesis, Shenyang Agricultural University, Shenyang, China, 2023; p. 47.
27. Zhang, J.; Yang, Y.; Zhang, Z.; Peng, L.; Lin, W.; Zhang, Y.; Xu, Z. Load spectrum extrapolation based on kernel density estimation based on KANN-DBSCAN bandwidth optimization. *Automot. Eng.* **2024**, *46*, 2100–2109.
28. Gazo, R.; Vanek, J.; Abdul-Massih, M.; Benes, B. A fast pith detection for computed tomography scanned hardwood logs. *Comput. Electron. Agric.* **2020**, *170*, 105107. [CrossRef]
29. Wei, Q.; Leblon, B.; La Rocque, A. On the use of X-ray computed tomography for determining wood properties: A review. *Can. J. For. Res.* **2011**, *41*, 2120–2140. [CrossRef]
30. Xie, Y.; Yang, T.; Wang, X.; Chen, X.; Pang, S.; Hu, J.; Wang, A.; Chen, L.; Shen, Z. Applying a portable backpack Lidar to measure and locate trees in a nature forest plot: Accuracy and error analyses. *Remote Sens.* **2022**, *14*, 1806. [CrossRef]
31. Cao, Y.; Wang, D.; Wang, Z.; Tian, L.; Zheng, C.; Tian, Y.; Liu, Y. Research on tree pith location in radial direction based on terrestrial laser scanning. *Forests* **2021**, *12*, 671. [CrossRef]
32. Sellier, D.; Ségura, R. Radial growth anisotropy and temporality in fast-growing temperate conifers. *Ann. For. Sci.* **2020**, *77*, 85. [CrossRef]
33. Szymański, N.; Wilczyński, S. Radial growth response of European larch provenances to interannual climate variation in Poland. *Forests* **2021**, *12*, 334. [CrossRef]
34. Gheyret, G.; Zhang, H.; Guo, Y.; Liu, T.; Bai, Y.; Li, S.; Schmid, B.; Bruelheide, H.; Ma, K.; Tang, Z. Radial growth response of trees to seasonal soil humidity in a subtropical forest. *Basic Appl. Ecol.* **2021**, *55*, 74–86. [CrossRef]
35. Huang, X.; Dai, D.; Xiang, Y.; Yan, Z.; Teng, M.; Wang, P.; Zhou, Z.; Zeng, L.; Xiao, W. Radial growth of Pinus massoniana is influenced by temperature, precipitation, and site conditions on the regional scale: A meta-analysis based on tree-ring width index. *Ecol. Indic.* **2021**, *126*, 107659. [CrossRef]
36. Pretzsch, H. Tree growth as affected by stem and crown structure. *Trees* **2021**, *35*, 947–960. [CrossRef]
37. Coomes, D.; Allen, R. Effects of size, competition and altitude on tree growth. *J. Ecol.* **2007**, *95*, 1084–1097. [CrossRef]
38. Wang, H.; Qin, J.; Hu, Y.; Guo, C. Asymmetric growth of belowground and aboveground tree organs and their architectural relationships: A review. *Can. J. For. Res.* **2023**, *53*, 315–327. [CrossRef]

39. Miodek, A.; Gizińska, A.; Klisz, M.; Wojda, T.; Ukalski, K.; Kojs, P. Direct exposure to solar radiation causes radial growth eccentricity at the beginning of the growing season in Robinia pseudoacacia. *IAWA J.* **2020**, *41*, 61–84. [CrossRef]
40. Simpson, H.; Denne, M. Variation of ring width and specific gravity within trees from unthinned Sitka spruce spacing trial in Clocaenog, North Wales. *Forestry* **1997**, *70*, 31–45. [CrossRef]

forests

MDPI

Article

Stand Age Affects Biomass Allocation and Allometric Models for Biomass Estimation: A Case Study of Two Eucalypts Hybrids

Runxia Huang, Wankuan Zhu, Apeng Du, Yuxing Xu * and Zhichao Wang

Research Institute of Fast-Growing Trees (RIFT), Chinese Academy of Forestry (CAF), State Key Laboratory of Efficient Production of Forest Resources, Zhanjiang 524022, China; runxia_huang@caf.ac.cn (R.H.); wzc2254@163.com (Z.W.)

* Correspondence: cercxuyx@caf.ac.cn

Abstract: We studied the effects of stand age on the allocation of biomass and allometric relationships among component biomass in five stands ages (1, 3, 5, 7, and 8 years old) of two eucalypts hybrids, including *Eucalyptus urophylla* × *E. grandis* and *E. urophylla* × *E. tereticornis*, in the Leizhou Peninsula, China. The stem, bark, branch, leaf, and root biomass from 60 destructively harvested trees were quantified. Allometric models were applied to examine the relationship between the tree component biomass and predictor variable (diameter at breast height, D, and height, H). Stand age was introduced into the allometric models to explore the effect of stand age on biomass estimation. The results showed the following: (1) Stand age significantly affected the distribution of biomass in each component. The proportion of stem biomass to total tree biomass increased with stand age, the proportions of bark, branch, and leaf biomass to total tree biomass decreased with stand age, and the proportion of root biomass to total tree biomass first decreased and then increased with stand age. (2) There were close allometric relationships between biomass (i.e., the components biomass, aboveground biomass, and total biomass per tree) and diameter at breast height (D), height (H), the product of diameter at breast height and tree height (DH), and the product of the square of the diameter at breast height and tree height (D^2H). The allometric relationship between biomass and measurement parameters (D, H, DH, D^2H) could be applied to the biomass assessment of eucalypts plantation. (3) Allometric equations that included stand age as a complementary variable significantly improved the fit and enhanced the accuracy of biomass estimates. The optimal independent variable for the biomass prediction model varied according to each organ. These results indicate that stand age has an important influence on biomass allocation. Allometric equations considering stand age could improve the accuracy of carbon sequestration estimates in plantations.

Keywords: eucalypts; stand age; biomass; allometric relationship; model simulation

1. Introduction

At present, global climate change, such as global warming, extreme weather happening frequently, glacier retreating, and global sea levels rising, is becoming more and more serious [1], and the mitigation of and adaptation to global climate change should not be delayed. Forest ecosystems are an important terrestrial carbon sink, with a high carbon storage capacity, globally accumulating about 2.4 ± 0.4 Pg C per year [2], and they play a critical role in mitigating global climate change and balancing the ecosystem carbon cycle. The area of plantations in China is 7954.28×10^4 ha, accounting for 36.45% of the national forest area, and the storage capacity of plantations is (33.88 million m^3), accounting for

19.29% of the national forest storage capacity [3]. Plantations have a high potential for carbon sink, absorbing and fixing CO_2 through photosynthesis conversion to the form of biomass in plants, which helps to reduce greenhouse gas emissions by increasing carbon sinks and plays an important role in improving the ecological environment and mitigating climate change [4].

Biomass is the amount of organic matter actually living in a unit area at a given point in time [5]. Tree biomass is an important indicator for evaluating forest carbon stocks [6]. The accumulation of biomass not only reflects tree growth and adaptation to the environment, but also the productivity and carbon sequestration capacity of a forest ecosystem, and its role as a carbon sink [7,8]. Biomass is influenced by biotic and abiotic factors, such as plant genetic characteristics, interspecific competition, stand age, precipitation, soil conditions, altitude, and annual average temperature [9–12]. For example, Zhang et al. [13] found that altitude had an important effect on above–belowground biomass distribution, and the altitude gradient pattern of forest above–belowground biomass allocation distribution was determined by climate factors. Dai et al. [14] revealed that the efficiency of plant photosynthesis decreases because of a reduction in temperature, which would lead to a reduction in carbohydrate production and potentially less growth investment in aboveground parts, so as to influence biomass distribution. Accurate biomass estimation of trees is important for quantifying forest carbon storage and for understanding the carbon cycle of a forest ecosystem in the context of the global carbon cycle. Estimating the biomass of forests may rely on different techniques at different scales, from remote sensing at the regional level to direct weighing at the local level [15,16]. Because direct tree biomass measurements are destructive and costly, this method is not suitable for estimating large-scale forest biomass [15]. Allometric equation, a statistical model that predicts tree biomass (stem, bark, branch, leaf, and root) from easily measured variables, has been widely used in biomass estimation [17–20]. The diameter at breast height (D), tree height (H), cross-sectional area at breast height, crown morphology (crown width and shape), and wood density are common independent variables in allometric models [7,21–23]. In previous research, allometric models have been shown to provide accurate and reliable estimates of biomass in forest ecosystems [24–27]. For example, McNicol et al. [24] and Peng et al. [25] used a power function model with D^2H as the predictive variable to estimate aboveground biomass. Yang et al. [26] found a good model fit using D and H as predictive variables ($W = c\ D^a\ H^b$) to estimate the total biomass of *Quercus acutissima* in middle-aged forests and using only D as a predictive variable ($W = c\ D^a$) to estimate the total biomass of *Q. acutissima* in young and mature forests. And Yang et al. [26] revealed the parameters of the allometric equation's dependence on the stand age.

Stand age provides a clear indication of cumulative biomass. Many studies have emphasized the importance of stand age for the assessment of forest biomass and carbon sink potential [28–31]. Stand age also has a huge impact on tree size, shape, and traits, alters biomass allocation among tree components, and consequently affects allometric relationships [19,26]. For example, in early periods of growth, more biomass is allocated to stems for height growth and to branches and leaves for canopy expansion in order to increase competitive advantage for light resources over their neighbors [19]. Peichl et al. [32] and Chen et al. [33] reported differences in tree biomass between different stand ages. Some studies proved that the influence of stand age was an important determinant of forest biomass [16,31,34]. Thus, there is a need to develop age-specific allometric equations for rapid and reliable estimation of tree component biomass at different developmental stages. Although previous studies have explored stand age-based biomass modeling, stand age-driven biomass modeling also varies by forest type and tree species because of species-specific physiological growth traits [35]. In addition, different environmental conditions, e.g., biogeographic region, climatic conditions, soil type, and stand category, may lead to differences in stand age-specific biomass modeling [10,34,36]. Therefore, the development of stand age-dependent biomass models is important for practical applications to reveal biomass changes in fast-growing plantation forests (e.g., eucalypts).

Eucalypts is an important species for afforestation in southern China due to its fast growth, short rotation period, and excellent material. Eucalypts plantations in China exceed 5.6 million ha, accounting for 2.57% of the country's forest area (218.22 million ha) and 7.04% of the country's plantation forest (79.54 million ha) [3,37]. The area of eucalypts plantation in China is second only to India and Brazil [37]. Eucalypts has the function of carbon sequestration. Accurately quantifying the biomass of eucalypts plantations is critical for understanding carbon stocks in forest ecosystems and mitigating global climate change. A great deal of research has been conducted on allometric models of eucalypts biomass [38,39], but the information about allometric models considering the effects of stand age is limited. Therefore, the objectives of this study were to (1) analyze the effect of stand age on biomass distribution among tree components (i.e., stems, bark, branch, leaf, and root); (2) examine if the predictive capacity of the allometric equations that include stand age as a complementary variable is improved.

2. Material and Methods

2.1. Study Site

This study was carried out in the Leizhou Peninsula (109°57′~110°8′ E, 20°55′~21°8′ N) in Zhanjiang, Guangdong Province, China. The study site was within one of the major production areas of eucalypts. The study area had a typical oceanic monsoon climate, with an annual mean temperature of 23.1 °C. The annual relative humidity was 80.4%, the mean annual precipitation was 1711.6 mm, and there was an annual mean of 2003.6 h of sunshine. The elevation was approximately 50 m, and the soil type was Rhodi-Udic Ferralosols. The asexual lines *E. urophylla* × *E. grandis* 3229 and *E. urophylla* × *E. tereticornis* LH1 were used, and there were minimal differences in the terrain or management measures.

2.2. Sample Selection and Biomass Measurement

During 2023, five *E. urophylla* × *E. grandis* and *E. urophylla* × *E. tereticornis* stands (1, 3, 5, 7, and 8 years old) in the Leizhou Peninsula were selected. Three 20 m × 20 m sample plots were established randomly within each stand, each plot was separated by 500 m, and the diameter at breast height (D, 1.3 m) and tree height (H) was measured (Table 1). According to the measurement data, two average trees (i.e., trees with the same diameter at breast height (D) and tree height as the average diameter and average tree height of the stand) in each sample plot were harvested. That is, six average trees were harvested per stand for destructive sampling. In total, 60 average trees were sampled. After felling, D and H were recorded. The trees were cut using a hand saw at 1.3 m, 2 m, 4 m, 6 m . . . 20 m, and 22 m of the tree height, and subsequently at 2 m intervals up to the apex. Bark was removed from the sampled stem. The fresh mass of the stem, bark, branches, and leaves was measured in the field. After the above measurements, the root systems of all sampled trees were excavated with a hook machine, soil clinging to the roots was removed, and the fresh mass of the roots was measured. Approximately 60 g subsamples of each component were selected and brought to the laboratory. The subsamples were oven-dried at 85 °C until they reached a constant weight. The dry mass (i.e., biomass) of all components was estimated according to the ratio of the fresh mass and dry mass of the subsamples.

Table 1. Characteristics *E. urophylla* × *E. grandis* and *E. urophylla* × *E. tereticornis* stands. (mean ± standard deviations).

Species	Stand Age (Year)	Regeneration Method	Stand Density (Trees/ha)	Average Tree Height (m)	Average DBH (cm)
E. urophylla × *E. grandis*	1	seedling	1200	5.550 ± 0.235	4.300 ± 0.245
	3	seedling	1400	12.467 ± 0.480	8.550 ± 0.164
	5	seedling	1250	19.167 ± 0.258	13.278 ± 0.441
	7	seedling	1150	20.483 ± 0.479	13.85 0 ± 0.197
	8	seedling	1100	21.717 ± 0.471	15.033 ± 0.186
E. urophylla × *E. tereticornis*	1	seedling	1350	5.750 ± 0.418	6.017 ± 0.426
	3	seedling	1400	11.342 ± 0.111	8.117 ± 0.075
	5	seedling	1300	18.550 ± 0.565	11.797 ± 0.259
	7	seedling	1000	21.983 ± 0.911	15.000 ± 0.379
	8	seedling	1000	22.450 ± 0.616	15.875 ± 0.395

2.3. Biomass Modeling

According to the methodology of McNicol et al. [24], allometric relationships between biomass components and D, H were often expressed using the following power function equation:

$$y_i = a_1 (x)^{b1} \tag{1}$$

where y_i is the biomass (kg) of the component (i.e., stem, bark, branch, leaf, root, above-ground) or the total biomass of a single tree; x is D, H, DH or D^2H; and a_1 and b_1 are equation parameters.

All data were logarithmically (base 10) transformed to ensure a linear relationship between variables and to meet the requirements of homogeneity [19,24]. Therefore, we established the following equation [24]:

$$\lg (y_i) = a_2 + b \lg (x) \tag{2}$$

where y_i is the biomass (kg) of component (i.e., stem, bark, branch, leaf, root, aboveground) or the total biomass of a single tree; x is D, H, DH or D^2H; and a_2 and b are equation parameters.

According to Xiang et al. [19], we introduced stand age as a complementary variable into the allometric equation to determine whether the equation was improved. The individual effect of stand age and the interactions between stand age and other predictor variables (i.e., D, H, DH, or D^2H) were considered. Then, we reconstructed the stand age-specific allometric equations by introducing the stand age factor based on Equation (2):

$$\lg (y_i) = a_2 + b \lg(x) + c (age) + d (age) \lg (x) \tag{3}$$

where y_i is the biomass (kg) of a tree component (i.e., stem, bark, branch, leaf, root, aboveground) or the total biomass of a single tree; x is D, H, DH or D^2H; age is the stand age; and a_2, b, c, and d are equation parameters.

2.4. Data Analysis

The coefficient of determination (R^2), total relative error (TRE), residual sum of squares (RSS), root mean squared error (RMSE), and Akaike information criterion (AIC) were used to assess the performance of all models:

$$TRE = \frac{\sum(y_i - \hat{y}_i)}{\sum \hat{y}_i} \times 100 \tag{4}$$

$$RSS = \sum_{i=1}^{n}(y_i - \hat{y}_i)^2 \quad (5)$$

$$RMSE = \sqrt{\frac{\sum_{1}^{n}(y_i - \hat{y}_i)^2}{n}} \quad (6)$$

$$AIC = 2k + n\ln\left(\frac{RSS}{n}\right) \quad (7)$$

where y_i is the observed biomass value, $\hat{y}_i$ is the predicted biomass value obtained by the allometric models, n is the number of samples, and k is the number of parameters in the model.

3. Results

3.1. Biomass Allocation Among Tree Components

The stem of *E. urophylla* × *E. grandis* accounted for the largest proportion of biomass, (42.17%–74% of total tree biomass) and increased with stand age. Root biomass represented 14.18%–25.84% of total tree biomass and first decreased with increasing stand age and then increased with increasing stand age. The proportions of bark, branch, and leaf biomass decreased with stand age. Crown biomass (branch + leaf) was higher in 1 year and 3 years plantations and represented 22.89% and 17.45%, respectively, and crown biomass decreased with stand age. The partitioning of biomass components in *E. urophylla* × *E. tereticornis* was equivalent to that in *E. urophylla* × *E. grandis*. Stem biomass accounted for 55%–73.68% of total tree biomass in the five stands and increased with stand age. Root biomass accounted for 11.90%–22.28% of the total tree biomass and first decreased as the stand age increased, then increased with stand age increased. Stem and root biomass accounted for 71.93%–90.69%, while crown biomass accounted for 4.38%–16.75%, with a higher contribution occurring in 1 year and 3 years (Figure 1).

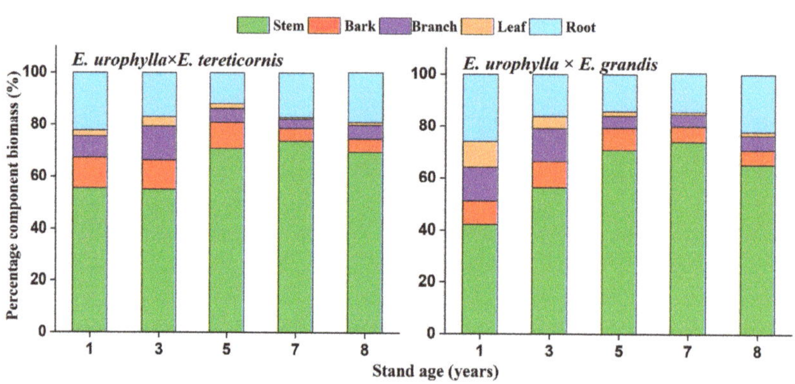

Figure 1. Changes in component biomass (%) relative to total tree biomass in relation to age (years) in two eucalypts hybrids.

3.2. Allometry Among Tree Biomass Components

Biomass was scaled allometrically in eucalypts. The stem, bark, roots, aboveground biomass, and total biomass were closely related to diameter at breast height, with an $R^2 > 0.917$. Branch and crown biomass had R^2 values = 0.794 and 0.733, respectively, and leaf biomass had an $R^2 = 0.390$ (Figure 2). The stem, bark, aboveground biomass, and total biomass were closely related to tree height, with an $R^2 > 0.968$ (Figure 3). Branch, crown, and root biomass had R^2 values = 0.865, 0.833, and 0.703, respectively, and leaf biomass had an R^2 = 0.534 (Figure 3). There were close allometric relationships among the components' biomasses. The R^2 between root and stem biomass was 0.786, the R^2 between crown and stem biomass was 0.675, and the R^2 between root and crown biomass was 0.675 (Figure 4). Our results demonstrate that an allometric equation was suitable for estimating the component biomass of eucalypts, and this relationship could be used to estimate biomass when diameter at breast height, tree height, or the biomass of one component is known.

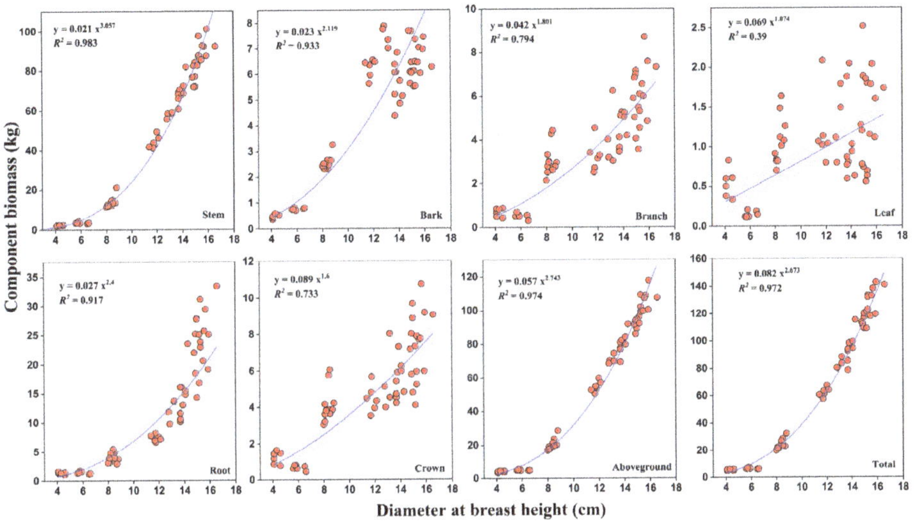

Figure 2. The allometric relationship between component biomass and diameter at breast height. Note: gray shaded areas are 95% confidence intervals.

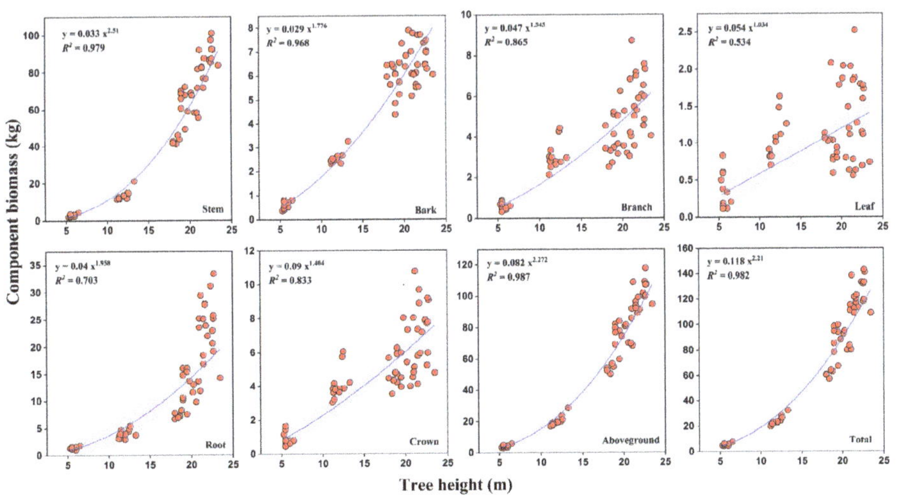

Figure 3. The allometric relationship between component biomass and tree height. Note: gray shaded areas are 95% confidence intervals.

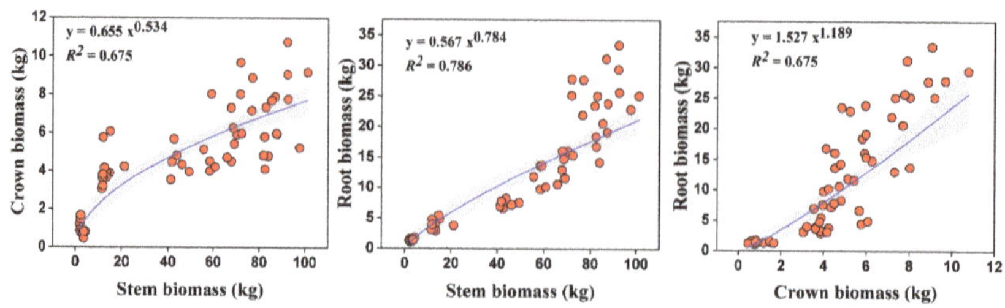

Figure 4. Allometric relationships among crown (branch + leaf), stem, and root biomass. Note: gray shaded areas are 95% confidence intervals.

3.3. Effect of Stand Age on Biomass Components

Stand age had a significant effect on the component biomass (including stem, bark, branch, leaf, root, aboveground, and crown) and total biomass of eucalypts ($p < 0.05$, Tables S1 and S2). However, tree variety had no significant effect on the component biomass or total biomass of eucalypts, except for stem and leaf (Tables S1 and S2). This study focused only on the effect of stand age on eucalypts biomass.

Introducing stand age, which was used to predict component (including stem, bark, branch, leaf, root, aboveground, and crown) biomass and total biomass, improved the fit of the allometric model. The R^2 of the stand age-specific allometric model was higher than that of a model that only considered D or H; in particular branch, leaf and root estimates were significantly improved. The R^2 values of stand age-specific allometric models based on D that were used to predict branch, leaf, and root biomass improved by 9.45%, 31.03%, and 4.69%, respectively (Table 2 and Figure 2). The R^2 values of stand age-specific allometric models based on H that were used to predict branch, leaf, and root biomass improved by 3.70%, 2.81%, and 37.56%, respectively (Table 2 and Figure 3). The RSS, RMSE, and AIC of stand age-specific allometric models based on H that predicted bark, branch, and leaf biomass were lower than those of other models. The RSS, TRE, RMSE, and AIC of the stand age-specific allometric model based on DH that predicted root and aboveground biomass were lower than those of other models. The RSS, RMSE, and AIC of the stand age-specific allometric model based on DH that predicted stem and total biomass were lower than those based on D or H (Table 2). The RSS, TRE, RMSE, and AIC of the stand age-specific allometric model based on D^2H that predicted stem biomass were lower than those of other models, while the RSS, RMSE, and AIC of the models that predicted root, aboveground, and total biomass were lower than those based on D or H.

The ratios among biomass components were significantly different across stand age (Table 3). The ratios of roots to stem biomass were 0.511, 0.301, 0.185, 0.218, and 0.310 across the stand age, and these ratios first decreased and then increased with stand age. The ratios of crown to stem biomass were 0.372, 0.311, 0.098, 0.069, and 0.099, and first decreased and then stabilized with stand age. The ratios of root to crown biomass were 1.688, 0.987, 1.963, 3.281, and 3.161, and first decreased and then increased with stand age. The ratio of belowground to aboveground biomass first decreased and then increased with stand age (Figure S1). Overall, the ratios between biomass components varied greatly in young stands (1 year and 3 years) and differed significantly from those in nearly mature stands (7 years and 8 years, Table 3). Therefore, large errors in the estimation of component biomass may occur if an allometric model does not consider eucalypts stand age.

Table 2. Allometric equations for tree biomass components and fitting statistics with stand age.

Independent Variable	Dependent Variable	a	b	c	d	R^2	p	RSS	TRE	RMSE	AIC
D	Stem biomass	−1.490	2.595	0.216	−0.143	0.986	0.000	2409.382	0.917	6.337	229.567
H		−1.033	1.969	−0.066	0.081	0.988	0.000	2628.201	1.369	6.618	234.783
DH		−1.401	1.283	−0.075	0.035	0.995	0.000	1766.375	−0.303	5.426	210.940
D^2H		−1.486	0.893	−0.014	0.008	0.994	0.000	1578.636	−0.889	5.129	204.198
D	Bark biomass	−2.146	2.503	0.412	−0.349	0.970	0.000	39.750	2.130	0.814	−16.705
H		−1.818	2.119	0.045	−0.054	0.974	0.000	20.893	1.089	0.590	−55.295
DH		−2.129	1.295	0.107	−0.059	0.983	0.000	26.621	−0.375	0.666	−40.760
D^2H		−2.185	0.883	0.185	−0.060	0.981	0.000	29.322	3.100	0.699	−34.960
D	Branch biomass	−1.124	0.806	0.737	−0.523	0.869	0.000	83.809	2.813	1.182	28.052
H		−1.468	1.302	0.559	−0.373	0.897	0.000	65.885	2.383	1.048	13.614
DH		−1.385	0.595	0.641	−0.222	0.883	0.000	71.773	1.895	1.094	18.750
D^2H		−1.315	0.362	0.680	−0.159	0.878	0.000	76.784	4.648	1.131	22.799
D	Leaf biomass	−0.941	0.025	0.845	−0.605	0.511	0.000	15.468	9.001	0.508	−73.333
H		−1.743	1.445	0.288	−0.227	0.549	0.000	12.080	7.468	0.449	−88.167
DH		−1.526	0.518	0.545	−0.201	0.502	0.000	13.348	9.924	0.472	−82.178
D^2H		−1.360	0.245	0.666	−0.161	0.496	0.000	14.106	8.879	0.485	−78.864
D	Root biomass	−0.543	0.880	0.054	0.049	0.960	0.000	452.262	1.159	2.745	129.195
H		−0.412	0.728	−0.094	0.146	0.967	0.000	492.147	3.277	2.864	134.266
DH		−0.553	0.482	−0.102	0.076	0.967	0.000	434.119	1.824	2.690	126.738
D^2H		−0.578	0.329	−0.067	0.043	0.965	0.000	434.716	2.447	2.692	126.821
D	Aboveground biomass	−1.045	2.189	0.310	−0.212	0.981	0.000	3423.626	1.035	7.554	250.647
H		−0.767	1.846	0.005	0.025	0.992	0.000	2313.416	−0.681	6.209	227.128
DH		−1.045	1.146	0.031	−0.004	0.994	0.000	1338.961	−1.481	4.724	194.318
D^2H		−1.095	0.782	0.097	−0.019	0.991	0.000	1400.379	−1.073	4.831	197.009
D	Total biomass	−0.736	1.941	0.264	−0.165	0.981	0.000	5349.070	1.002	9.442	277.420
H		−0.491	1.648	−0.021	0.052	0.993	0.000	3592.137	−0.432	7.738	253.529
DH		−0.743	1.026	0.002	0.012	0.993	0.000	2171.750	−0.885	6.016	223.337
D^2H		−0.787	0.699	0.065	−0.007	0.990	0.000	2768.508	−2.397	6.793	237.903

Notes: D is the diameter at breast height, H is the tree height, DH is the product of diameter at breast height and tree height, D^2H is the product of square of the diameter at breast height and tree height, R^2 is the coefficient of determination, RSS is the residual sum of squares, TRE is total relative error, RMSE is the root mean squared error, AIC is Akaike information criterion, and a, b, c, and d are model parameters.

Table 3. Component biomass ratios in different stand age classes and species classes. (mean ± standard deviations).

Factors	Roots/Stem Ratio	Crown/Stem Ratio	Roots/Crown Ratio	Belowground/Aboveground Ratio
		Stand age classes		
1 year	0.511 ± 0.138 a	0.372 ± 0.218 a	1.688 ± 0.643 b	0.319 ± 0.055 a
3 years	0.301 ± 0.066 b	0.311 ± 0.073 a	0.987 ± 0.219 c	0.199 ± 0.040 c
5 years	0.185 ± 0.029 c	0.098 ± 0.022 b	1.963 ± 0.469 b	0.150 ± 0.023 d
7 years	0.218 ± 0.039 c	0.069 ± 0.013 b	3.281 ± 0.885 a	0.191 ± 0.036 c
8 years	0.310 ± 0.055 b	0.099 ± 0.018 b	3.161 ± 0.389 a	0.262 ± 0.043 b
		Species classes		
EG 1 year	0.620 ± 0.108 a	0.555 ± 0.150 a	1.175 ± 0.304 b	0.351 ± 0.058 a
ET	0.403 ± 0.042 b	0.189 ± 0.036 b	2.200 ± 0.434 a	0.287 ± 0.031 b
EG 3 years	0.293 ± 0.08	0.318 ± 0.104	0.952 ± 0.253	0.194 ± 0.047
ET	0.309 ± 0.055	0.305 ± 0.028	1.023 ± 0.196	0.205 ± 0.035

Table 3. *Cont.*

Factors		Roots/Stem Ratio	Crown/Stem Ratio	Roots/Crown Ratio	Belowground/Aboveground Ratio
EG	5 years	0.201 ± 0.03 a	0.093 ± 0.025	2.233 ± 0.480 a	0.166 ± 0.022 a
ET		0.168 ± 0.016 b	0.102 ± 0.019	1.694 ± 0.282 b	0.135 ± 0.012 b
EG	7 years	0.203 ± 0.032	0.078 ± 0.010 a	2.643 ± 0.48 b	0.176 ± 0.028
ET		0.232 ± 0.043	0.06 ± 0.008 b	3.919 ± 0.719 a	0.206 ± 0.039
EG	8 years	0.341 ± 0.038 a	0.106 ± 0.018	3.261 ± 0.379	0.285 ± 0.028
ET		0.278 ± 0.054 b	0.091 ± 0.016	3.060 ± 0.407	0.239 ± 0.045

Note: different lowercase letters indicate significant differences ($p < 0.05$). ET is *E. urophylla* × *E. tereticornis*, EG is *E. urophylla* × *E. grandis*.

3.4. Biomass Prediction Comparisons Among Tree Components

Except for branches and leaves, the biomass values predicted by the stand age-specific allometric models for stem, bark, root, aboveground, and total tree biomass were comparable with the observed biomass values (Figure 5). These results indicate that stand the age-specific allometric models based on D, H, DH, and D^2H were able to efficiently estimate stem, bark, root, aboveground, and total tree biomass with a high degree of precision. The R^2 values for observed biomass (bark, branch, and leaf) versus predicted biomass (using H) from the stand age-specific allometric model were 0.945, 0.753, and 0.375, respectively, which were better than those from models based on D, DH, or D^2H (Figure 5). For bark, branches and leaves, the H-based stand age-specific allometric model was more precise, and the RSS, RMSE, and AIC values using H as a predictor variable were smaller than those from other models (Table 2). The R^2 values for observed stem biomass values predicted biomass (using D^2H) from the stand age-specific allometric model to be 0.979, and that value was higher than those from the other allometric models (Figure 5). For stem, the stand age-specific allometric model using D^2H as a predictor variable was better than models based on D, H, or DH, and the TRE, RSS, RMSE, and AIC of the based D^2H stand age-specific allometric model were smaller than those from other allometric models (Table 2). For roots, aboveground biomass, and total tree biomass, the values predicted by the DH- and D^2H-based stand age-specific allometric models were close to the observed values, with R^2 values of 0.920 and 0.921 (root biomass), 0.986 and 0.985 (aboveground biomass), and 0.984 and 0.982 (total biomass), which were higher than those from the models using D or H as predictor variables (Figure 5). These results were consistent with the RSS, RMSE, and AIC values from the DH- and D^2H-based stand age-specific allometric models, which were smaller than those from other models (Table 2).

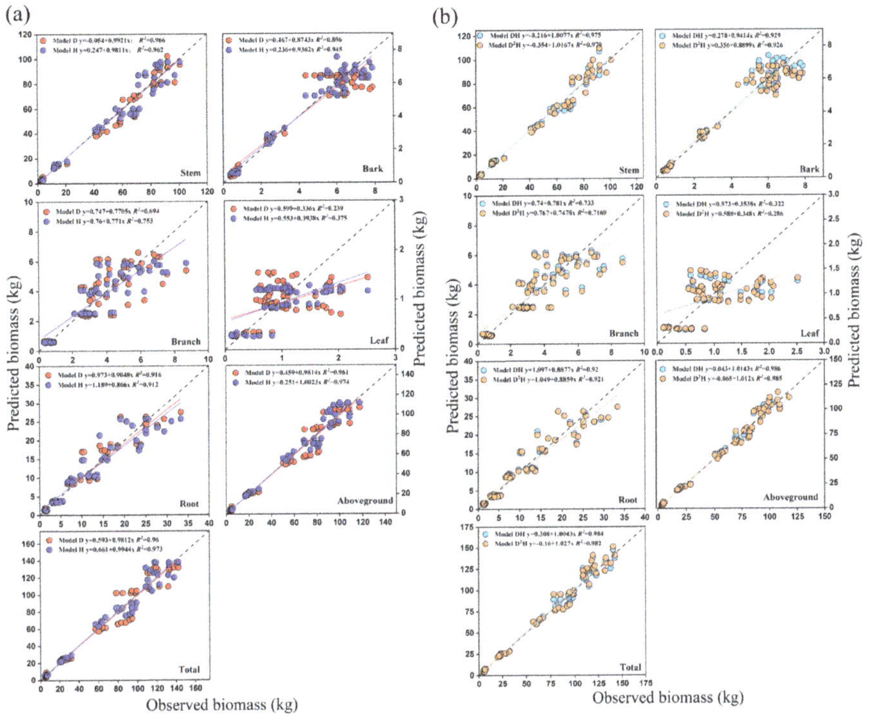

Figure 5. Comparison of the predicted component biomass versus the observed biomass, with (**a**) D-, H-, (**b**) DH- and D^2H-based models. Note: the red dots represent stand age-specific allometric models based on D, the blue dots represent models based on H, the azure dots represent models based on DH, the light brown dots represent models based on D^2H, and the black dashed line represents the function y = x.

4. Discussion

4.1. Biomass Allocation Characteristics

Biomass production is a primary function of forest ecosystems and is influenced by numerous processes: roots capture nutrients from the soil; stems and branches provide mechanical support and conduct water and nutrients; and leaves fix carbon [40]. Balancing the allocation of biomass to different organs enables trees to perform necessary physiological activities and achieve normal growth, thus reflecting their resource use efficiency and environmental adaptability [41,42]. The effect of stand age on biomass allocation is a result of tree growth and the balance of resource allocation among biomass components in a way that maintains the physiological activities and functions of each organ [22]. Our results show that except for 1 year *E. urophylla* × *E. grandis* (42.17%) and *E. urophylla* × *E. tereticornis* (55.43%) and 3 years *E. urophylla* × *E. grandis* (56.24%) and *E. urophylla* × *E. tereticornis* (55%), the proportions of stem biomass to total tree biomass in *E. urophylla* × *E. grandis* (65.18%–74%) and *E. urophylla* × *E. tereticornis* (69.62%–73.68%) (Figure 1) were higher than the average stem biomass ratio for forest ecosystems in China (58.20%) [43]. The proportions of stem biomass to total tree biomass in *E. urophylla* × *E. grandis* and *E. urophylla* × *E. tereticornis* from all stand ages were higher than the biomass of 1 year–7 years *E. urophylla* × *E. grandis* (24.91%–66.78%) measured by Zhang et al. [44] in Guangdong, but lower than the biomass of 1 year–8 years *E. urophylla* × *E. grandis* (51.07%–98.48%) measured by Fu et al. [45] in Guangxi. In this study, the proportion of leaf biomass to total tree biomass of eucalypts (0.91%–10.01%) at all stand ages was smaller than that measured by Zheng et al. [39] (2.37%–23.63%) and Fu et al. [45] (2.17%–21.01%).

Stand age significantly affected biomass allocation in each organ of eucalypts. We observed an increase in the proportion of stem biomass to total tree biomass and a decrease in the proportion of bark, branch, and leaf biomass to total tree biomass as stand age increased (Figure 1), which was consistent with other studies [19,26]. Changes in biomass allocation with age may be explained by the strategies that trees employ during stand development [19]. Stem biomass is an important component of tree biomass, which represents a net accumulation process with stand age [26]. In order to occupy the upper canopy space and obtain more lights, the branches and leaves of young eucalypts would undergo rapid expansion, which is conducive to photosynthesis and transpiration and crucial for the survival of young eucalypts [32,46]. The proportions of branch and leaf biomass to total tree biomass in eucalypts were higher in young stands, demonstrating that more carbohydrates may be allocated to branches and leaves to help eucalypts grow rapidly [22]. Trees will adjust their biomass allocation strategy according to competitive demands at different stages [26]. Therefore, changes in biomass distribution during stand development must be considered to obtain accurate biomass predictions.

Roots play an important role in material cycling and energy flow in forest ecosystems [47]. Roots not only absorb and transport water and nutrients but also input organic matter into the soil, which maintains soil fertility, and fix carbon [23,47]. In this study, the proportions of root biomass to total tree biomass in eucalypts of all stand ages were 11.90%–25.84% (Figure 1), which indicates the importance of root biomass. A well-developed root system improves tree competition for water and nutrients, which is conducive to growth [26]. We also found that the proportion of root biomass to total tree biomass first decreased and then increased as stand age increased (Figure 1), which is not consistent with previous studies [45,47]. The proportion of root biomass to total tree biomass decreased as stand age increased in Fu et al. [45], while an increase with stand age was reported by Han et al. [47], and an initial increase and then a decrease with stand age was found by Zhang et al. [44]. However, Chen et al. [33] showed that the proportion of root biomass to total tree biomass gradually stabilized as stand age increased. These differences may result from different tree species and study sites. Root biomass is also related to soil moisture, nutrients, stand structure, and community composition [47]. In addition, the root biomass may be estimated based on the relationship between root biomass and the biomass of other components [48]. In this study, root biomass was closely related to stem biomass and crown biomass (R^2 = 0.786 and 0.675, respectively) (Figure 4), which was similar to the results of other studies [49,50]. The ratio of belowground to aboveground biomass of eucalypts ranged from 0.150 to 0.319 (Table S2), which was close to that of other forest types in China (0.233) [51]. The ratio of aboveground biomass to root biomass helps us estimate root biomass from the more easily measured aboveground biomass and also reflects investment in aboveground organs versus roots in response to environmental change [23,52].

4.2. Analysis of Biomass Allometric Equations

Stand age had a significant effect on component biomass in eucalypts. Other studies have also reported an effect of stand age on the relationship between biomass and measurement factors [19,26]. Hong et al. [53] showed that the introduction of a stand age factor improved the fit and the prediction accuracy of their biomass model. Similarly, Wang et al. [50] reported that there were differences in aboveground and belowground biomass allocation by *Abies nephrolepis* at different growth stages, and a biomass estimation without considering stand age could be slightly biased. Our results show that stand age-specific allometric equations for component biomass in eucalypts improved the estimates of biomass, especially in branch and leaf biomass equations (Figures 2 and 3, Table 2), which was similar to the results of previous studies [32,54]. These are the results of changes in tree growth and a combination of stand age and tree social status [19]. Therefore, we should consider the effect of stand age on biomass to improve the accuracy of biomass estimation. Similar findings have been applied at other species, such as Chinese fir (*Cunninghamia lanceolata*), Mongolian pine (*Pinus sylvestris* L. var. *mongolica* Litv.), and White pine (*Pinus strobus*) [4,19,32].

Using D, H, or combinations of D and H as predictor variables in allometric equations is a simple and effective method for estimating tree biomass. In this study, we used *E. urophylla* × *E. grandis* and *E. urophylla* × *E. tereticornis* from different-aged stands and constructed and screened an optimal biomass model, which was reliable for estimating the biomass of eucalypts in the Leizhou Peninsula of Guangdong Province (Table 2). Stand age-specific allometric models for stem, bark, root, aboveground, and total biomass per tree based on D, H, DH, or D^2H explained more than 96.5% of variation in the data (Table 2). Stand age-specific allometric models for branch biomass based on D, H, DH, or D^2H explained more than 86.9% of variation in the data but stand age-specific allometric models for leaf biomass explained only 49.6%–54.9% of variation in the data (Table 2). Except for leaf biomass, the stand age-specific allometric models for other components (stem, bark, branch, root, aboveground, and total biomass) based on D, H, DH, or D^2H had high explanatory power and provided good estimates of biomass (Table 2). Similar trends in leaf biomass have been reported in other studies [20,55] and may be due to the felling process, which causes a loss of leaves that results in a large error. Leaf biomass is strongly influenced by canopy traits and tree growth because of canopy competition [56]. In addition, leaf biomass is influenced by light, moisture, temperature, soil nutrients, and interspecific competition [20]. The fit of the leaf biomass model was lower than that of other biomass components (Table 2). However, in general, stand age-specific allometric models fit better (Figures 2 and 3; Table 2).

The D and H, which are accessible and accurate measurements, are widely used in allometric models [21]. Our study confirmed a strong correlation between component biomass and D in eucalypts, and a stronger correlation between branch and leaf biomass and H (Table 2). It is possible that branches and leaves were more responsive to competition and the environment [57], and that competition for canopy dominance may have resulted in a greater allocation of photosynthate to height than to radial growth [46]. H was added into the biomass allometric model as the second parameter to improve its accuracy [24,25,58]. Our results showed that stem biomass predicted by the stand age-specific allometric model based on D^2H fitted well and had high explanatory power. The R^2 value for this model was higher than that of other models, and the RSS, TRE, RMSE, and AIC were lower than those of other models (Table 2), demonstrating that D^2H was strongly intrinsically linked to stem biomass. The stem biomass predicted with D^2H was comparable to the observed data (Figure 5). The DH and D^2H were reliable predictors of root, aboveground, and total tree biomass, and those models had higher R^2 values and lower RSS, TRE, RMSE, and AIC values (Table 2), indicating that DH and D^2H were both strongly intrinsically related to root, aboveground, and total tree biomass. Root, aboveground, and total tree biomass predicted with DH or D^2H were comparable to the observed data (Figure 5), indicating that DH and D^2H improved the accuracy of the models. This may be because DH and D^2H captured the variation in tree structure and avoided the problem of multicollinearity between D and H as separate variables in an allometric model [36,59]. These models are suitable for predicting the component biomass of eucalypts in Leizhou Peninsula, Guangdong province, China. However, when using these models to predict the biomass of eucalypts plantations in other place of China, they should preferably be calibrated to improve the precision and accuracy of biomass estimation because biomass is affected by light, moisture, temperature, soil nutrients, altitude, etc. [13,20,57], which affect the relationship between component biomass and measuring factor (i.e., D and H).

5. Conclusions

There was no significant effect of different eucalypts varieties on biomass except for stem and leaf biomass (Figure 1). Stand age significantly affected the allocation of biomass in eucalypts. With increasing stand age, the proportion of stem biomass to total tree biomass increased, the proportion of bark, branch, and leaf biomass to total tree biomass decreased, and the proportion of root biomass to total tree biomass first decreased and then increased (Figure 1).

This study confirms that organ biomass, aboveground biomass, total biomass, and D, H, DH, and D^2H were allometrically consistent and can be used to accurately estimate the biomass of eucalypts plantation (Figures 2 and 3, Table 2).

Stand age-specific allometric equations that include stand age as a complementary predictor variable, significantly improved the prediction effect and the accuracy of biomass assessment (Table 2). The optimal independent variables in the biomass prediction models differed according to each organ. Bark, branch, and leaf biomass were best predicted with H as the independent variable, stem biomass was best predicted with D^2H as the independent variable, and root, aboveground, and total biomass were best predicted with DH or D^2H (Table 2 and Figure 5). Therefore, stand age-specific allometric equations are recommended to improve the estimates of component biomass in eucalypts plantations.

Supplementary Materials: The following supporting information can be downloaded at https://www.mdpi.com/article/10.3390/f16020193/s1, Figure S1: Changes in component biomass ratios across stand age. Table S1: Analysis of variance of component biomass by stand age and species variety. Table S2: Stand age classes and species classes affects the component biomass.

Author Contributions: R.H.: Writing—original draft, methodology, formal analysis, investigation. Y.X.: Conceptualization, visualization, supervision, investigation. W.Z.: Data curation, visualization, investigation. A.D.: funding acquisition, conceptualization. Z.W.: Writing—review and editing, supervision, methodology, investigation. All authors have read and agreed to the published version of the manuscript.

Funding: This study was funded by the Guangxi Key Research and Development Projects (Guike AB23026010); China National Key R&D Program during the 14th Five-year Plan Period (2023YFD2201005); Forestry Ecological Monitoring Network Platform Construction Project (2024CG232); and the Operation Project for Guangdong Zhanjiang Eucalyptus Forest Ecosystem National Positioning Observation and Research Station (KS2024160017).

Data Availability Statement: Data are contained within the article.

Conflicts of Interest: The authors declare that they have no known competing financial interests or personal relationships that could have appeared to influence the work reported in this paper.

References

1. IPCC. *Climate Change 2021: The Physical Science Basis. Contribution of Working Group I to the Sixth Assessment Report of the Intergovernmental Panel on Climate Change*; Cambridge University Press: Cambridge, MA, USA, 2021.
2. Andrzej, M.J.; Marcin, K.D.; Kamil, G.; Paweł, H.; Agnieszka, C.; Sylwia, W.; Karol, M. How do Tree Stand Parameters Affect Young Scots Pine Biomass?—Allometric Equations and Biomass Conversion and Expansion Factors. *For. Ecol. Manag.* **2018**, *409*, 74–83. [CrossRef]
3. Sun, C. Exploration on the Development Status of Forest Resources in China. *For. Investig. Design* **2020**, *49*, 22–24+42. (In Chinese)

4. Zhang, X.; Zhang, X.L.; Han, H.; Shi, Z.J.; Yang, X.H. Biomass Accumulation and Carbon Sequestration in an Age-Sequence of Mongolian Pine Plantations in Horqin Sandy Land, China. *Forests* **2019**, *10*, 197. [CrossRef]
5. Sarmiento, G. The Ecology of Neotropical Savannas. Ph.D. Thesis, Harvard University Press, Cambridge, MA, USA, 1984. [CrossRef]
6. Du, X.L.; Car, S.F.; Luo, S.Z.; Liu, Y.H.; Pei, Y.; Huang, J.X.; Yang, Z.J.; Xiong, D.C. Effects of warming on biomass and allocation characteristic of forest tree: A review. *Chin. J. Ecol.* **2024**, *43*, 3180–3190. (In Chinese) [CrossRef]
7. Smith-Martin, C.M.; Xu, X.; Medvigy, D.; Schnitzer, S.A.; Powers, J.S. Allometric Scaling Laws Linking Biomass and Rooting Depth Vary across Ontogeny and Functional Groups in Tropical Dry Forest Lianas and Trees. *New Phytol.* **2020**, *226*, 714–726. [CrossRef]
8. Skarpaas, O.; Meineri, E.; Bargmann, T.; Pötsch, C.; Töpper, J.; Vandvik, V. Biomass Partitioning in Grassland Plants along Independent Gradients in Temperature and Precipitation. *Perspect. Plant Ecol.* **2016**, *19*, 1–11. [CrossRef]
9. Genet, A.; Wernsdörfer, H.; Jonard, M.; Pretzsch, H.; Rauch, M.; Ponette, Q.; Nys, C.; Legout, A.; Ranger, J.; Vallet, P.; et al. Ontogeny Party Explains the Apparent Heterogeneity of Published Biomass Equations for *Fagus sylvatica* in Central Europe. *For. Ecol. Manag.* **2011**, *261*, 1188–1202. [CrossRef]
10. Gargaglione, V.; Peri, P.L.; Rubio, G. Allometric Relations for Biomass Portioning of *Nothofagus antarctica* Trees of Different Crown Classes Over a Site Quality Gradient. *For. Ecol. Manag.* **2010**, *259*, 1118–1126. [CrossRef]
11. Poorter, H.; Jagodzinski, A.M.; Ruiz-Peinado, R.; Kuyah, S.; Luo, Y.; Oleksyn, J.; Usoltsev, V.A.; Buckley, T.N.; Reich, P.B.; Sack, L. How Does Biomass Distribution Change with Size and Differ among Species? An Analysis for 1200 Plant Species from Five Continents. *New Phytol.* **2015**, *208*, 736–749. [CrossRef] [PubMed]
12. Wang, Z.; Huang, X.; Li, F.; Chen, D.; Xu, X. Global Patterns of Allometric Model Parameters Prediction. *Sci. Rep.* **2023**, *13*, 1550. [CrossRef]
13. Zhang, X.; Wang, Y.; Wang, J.F.; Yu, M.Y.; Zhang, R.Z.; Mi, Y.L.; Xu, J.L.; Jiang, R.F.; Gao, J. Elevation Influences Belowground Biomass Proportion in Forests by Affecting Climatic Factors, Soil Nutrients and Key Leaf Traits. *Forests* **2024**, *13*, 674. [CrossRef] [PubMed]
14. Dai, L.C.; Ke, X.; Guo, X.W.; Du, Y.G.; Zhang, F.W.; Li, Y.K.; Li, Q.; Lin, L.; Peng, C.J.; Shu, K.; et al. Responses of Biomass Allocation across Two Vegetation Types to Climate Fluctuations in the Northern Qinghai–Tibet Plateau. *Ecol. Evol.* **2019**, *9*, 6105–6115. [CrossRef] [PubMed]
15. Zhao, H.B.; Zhou, G.Y.; Li, Y.J.; Qiu, Z.J.; Wu, Z.M.; Wang, X. Biomass Allocation and Allometric Growth Models of Four Common Tree Species in Southern Subtropical Evergreen Broad-Leaved Forest. *Sci. Silvae Sin.* **2022**, *58*, 23–31. (In Chinese)
16. Tompalski, P.; Wulder, M.A.; White, J.C.; Hermosilla, T.; Riofrío, J.E.; Kurz, W.A. Developing Aboveground Biomass Yield Curves for Dominant Boreal Tree Species from Time Series Remote Sensing Data. *For. Ecol. Manag.* **2024**, *561*, 121894. [CrossRef]
17. Peichl, M.; Leava, N.A.; Kiely, G. Above- and Belowground Ecosystem Biomass, Carbon and Nitrogen Allocation in Recently Afforested Grassland and Adjacent Intensively Managed Grassland. *Plant Soil.* **2012**, *350*, 281–296. [CrossRef]
18. Preece, N.D.; Crowley, G.M.; Lawes, M.J.; Oosterzee, P.V. Comparing Above-Ground Biomass among Forest Types in the Wet Tropics: Small Stems and Plantation Types Matter in Carbon Accounting. *For. Ecol. Manag.* **2012**, *264*, 228–237. [CrossRef]
19. Xiang, W.; Li, L.; Ouyang, S.; Xiao, W.; Zeng, L.; Chen, L.; Lei, P.; Deng, X.; Zeng, Y.; Fang, J.; et al. Effect of Stand Age on Tree Biomass Partitioning and Allometric Equations in Chinese Fir (*Cunninghamia lanceolata*) Plantations. *Eur. J. For. Res.* **2021**, *140*, 317–332. [CrossRef]
20. Liu, B.; Bu, W.; Zang, R. Improved Allometric Models to Estimate the Aboveground Biomass of Younger Secondary Tropical Forests. *Glob. Ecol. Conserv.* **2023**, *41*, e02359. [CrossRef]
21. Chave, J.; Réjou-Méchain, M.; Búrquez, A.; Chidumayo, E.; Colgan, M.S.; Delitti, W.B.C.; Duque, A.; Eid, T.; Fearnside, P.M.; Goodman, R.C.; et al. Improved Allometric Models to Estimate the Aboveground Biomass of Tropical Trees. *Glob. Change Biol.* **2014**, *20*, 3177–3190. [CrossRef] [PubMed]
22. Mensah, S.; Kakaï, R.L.G.; Seifert, T. Patterns of Biomass Allocation between Foliage and Woody Structure: The Effects of Tree Size and Specific Functional Traits. *Ann. For. Res.* **2016**, *59*, 49–60. [CrossRef]
23. Waring, B.G.; Powers, J.S. Overlooking What is Underground: Root:Shoot Ratios and Coarse Root Allometric Equations for Tropical Forests. *For. Ecol. Manag.* **2017**, *385*, 10–15. [CrossRef]
24. McNicol, I.M.; Berry, N.; Bruun, T.B.; Hergoualc'h, K.; Mertz, O.; Neergaard, A.d.; Ryan, C.M. Development of Allometric Models for Above and Belowground Biomass in Swidden Cultivation Fallows of Northern Laos. *For. Ecol. Manag.* **2015**, *357*, 104–116. [CrossRef]
25. Peng, S.; He, N.; Yu, G.; Wang, Q. Aboveground Biomass Estimation at Different Scales for Subtropical Forests in China. *Bot. Stud.* **2017**, *58*, 45. [CrossRef]
26. Yang, B.; Xue, W.; Yu, S.; Zhou, J.; Zhang, W. Effects of Stand Age on Biomass Allocation and Allometry of *Quercus Acutissima* in the Central Loess Plateau of China. *Forests* **2019**, *10*, 41. [CrossRef]

27. Zhou, H.; Yeerjiang, B.; He, H.J.; Zhang, C.Y.; Zhao, X.H.; Han, M.H. Biomass Distribution Characteristics and Species-Specific Allometric Equations for Afforestation Species in Northeast China. *Sci. Silvae Sin.* **2023**, *59*, 23–32. (In Chinese)
28. Peichl, M.; Arain, M.A. Above- and Belowground Ecosystem Biomass and Carbon Pools in an Age-Sequence of *Temperate Pine* Plantation Forests. *Agric. For. Meteorol.* **2006**, *140*, 51–63. [CrossRef]
29. Zhu, k.; Zhang, J.; Niu, S.L.; Zhu, C.J.; Luo, Y.Q. Limits to Growth of Forest Biomass Carbon Sink under Climate Change. *Nat. Commun.* **2018**, *9*, 2709. [CrossRef] [PubMed]
30. Besnard, S.; Carvalhais, N.; Arain, M.A.; Black, A.; de Bruin, S.; Buchmann, N.; Cescatti, A.; Chen, J.; Clevers, J.G.P.W.; Desai, A.R.; et al. Quantifying the Effect of Forest Age in Annual Net Forest Carbon Balance. *Environ. Res. Lett.* **2018**, *13*, 124018. [CrossRef]
31. Tian, H.L.; Zhu, J.H.; Lei, X.D.; Jian, Z.J.; Chen, X.Y.; Zeng, L.X.; Huang, G.S.; Liu, C.F.; Xiao, W.F. Models Considering the Theoretical Stand Age Will Underestimate the Future Forest Carbon Sequestration Potential. *For. Ecol. Manag.* **2024**, *562*, 121982. [CrossRef]
32. Peichl, M.; Arain, M.A. Allometry and Partitioning of Above– and Belowground Tree Biomass in an Age-Sequenee of White Pine Forests. *For. Ecol. Manag.* **2007**, *253*, 68–80. [CrossRef]
33. Chen, D.S.; Sun, X.M.; Jin, Y.B.; Zhang, Y.; Zhang, S.G. Effects of Stand Age and Competition on Allometric Relationships for Biomass Partitioning in *Larixkaempferi* Plantation. *Acta Ecol. Sinica.* **2020**, *40*, 843–853. (In Chinese) [CrossRef]
34. Repo, A.; Rajala, T.; Henttonen, H.M.; Lehtonen, A.; Peltoniemi, M.; Heikkinen, J. Age-Dependence of Stand Biomass in Managed Boreal Forests Based on the Finnish National Forest Inventory Data. *For. Ecol. Manag.* **2021**, *498*, 119507. [CrossRef]
35. Yao, Y.T.; Piao, S.L.; Wang, T. Future Biomass Carbon Sequestration Capacity of Chinese Forests. *Sci. Bull.* **2018**, *63*, 1108–1117. [CrossRef] [PubMed]
36. Dutcă, I.; Mather, R.; Blujdea, V.N.B.; Ioraș, F.; Olari, M.; Abrudan, I.V. Site-Effects on Biomass Allometric Models for Early Growth Plantations of Norway Spruce (*Picea abies* (L.) Karst.). *Biomass Bioenerg.* **2018**, *116*, 8–17. [CrossRef]
37. Ouyang, L.; Zhao, P.; Zhou, G.; Zhu, L.; Huang, Y.; Zhao, X.; Ni, G. Stand–Scale Transpiration of a *Eucalyptus Urophylla* × *Eucalyptus Grandis* Plantation and Its Potential Hydrological Implication. *Ecohydrology* **2018**, *11*, e1938. [CrossRef]
38. Magalhães, T.M.; Cossa, V.N.; Guedes, B.S.; Saraiva, A.; Fanheiro, M. Species-Specific Biomass Allometric Models and Expansion Factors for Indigenous and Planted Forests of the Mozambique Highlands. *J. For. Res.* **2021**, *32*, 1047–1065. [CrossRef]
39. Zheng, X.M.; Weng, X.; Ou, L.L.; Ren, Y. Allometric Equation and Biomass Estimation of Eucalyptus in Fujian. *J. Univ. Chin. Acad. Sci.* **2024**, *41*, 321–333. (In Chinese) [CrossRef]
40. Poorter, H.; Niklas, K.J.; Reich, P.B.; Oleksyn, J.; Poot, P.; Mommer, L. Biomass Allocation to Leaves, Stemsand Roots: Meta-Analyses of Interspecific Variation and Environmental Control. *New Phytol.* **2012**, *193*, 30–50. [CrossRef] [PubMed]
41. Xie, X.; Hu, Y.; Pan, X.; Liu, F.; Song, Y.; Dong, M. Biomass Allocation of Stoloniferous and Rhizomatous Plant in Response Toresource Availability: A Phylogenetic Meta-Analysis. *Front. Plant Sci.* **2016**, *7*, 603. [CrossRef] [PubMed]
42. Sun, Y.; He, H.J.; Li, L.; Song, C.M.; Wang, F.J.; Xia, F.C. Biomass Allocation and Biomass Allometric Models of Six Early-Summer Herbs under the Canopy of Broad-Leaved Korean Pine Forest during Different Growth Periods in Jiaohe, Jilin Province. *Acta Ecol. Sin.* **2017**, *37*, 6523–6533. (In Chinese) [CrossRef]
43. Luo, Y.J. Biomass and Allocation of China Forest Ecosystems. Ph.D. Thesis, Research Center for Eco-Environmental Sciences Chinese Academy of Sciences, Beijing, China, 2012. (In Chinese).
44. Zhang, L.L.; Wang, Z.C.; Chen, S.X.; Zhu, W.K.; Du, A.P. Biomass Allocation Pattern of *Eucalyptus urophylla* × *Eucalyptus grandis* Plantation at Different Ages. *J. Northwest A&F Univ. (Nat. Sci. Ed.)* **2017**, *45*, 61–68. (In Chinese) [CrossRef]
45. Fu, W.B.; Peng, W.X.; Song, T.Q.; Zeng, F.P.; Du, H.; Wen, Y.G.; Xu, H.F. Biomass and Its Allocation Characteristics of *Eucalyptus Urophylla* × *E. Grandis* Plantations at Different Stand Ages. *Acta Ecol. Sin.* **2014**, *34*, 5234–5241. (In Chinese) [CrossRef]
46. Fatemi, F.R.; Yanai, R.D.; Hamburg, S.P.; Vadeboncoeur, M.A.; Arthur, M.A.; Briggs, R.D.; Levine, C.R. Allometric Equations for Young Northern Hardwoods: The Importance of Age–Specific Equations for Estimating Aboveground Biomass. *Can. J. Forest Res.* **2011**, *41*, 881–891. [CrossRef]
47. Han, C.; Song, M.; Du, H.; Zeng, F.P.; Peng, W.X.; Wang, H.; Chen, L.; Su, L. Biomass and Carbon Storage in Roots of *Cunninghamia lanceeolata* and *Pinus massoniana* Plantations at Different Stand Ages in Guangxi. *Acta Ecol. Sin.* **2017**, *37*, 2282–2289. (In Chinese) [CrossRef]
48. Lim, H.; Lee, K.H.; Lee, K.H.; Park, I.H. Biomass Expansion Factors and Allometric Equations in an Age Sequence for Japanese Cedar (*Cryptomeria japonica*) in Southern Korea. *J. For. Res.* **2013**, *18*, 316–322. [CrossRef]
49. Brown, S. Measuring Carbon in Forests: Current Status and Future Challenges. *Environ. Pollut.* **2002**, *116*, 363–372. [CrossRef] [PubMed]
50. Wang, J.S.; Zhang, C.Y.; Fan, X.H.; Zhao, Y.Z. Biomass Allocation Patterns and Allometric Models of *Abies nephrolepis* Maxim. *Acta Ecol. Sin.* **2011**, *31*, 3918–3927. (In Chinese)
51. Luo, Y.; Wang, X.; Zhang, X.; Booth, T.H.; Lu, F. Root:Shoot Ratios Across China's Forests: Forest Type and Climatic Effects. *For. Ecol. Manag.* **2012**, *269*, 19–25. [CrossRef]

52. Zhou, X.; Zhang, Y.; Niklas, K.J. Sensitivity of Growth and Biomass Allocation Patterns to Increasing Nitrogen: A Comparison Between Ephemerals and Annuals in the Gurbantunggut Desert, North-Western China. *Ann. Bot.* **2014**, *113*, 501–511. [CrossRef] [PubMed]
53. Hong, Y.F.; Chen, D.S.; Shen, J.P.; Sun, X.M.; Zhang, S.G. Compatible Biomass Models for *Larix Olgensis* Plantation Based on Tree-Level and Stand−Level. *For. Res.* **2019**, *32*, 33–40. (In Chinese) [CrossRef]
54. King, J.S.; Giardina, C.P.; Pregitzer, K.S.; Friend, A.L. Biomass Partitioning in Red Pine (*Pinus resinosa*) along a Chronosequence in the Opper Peninsula of Michigan. *Can. J. For. Res.* **2007**, *37*, 93–102. [CrossRef]
55. Chan, N.; Takeda, S.; Suzuki, R.; Yamamoto, S. Establishment of Allometric Models and Estimation of Biomass Recovery of Swidden Cultivation Fallows in Mixed Deciduous Forests of the Bago Mountains, Myanmar. *For. Ecol. Manag.* **2013**, *304*, 427–436. [CrossRef]
56. Shen, Y.Z.; Sun, X.M.; Zhang, J.T.; Du, Y.C.; Ma, J.W. Study on the Individual Tree Biomass of *Larix kaempferi* Plantation in Xiaolong Mountain, Gansu Province. *For. Res.* **2011**, *24*, 517–522. (In Chinese) [CrossRef]
57. Xiang, W.; Zhou, J.; Ouyang, S.; Zhang, S.; Lei, P.; Li, J.; Deng, X.; Fang, X.; Forrester, D.I. Species-Specific and General Allometric Equations for Estimating Tree Biomass Components of Subtropical Forests in Southern China. *Eur. J. For. Res.* **2016**, *135*, 963–979. [CrossRef]
58. Larreta, B.V.; López-Sánchez, C.A.; Corral-Rivas, J.J.; López-Martínez, J.O.; Aguirre-Calderón, C.G.; Álvarez-González, J.G. Allometric Equations for Estimating Biomass and Carbon Stocks in the Temperate Forests of North-Western Mexico. *Forests* **2017**, *8*, 269. [CrossRef]
59. Rojas-García, F.; Jong, B.H.J.D.; Martínez-Zurimendí, P.; Paz-Pellat, F. Database of 478 Allometric Equations to Estimate Biomass for Mexican Trees and Forests. *Ann. For. Sci.* **2015**, *72*, 835–864. [CrossRef]

forests

MDPI

Article

Quantifying the Effects of Carbon Growth Grade and Structural Diversity on Carbon Sinks of Natural Coniferous–Broadleaved Mixed Forests Across the Jilin Province of China

Xiao He [1], Hong Guo [1,*], Xiangdong Lei [1], Wenqiang Gao [1] and Yutang Li [2]

[1] State Key Laboratory of Efficient Production of Forest Resources, Key Laboratory of Forest Management and Growth Modelling, National Forestry and Grassland Administration, Institute of Forest Resource Information Techniques, Chinese Academy of Forestry, Beijing 100091, China; hexiao@ifrit.ac.cn (X.H.); xdlei@ifrit.ac.cn (X.L.); gwq9975@ifrit.ac.cn (W.G.)

[2] Jilin Forestry Inventory and Planning Institute, Changchun 130022, China; yutanglee@163.com

* Correspondence: hongguo@ifrit.ac.cn

Abstract: Natural mixed forests' carbon sequestration capacity is crucial for mitigating climate change and maintaining ecological balance. However, most of the current studies only consider the role of forest age, ignoring the influence of carbon growth grade and stand structural diversity, which leads to an increase in uncertainty in large-scale forest carbon sink assessment. The aim of this study was to quantify the effects of carbon growth grade and stand structure diversity on the carbon sink of natural mixed forests and to establish a more accurate stand carbon growth model. Based on sample data from the National Forest Inventory (NFI) of China, the stand carbon growth model was established based on Gompertz and Logistic theoretical growth models, and the forest carbon sink at the regional scale was predicted. It was found that the stand carbon growth model considering only the stand age as a single variable often had poor results, with R^2 less than 0.36, while R^2 values of the optimal model introducing carbon growth grade and stand structural diversity were 0.87 and 0.48, respectively, which significantly improved the prediction accuracy of the model, and both had significant effects on stand carbon stocks. By predicting the future forest carbon sink, it was found that the forest carbon sink of the natural coniferous–broadleaved mixed forests in Jilin Province would reach 791 (781–801) t c/a and 843 (833–852) t c/a in 2030 and 2060, respectively, which were 17% lower and 51% higher than that of the forest carbon sink estimated by considering only the age. Moreover, the model considering structural diversity predicted a more positive carbon sink trend, indicating that forest carbon stocks could be more effectively maintained and carbon sinks increased by increasing the complexity of stand diameter at breast height structure, which has important guiding significance for future forest carbon sink management. This study provides scientific support for achieving the goal of "carbon neutrality" proposed by China.

Keywords: carbon sink; structural diversity; growth grade classification; mixed forests

1. Introduction

Climate change is one of the most serious challenges facing the world today. Through its biomass and carbon storage in soil, forests effectively absorb CO_2 in the atmosphere, thus playing a vital role in the process of restraining global warming [1]. The United Nations Framework Convention on Climate Change (UNFCCC) emphasizes the necessity of annual monitoring of forest carbon sinks. In response to global climate change, China has set a goal to "strive to reach a peak in CO_2 emissions before 2030 and strive to achieve carbon neutrality before 2060". Forests' carbon sequestration capacity is crucial for mitigating climate change and maintaining ecological balance. Mixed forests, due to their complex structure and biodiversity, play a key role in providing ecosystem services, especially in carbon storage. Accounting for forest carbon sinks requires a carbon storage growth model with high accuracy and low uncertainty. Different levels of biomass estimation models have a significant impact on local, regional, and even global biomass estimation [2]. On a large scale, a direct and effective method is to establish the relationship between forest age and carbon sinks on the same spatial regional scale. Compared to monoculture forests with a single tree species, natural and mixed forests exhibit stronger environmental adaptability and resistance to disturbances, as well as higher biomass [3,4] and greater stability of carbon capture [5], and natural forests' regrowth has great potential for climate mitigation [6]. Therefore, accurately assessing the carbon storage and growth of mixed forests is essential for understanding and enhancing their role in the global carbon cycle. However, the accuracy of mixed forest models is generally lower than that of pure forests because the diversity of stand structure is less considered in the model, which causes challenges to the simulation of forest carbon sinks.

Using plot survey data from the National Forest Inventory (NFI) to establish forest growth models is a reliable method for predicting forest carbon sinks in a region [7]. Because stand age is the main determinant of forest biomass carbon storage [8], theoretical growth equations such as Gompertz, Logistic, and Richards are commonly employed to establish forest growth models [9–13]. These models have different functional forms, but all have an asymptotic limit, indicating the ultimate limit of forest growth, and are often used to simulate the distribution of stand parameters (such as diameter at breast height, basal area (*BA*), and accumulation, etc.) over time. The biomass growth models for various dominant tree species were generally well fitted. However, the best carbon storage growth model might vary for different regions and stand types [9,10]. Although the growth model is one of the most directly and widely used methods for estimating forest carbon storage, considering only age as a single variable in stand carbon growth models and neglecting key factors such as carbon growth grades and structural diversity can limit the prediction of carbon sink potential. These key factors were especially important for accurately estimating forest carbon sinks in natural coniferous–broadleaved forests.

Carbon growth grades are an important indicator describing the relationship between stand carbon and age, reflecting the potential differences in the forest's ability to capture CO_2. The carbon growth grade is determined by categorizing the stand carbon growth process based on the relationship between stand carbon and age, usually represented by an ordered categorical variable, and stand carbon storage generally increases with the increase in carbon fixation level [14]. However, due to the influence of environmental factors such as climate, soil, and topography, there can be significant differences in carbon storage of the same stand type at the same age [12]. Relying solely on age as a predictor makes it challenging to achieve accurate forest carbon prediction results. Although the elements such as altitude, temperature, precipitation, soil pH, and bulk density of the fine earth fraction can be directly added into the carbon growth model to improve the estimation accuracy of the models [15,16], it is very difficult to apply growth models under different conditions to forests on a regional scale due to high-dimension predictors. Therefore, it is convenient to describe variations in carbon at the same age by dividing the complex, comprehensive factors affecting carbon growth into carbon growth grades; however, carbon growth grades are rarely considered in existing models.

One of the important characteristics of mixed forests is structural diversity, which is influenced by different tree species and size combinations. Structural diversity refers to the diversity of spatial and morphological characteristics of trees within the forests, an emerging indicator usually measured by DBH size differentiation based on the Shannon index, Gini coefficient, etc. [17–20]. The higher the structural diversity, the greater the difference in tree sizes within the forest. Compared to species diversity, structural diversity is more closely related to forest productivity [19,21]. Forests with complex structures can usually make effective use of resources and show better adaptability. Therefore, structural diversity is one of the important factors affecting forest carbon sinks [22]. However, previous studies mostly used linear models to detect the relationship between the two and rarely included structural diversity in the carbon storage growth model. At the same time, there are many indicators for measuring structural diversity, and it is not clear which indicator has the best predictive effect, leading to increasing uncertainty in forest carbon sink prediction results [23].

Accurately predicting the future changes in carbon sinks is essential for effective forest carbon sink management. Previous studies have primarily focused on age; however, two important factors, carbon growth grades and structural diversity, were not considered for mixed forests; hence, it has been challenging to quantify the impact of these factors on forest carbon sink estimation, and there is a large uncertainty in the prediction results.

Jilin Province is a key state-owned forest region and natural forest area in China, with natural forests accounting for a large proportion (~78%) of the area. Natural coniferous–broadleaved mixed forests are one of the most important forest types that play a very important role in safeguarding national ecological security and timber supply. Natural coniferous–broadleaved mixed forests are vital for Jilin Province, China, and play an important role in providing ecological services, especially carbon capture. Consequently, this study took natural coniferous and broadleaved mixed forests in Jilin Province, China, as the research object. We aimed to develop stand carbon growth models with carbon growth grade and structural diversity, which improve and optimize the stand carbon growth model structure, to increase the accuracy of the forest carbon sink and to explore the impact of carbon growth grade and structural diversity. The specific objectives were to (1) develop stand carbon growth models based on theoretical growth equation; (2) compare the results of carbon sequestration classification of different growth models; (3) select structural diversity variables that influence the carbon storage of stands; and (4) quantify the varying contributions of carbon growth grade and structural diversity to forest carbon sinks. Through this research, we hoped to better understand how carbon growth grades and structural diversity affect the carbon storage capacity of forests and provide scientific support for future carbon sink forecasts and achievement of China's proposed "carbon neutrality" goals.

2. Materials and Methods

2.1. Study Area

The study area is situated in Jilin Province, located in Northeast China, spanning the coordinates from 121°38′ E to 131°19′ E longitude and from 40°50′ N to 46°19′ N latitude. This region experiences a continental monsoon climate characterized by a mean air temperature of approximately 2.43 °C and a mean annual precipitation of around 199.37 mm [24]. The predominant soil types include dark brown soil, mollisols, chernozems, albic soils, and meadow soils. The total area is 18.74 million hectares, with a forest area of 7.8487 million hectares and a forest coverage rate of 41.49% in 2014. Natural coniferous–broadleaved mixed forests are one of the main forest types.

2.2. Sample Plot Data and Stand Carbon Calculation

Natural coniferous–broadleaved mixed forest permanent sample plot data were obtained from the 5th~9th (1994, 1999, 2004, 2009, 2014) NFI in Jilin Province of China. The permanent sample plots were laid over a grid of 4 km × 8 km, and the area of a sample plot was 0.06 ha. Data for each plot were collected over a survey period of 5 years and contained the plot information such as forest origin, stand average age, average tree height, and so on [25,26]. Living trees with a *dbh* ≥ 5 cm were measured and used to deduce the stand quadratic mean diameter (*Dg*), basal area (*BA*), and stem density (*N*) within each plot. Stand average age was obtained by measuring 3 to 5 medium trees by the tree ring in the sample plot. Dominant tree species of natural coniferous–broadleaved mixed forests are listed in Table 1. Plots with cutting intensities of less than 5% were used for analysis, with a total of 404 sample plots and 1 to 5 repeated observations. A total of 1013 measurements were used for this study (Figure 1).

Table 1. Dominant tree species in natural coniferous–broadleaved mixed forests in Jinlin Province.

Tree Types	Tree Species
Coniferous	*Pinus koraiensis* Siebold & Zucc., *Picea asperata* Mast., *Abies fabri* (Mast.) Craib, *Larix olgensis* A. Henry, *Pinus densiflora* Siebold & Zucc.
Broad-leaved	*Quercus mongolica* Fisch. ex Ledeb., *Fraxinus mandshurica* Rupr., *Phellodendron amurense* Rupr., *Juglans mandshurica* Maxim., *Betula* L., *Tilia amurensis* Rupr., *Acer pictum* subsp. mono (Maxim.) Ohashi, *Acer tegmentosum* Maxim., *Ulmus pumila* L., *Populus* L.

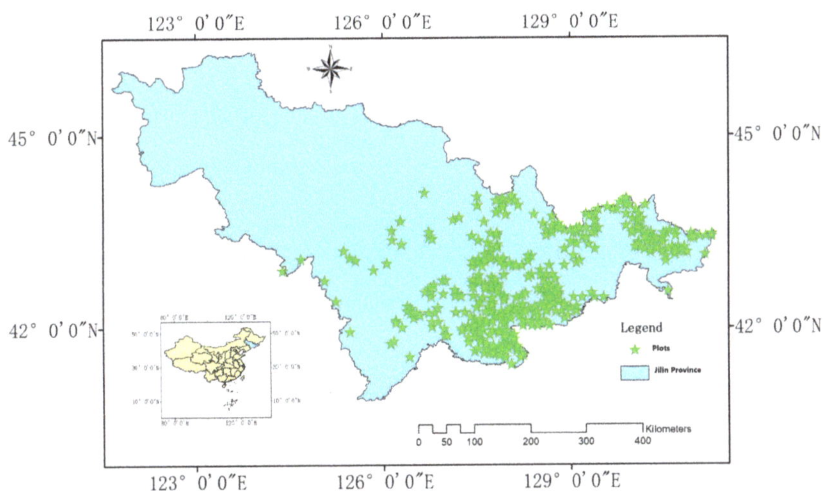

Figure 1. The study area and distribution of sample plots of natural coniferous–broadleaved mixed forests across Jilin Province of China.

The allometric models for tree-level species and site-specific models came from reference [27], and species-level carbon factors were utilized to estimate stand-level total carbon (sum of aboveground and root). Table 2 shows the basic descriptive statistics of stand factors of natural coniferous–broadleaved mixed forests in the present study.

Table 2. Basic descriptive statistics of natural coniferous–broadleaved mixed forest plots (n = 1013).

Variables	Mean	Min.	Max.	S.D.
N (trees/ha)	1033	333	3917	523
Dg (cm)	16.9	5.7	33.9	5.1
BA (m^2/ha)	23.30	1.31	62.81	10.20
Age (a)	79	4	202	42
Carbon (t/ha)	53.01	1.12	201.24	32.77
Mean air temperature (°C)	2.43	−1.40	6.10	1.15
Mean annual precipitation (mm)	199.37	110.11	333.39	30.56

2.3. Stand Structural Diversity

Stand structural diversity was quantified by the Shannon–Weaver, Simpson, and GiniSimpson index in terms of DBH class abundance (Table 3). The DBH width class is equal to 2 cm or 4 cm in this study and is mainly based on the actual needs of forestry management and the customary consideration of data processing [17].

Table 3. Summary of formulas for stand structural diversity.

Variables	Formula
N_Shannon_Size2_Div and *N_Shannon_Size4_Div*	$N\_Shannon\_Size\_Div = -\sum_{i=1}^{d_C} \frac{n_i}{n} \ln(\frac{n_i}{n})$
N_Simpson_Size2_Div and *N_Simpson_Size4_Div*	$N\_Simpson\_Size\_Div = \frac{1}{\sum_{i=1}^{d_C} (\frac{n_i}{n})^2}$
N_GiniSimpson_Size2_Div and *N_GiniSimpson_Size4_Div*	$N\_GiniSimpson\_Size\_Div = 1 - \sum_{i=1}^{d_C} (\frac{n_i}{n})^2$
BA_Shannon_Size2_Div and *BA_Shannon_Size4_Div*	$BA\_Shannon\_Size\_Div = -\sum_{i=1}^{d_C} \frac{BA_i}{BA} \ln(\frac{BA_i}{BA})$
BA_Simpson_Size2_Div and *BA_Simpson_Size4_Div*	$BA\_Simpson\_Size\_Div = \frac{1}{\sum_{i=1}^{d_C} (\frac{BA_i}{BA})^2}$
BA_GiniSimpson_Size2_Div and *BA_GiniSimpson_Size4_Div*	$BA\_GiniSimpson\_Size\_Div = 1 - \sum_{i=1}^{d_C} (\frac{BA_i}{BA})^2$

Note: d_C represents the number of DBH classes in the plot; n_i denotes the stem number of the *i*-th DBH class in the plot; *n* is the stem number in the plot; BA_i is the stand basal area of the *i*-th DBH class in the plot; *BA* is the stand basal area in the plot. "2" and "4", which represent DBH width class, are equal to 2 cm and 4 cm, respectively.

2.4. Development of a Carbon Growth Model with Grade or Stand Structural Diversity

Two models, the Logistic model and the Gompertz model, were used to fit the stand carbon growth equations (Equations (1) and (2)).

$$Carbon = \frac{a}{(1 + c \cdot e^{-r \cdot Age})} \tag{1}$$

$$Carbon = a \cdot e^{-c \cdot e^{-r \cdot Age}} \tag{2}$$

where *Carbon* represents stand carbon; *Age* represents stand average age; and *a*, *r*, and *c* are the parameters of the stand carbon growth model.

Due to stand carbon growth being different at the same age, we utilized the parameter stand carbon growth grade (*k*) to describe stand carbon growth trends; the *k* of optimal grade value was defined as "1" (Equation (3)) and was achieved using equal model parameter gap spacing. For example, there is a parameter β that is transformed into a set of arithmetic sequence parameters (β_1~β_k) with a tolerance of *d* after considering *k* (Equation (4)) [12,14]. The *k* value of each plot is also obtained through iteration, i.e., the *k* value between two iterations no longer changes. The *k* value was primarily determined based on the model performances, stand characteristics (bscheffe.test), the number of plots, and convenience of application.

$$k = \begin{cases} 1,\ Carbon \geq f(Age|\beta_1) \\ L_i,\ f(\text{Age}|\beta_{i+1}) \leq Carbon < f(Age|\beta_i),\ i = 1,2,3,\ldots,k \\ k,\ Carbon \leq f(Age|\beta_k) \end{cases}$$
$$L_i = i + \text{int}\left[\frac{f(Age|\beta_i)}{f(Age|\beta_i) - f(Age|\beta_{i+1})}\right], i = 1,2,3,\ldots,k \tag{3}$$

$$\beta \Rightarrow \beta + \sum_{k=1}^{1} (k-1)\cdot d\cdot L_k \tag{4}$$

where f represents the theoretical growth model; $\beta_1 \sim \beta_k$ are the parameters of the i-th level stand carbon growth model, and any combination of parameters in Equations (1) and (2) can be selected, and parameter $\boldsymbol{a}$ is usually preferred; int is an integer function; d is the difference between the adjacent stand carbon growth grade parameter estimates; L_k is a dummy variable of stand carbon growth grade consisting of 0 and 1; and the other variables are the same as described above.

Because of the different model selections, the carbon sequestration grade of each plot will be different under the condition of a given classification number k. The Kappa value was used to compare the differences between the Logistic model and the Gompertz model (Equation (5)).

$$\text{Kappa} = \frac{P_1 - P_2}{1 - P_2}$$
$$\text{where } P_1 = \frac{\sum_i^k \mathbf{I}(n_i^L = n_i^C)}{n} \text{ and } P_2 = \frac{\sum_i^k n_i^L \cdot n_i^C}{n^2}. \tag{5}$$

where n_i^L and n_i^C are the sample numbers of the Logistic grading model and the Gompertz grading model at the i-th carbon growth grade, respectively; $\mathbf{I}()$ is the indicator function, and its value is 1 if the conditions in brackets are met, otherwise it is 0; n is the number of samples for training data.

Due to the complex structure of mixed forests, we introduced the stand structural diversity in the stand carbon growth model. There were 12 candidate variables of stand structural diversity. In order to select the most suitable predictive variable of stand structural diversity for carbon storage estimation, we used a comprehensive feature selection method for multi-variable input, which consists of 8 widely used methods, namely Pearson correlation analysis, principal component analysis (PCA, including the principal component regression judgment method (PCAreg) and the sum of squared loadings (PCALss)), stepwise regression, redundancy analysis (RDA), generalized additive model (GAM), random forest (RF), and boosted regression tree (BRT). This method has been well applied in biomass prediction [28]. The screened variables were used to develop the carbon storage growth model with stand structural diversity by the reparameterization method, which includes both multiplicative and additive forms. For example, the stand carbon growth models with multiplicative and additive parameters $\boldsymbol{c}$ in the Logistic model were as follows (Equations (6) and (7)). All models developing work were done in R statistical software (v4.2.2).

$$Carbon = \frac{a}{\left(1 + (c_1 \cdot SS^{c_2}) \cdot e^{-r \cdot Age}\right)} \tag{6}$$

$$Carbon = \frac{a}{\left(1 + (c_1 + c_2 \cdot SS) \cdot e^{-r \cdot Age}\right)} \tag{7}$$

where SS is the stand structural diversity variable; other variables are the same as described above.

2.5. Model Evaluation

The goodness-of-fit statistics for evaluating the performance of the stand carbon growth models were the coefficient of determination (R^2) and root mean square error ($RMSE$), calculated according to Equations (8) and (9).

$$R^2 = 1 - \frac{\sum\limits_{i=1}^{n} (C_i - \hat{C}_i)^2}{\sum\limits_{i=1}^{n} (C_i - \overline{C})^2} \quad (8)$$

$$RMSE = \sqrt{\frac{\sum\limits_{i=1}^{n} (C_i - \hat{C}_i)^2}{n}} \quad (9)$$

where C_i and $\hat{C}_i$ represent the observed and predicted stand carbon values, respectively; $\overline{C}$ is the observed mean stand carbon; and n is the number of samples for training data.

2.6. Modified Carbon Sink Estimates

The carbon sink per unit area can be determined by deriving the carbon growth model. Considering the stand carbon growth grade and structural diversity, and assuming that the future forest area will remain unchanged, Equation (10) can be used to estimate the dynamic change in the regional carbon sink of natural arbor forests in Jilin Province.

$$CST_q = \frac{\sum_{i=1}^{N} f'(Age_q) \cdot A}{N} \cdot S \quad (10)$$

where CST is the regional carbon sink; f' is the derivative function of the carbon growth model; Age_q is stand average age at a time node q; A is the sample plot area; N is the number of sample plots in 2014 (N = 265); and S is the expansion coefficient, which is 43,200 ($4 \times 8 \times 100 \times 0.9 \times 15$) in Jilin province.

3. Results

3.1. Carbon Growth Models' Development

The goodness-of-fit values of carbon growth models were very bad when only considering stand age, with the R^2 of the basic stand carbon growth models being only between 0.31 and 0.36 (Table 4).

Table 4. The parameters and goodness-of-fit values (R^2 and RMSE) of the stand carbon growth model with stand age only.

Model	Parameter			R^2	RMSE (t/ha)
	a	*c*	*r*		
Logistic	76.666	2.451	0.0275	0.31495	27.126
Gompertz	76.666	2.451	0.0274	0.35675	26.285

3.2. Carbon Growth Grade Classification and Models' Development

Using Logistic and Gompertz equations to obtain carbon growth grades, the R^2 values of the stand carbon storage growth models were all above 0.8, and the RMSEs were less than 14 t/ha. There was little difference in the evaluation indices of the two growth models (Table 5). However, models with varying numbers of carbon growth grades had different performances; as the number of carbon growth grades increased, the model performance became better (Table 5). R^2 had the largest rate of increase when the number of carbon growth grades was 5 (6.91% for the Logistic equation and 6.81% for the Gompertz equation). At the same time, the Kappa coefficients of the two types of growth models were above 0.97, and if the number of carbon growth grade classifications was 3, the Kappa coefficients of both types of growth models reached 0.99 (Table 6), indicating that both types of growth models are highly consistent with the carbon growth grade.

Table 5. The goodness-of-fit values (R^2 and RMSE) for three different models among four grade classifications.

Model	No. of Classifications			
	3	5	7	9
Logistic	R^2 = 0.81737, RMSE = 13.999 (t/ha)	R^2 = 0.87383, RMSE = 11.635 (t/ha)	R^2 = 0.89182, RMSE = 10.774 (t/ha)	R^2 = 0.89998, RMSE = 10.360 (t/ha)
Gompertz	R^2 = 0.81833, RMSE = 13.962 (t/ha)	R^2 = 0.87405, RMSE = 11.625 (t/ha)	R^2 = 0.89251, RMSE = 10.740 (t/ha)	R^2 = 0.90055, RMSE = 10.330 (t/ha)

Table 6. The Kappa values of carbon growth grade classifications for two different models.

Kappa		Logistic			
		3	5	7	9
Gompertz	3	0.9922	-	-	-
	5	-	0.9739	-	-
	7	-	-	0.9820	-
	9	-	-	-	0.9731

According to the multiple comparisons of the different grades, the mean values of stand carbon and productivity had significant differences when the grades were separately divided into levels 5 (Table 7) and 3 (not listed). And the greater the mean value of stand carbon and productivity, the higher the grade level. However, when the grades were divided into levels 7 and 9, the results of multiple comparisons were not significant. At the same time, considering the convenience of the application of the model, the number of carbon growth grade classifications was finally chosen to be 5; the model expression with carbon growth grade and its simulation diagram are shown in Equation (11) and Figure 2, respectively. The model containing carbon growth grade can describe the growth process of stand carbon well.

$$Carbon = \begin{cases} 242.84 \cdot e^{-1.887 \cdot e^{-0.009572 \cdot Age}}, & Grade\ 1 \\ 195.17 \cdot e^{-1.887 \cdot e^{-0.009572 \cdot Age}}, & Grade\ 2 \\ 147.49 \cdot e^{-1.887 \cdot e^{-0.009572 \cdot Age}}, & Grade\ 3 \\ 99.82 \cdot e^{-1.887 \cdot e^{-0.009572 \cdot Age}}, & Grade\ 4 \\ 52.14 \cdot e^{-1.887 \cdot e^{-0.009572 \cdot Age}}, & Grade\ 5 \end{cases} \quad (11)$$

Table 7. Multiple comparisons between two different models when carbon growth grades were classified as 5.

Model	Indicator	F	Grade	The Number of Sample Plots	Mean
Logistic	Stand carbon (t/ha)	2.38	Grade 1	67	108.21 a
			Grade 2	150	82.02 b
			Grade 3	262	60.74 c
			Grade 4	347	42.18 d
			Grade 5	187	19.25 e
	Carbon productivity (t/ha/a)	2.38	Grade 1	67	1.45 a
			Grade 2	150	1.06 b
			Grade 3	262	0.80 c
			Grade 4	347	0.60 d
			Grade 5	187	0.40 e

Table 7. *Cont.*

Model	Indicator	F	Grade	The Number of Sample Plots	Mean
Gompertz	Stand carbon (t/ha)	2.38	Grade 1	68	107.31 a
			Grade 2	145	83.83 b
			Grade 3	259	60.76 c
			Grade 4	350	42.33 d
			Grade 5	191	19.36 e
	Carbon productivity (t/ha/a)	2.38	Grade 1	68	1.46 a
			Grade 2	145	1.05 b
			Grade 3	259	0.81 c
			Grade 4	350	0.61 d
			Grade 5	191	0.40 e

Note: Carbon productivity is obtained by dividing stand carbon by stand average age; different lowercase letters represent significant differences at the 0.05 level.

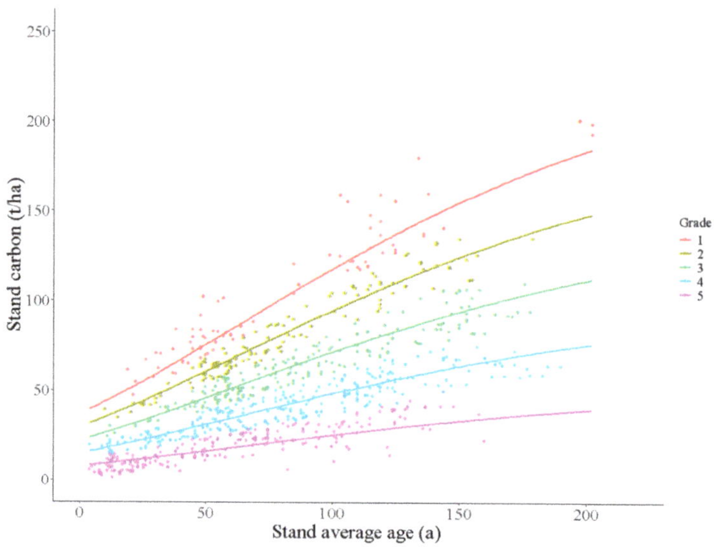

Figure 2. Simulation of the stand carbon growth process based on the Gompertz equation with carbon growth grade 5 level.

3.3. Significant Structural Diversity Variables Affecting Stand Carbon

Based on the comprehensive feature selection method for multivariable input, *N_Shannon_Size4_Div* was selected seven times, followed by *BA_GiniSimpson_Size2_Div*, which was selected five times. The remaining nine stand structure diversity variables were selected no more than four times; hence, *N_Shannon_Size4_Div* was chosen as the representative variable for stand structure diversity when establishing the stand carbon storage growth model (Table 8).

3.4. Stand Carbon Growth Model with Stand Structural Diversity

The goodness-of-fit values of different theoretical growth models were very different when only considering stand age and introducing structural diversity. After introducing structural diversity, the R^2 value of both Logistic and Gompertz models was improved (Tables 4 and 9). The Additive reparameterization method based on Gompertz further reduced the RMSE, showing that the structural diversity (*N_Shannon_Size4_Div*) significantly improved the prediction accuracy of the models. Compared with the basic Gompertz model, the curve predicted and simulated by the stand carbon storage growth model considering structural diversity was closer to the actual stand carbon growth process (Figure 3), which further verified the role of structural diversity in improving the prediction accuracy of the model. These findings emphasize the importance of considering the diversity of stand structural for forest carbon sink assessment.

Table 8. Results of stand structural diversity variables based on comprehensive multivariable input feature selection.

Variables	Frequency	Approach
N_Shannon_Size4_Div	7	Corr, Step, RDA, PCAReg, GAM, RF, BRT
BA_GiniSimpson_Size2_Div	5	Step, PCAReg, GAM, RF, BRT
BA_Shannon_Size2_Div	4	PCALss, PCAReg, GAM, RF
N_Shannon_Size2_Div	4	RDA, GAM, RF, BRT
N_Simpson_Size2_Div	4	Step, RDA, PCAReg, GAM
N_Simpson_Size4_Div	4	Step, RDA, PCAReg, GAM
BA_Shannon_Size4_Div	3	Step, RDA, BRT
BA_Simpson_Size2_Div	2	PCAReg, RF
BA_Simpson_Size4_Div	2	PCAReg, GAM
N_GiniSimpson_Size2_Div	2	Step, RDA
N_GiniSimpson_Size4_Div	1	RDA

Note: Corr represents correlation analysis; Step represents stepwise regression; RDA represents redundancy analysis; PCAReg represents the principal component regression judgment method; PCALss represents the sum of squared loadings method; GAM represents generalized additive model; RF represents random forest; BRT represents boosted regression tree.

Table 9. The parameters and goodness-of-fit values (R^2 and RMSE) of the stand carbon growth model considering stand structural diversity variables.

Model		Parameter				R^2	RMSE (t/ha)
		a	r	c_1	c_2		
Logistic	Additive	86.915	0.0160	9.670	−4.471	0.45736	24.142
	Multiplicative	600.000	0.0081	30.693	−0.836	0.46445	23.984
Gompertz	Additive	327.503	0.0032	3.967	−0.964	0.48359	23.551
	Multiplicative	600.000	0.0033	3.513	−0.202	0.44745	24.361

Note: The parameter a in all models is limited to a range less than 600 in order to ensure the rationality of the model.

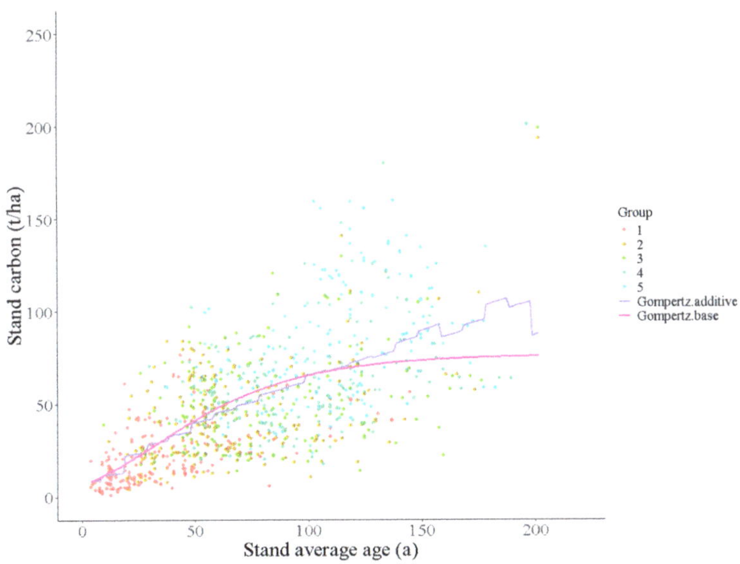

Figure 3. Simulation of the stand carbon growth process based on the basic Gompertz model (pink line) and considering stand structural diversity (purple line). The colors of points represent stand structure diversity divided into five uniform groups according to the numerical value.

3.5. Future Carbon Sink Estimation Based on Different Models

The changing trends of carbon sinks predicted by three different models were very different (Figure 4). The forest carbon sink predicted by the basic model decreased from ~1200 t/a in 2014 to ~550 t/a in 2060. The carbon sink predicted by the model considering carbon growth grade also showed a downward trend, but the decline was small, from ~1100 t/a in 2014 to ~970 t/a in 2060. However, the model considering structural diversity showed a more positive trend, with forest carbon sinks gradually increasing from ~760 t/a in 2014 to ~840 t/a in 2060. These results show that enhancing the complexity of the stand DBH structure can maintain the forest carbon more effectively and realize the increase in carbon sinks. This has important guiding significance for forest management and carbon sink strategy formulation.

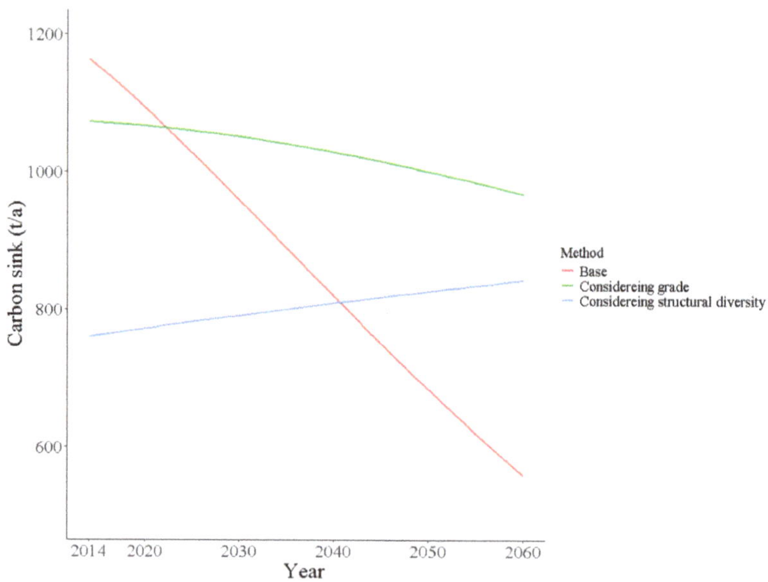

Figure 4. Comparison of trends of forest carbon sinks from 2014 to 2060. The red line represents the predicted value of the forest carbon sink based on the Gompertz basic model, the green line represents the predicted value of the forest carbon sink considering carbon growth grade in the Gompertz model, and the blue line represents the predicted value of the forest carbon sink considering structural diversity in the Gompertz model.

4. Discussion

4.1. The Performance of Stand Carbon Growth Models

The data from the National Forest Inventory can provide an effective measurement of carbon storage, which makes it possible to establish the forest carbon storage growth model and greatly improves the accuracy of forest carbon sink estimation [29]. Carbon growth model development and carbon sink estimation analysis can be used as a decision-making tool for forest management. Previous studies usually estimated forest carbon using the biomass–volume conversion expansion factor or a stand biomass model, which related biomass to stand basal area and average height [30–32]; however, there is little research on stand carbon growth models, especially those considering the effects of carbon growth grade classification and stand structural diversity in mixed forests. This study used the theoretical growth models, named the Logistic equation and the Gompertz equation, to develop stand carbon growth models. The results of the present study, which analyzed

R^2 and RMSE values of the stand carbon growth models, showed that the Gompertz models were better than the Logistic models (Tables 4, 5 and 9). This is different from previous studies by Peng et al. [11] and Xu et al. [9], and the study by Qin et al. [12], which stated that the Logistic equation and the weighted mean of Richard, Hossfeld, and Korf equations could fit the forest carbon growth process and dynamic balance well, respectively. Compared with the model from Zhang et al. [33] for mixed coniferous–broadleaved forests (Carbon = $0.5 \times 326.4180/(1 + \exp(3.3010 - 0.1181 \times \text{Age}))$, $R^2 = 0.79$, N = 15), the results of our data prediction were quite different, mainly due to the small number of samples of Zhang et al. [33]. In addition, because of the different tree species and environmental factors, the carbon growth model forms and parameters are different in different forest types, and the saturation time of natural forest growth varies greatly [34]. Thus, the model based on age alone cannot fully capture the large dispersion of carbon growth [13]. At present, it is difficult to use a unified model form to establish the relationship between carbon storage and age, and it was the best method for establishing the optimal carbon growth model through comparative testing [33]. This study showed that the Gompertz equation fitted the growth process well when stand age increased in the stand carbon in the natural coniferous–broadleaved mixed forests in Jilin Province of China, and it was observed that the natural coniferous and broadleaved mixed forests reached carbon saturation later (Figures 3 and 4), which is similar to the results of the research by Zhang et al. [34].

In addition, models considering the carbon growth grade classification and stand structural diversity variables were stable overall and the model parameters were significant (at 0.05 level) and reliable. The models explained 87% and 48% of the variations in stand carbon, respectively, and the RMSEs were less than those of the basic model (Tables 4, 5 and 9), which illustrates that the two models were superior to the basic model; thus, the effects of carbon growth grade classification and stand structural diversity on stand biomass were significant. The carbon growth grade classification improved the stand carbon model accuracy, which was consistent with the results of Li and Zhao [35]. Of all four grades, the grade with 5 levels best improved the accuracy of the stand carbon growth model (54%) and there were significant differences in stand carbon and carbon productivity [36,37]. Different structural diversity indexes reflect the distributions of different diameters, and the Shannon index is considered the most potential structural diversity variable to estimate the carbon storage of stands, which is consistent with our previous research [28]. Furthermore, the fitting effect of 4 cm width was better than that of 2 cm, which was consistent with previous research [17].

4.2. Forest Carbon Estimation Based on the Stand Carbon Growth Model

Forest carbon sink estimation in our results showed that the forest carbon sink of the natural coniferous and broadleaved mixed forests in Jilin Province of China was expected to reach 791 (781~801) t c/a in 2030 and 843 (833~852) t c/a in 2060, which was separately 17% lower and 51% higher than forest carbon sinks estimated by considering only stand age. These different estimations of future forest carbon sinks stem from the calibration of key parameters (stand structural diversity) of mixed forest growth with long-term datasets when we built the model, which enables our model to simulate forest carbon storage on a large scale [38,39], as stand age is a one-dimensional variable and shows a weaker correlation with stand carbon compared to other stand structural diversity [40]. The carbon growth grade classification can address the variations in carbon storage when the stand age is the same since the carbon growth grade classification model has good prediction accuracy [12].

Although environmental factors can also be directly added to the model, it is difficult to obtain these indicators directly, especially on a large scale. Previous studies mainly obtained these by interpolation, which introduces uncertainties. These environmental factors can be represented by carbon sequestration grade, which can simplify the model and facilitate application; however, carbon prediction needs to determine the carbon growth grade of the sample plot before it can be introduced into the model to estimate the carbon storage of the stand. We showed a method for determining carbon growth grade classification here:

(1) If the study site is a participatory modeling site:

The carbon growth grade classification level has already been established and only the age of the stand needs to be measured in order to apply the model for estimating forest carbon storage and carbon sink.

(2) If a plot was not included in the modeling data and the age is known:

The carbon stock of the stand can be calculated based on the volume and carbon coefficient, and then the carbon stock of the stand can be estimated using different carbon grade models based on stand age. Finally, the estimated error of the carbon stock of the stand is utilized to determine the carbon growth grade classification.

(3) If a plot was not included in the modeling data and the age is unknown:

Carbon growth grade classification can be used to determine stand carbon surveys over two periods. Specifically, the stand age in two periods can be calculated by stand carbon models, and the grade from the two periods closest to the survey interval period can then be used as the final carbon growth grade of the plot.

This method of establishing a model by dividing carbon growth grades and estimating carbon storage can also be used for other tree species, but the number of grades needs to be further studied according to the sample number and model accuracy [35].

4.3. Positive Impact of Stand Structural Diversity on Forest Carbon Sequestration

Our research results show that the diversity of stand structures has a positive impact on forest carbon sinks, which is consistent with previous research results. This can be explained as follows: forests with high structural diversity can increase complementary effects; improve the efficiency of light, water, and nutrient utilization; and enhance their adaptability to environmental changes [41,42]. Specifically, trees of different sizes and structures occupy different horizontal and vertical spaces within the ecosystem and ecological niches, complex forest structures allow light to reach different depths more efficiently, and also better capture rainwater and retain water and nutrients in the soil, avoiding resource competition. Thus, the ability to convert more resources into growth is improved [43] and forest carbon sinks are enhanced.

Structural diversity is one of the aspects of forest control, and growth models of forest carbon storage considering structural diversity can provide a supplementary method for improving the forest carbon capture management model. Based on the positive influence of forest structure diversity on forest carbon sinks, we propose that adjusting DBH class diversity and promoting the formation of a complex DBH class structure in the forest can effectively improve forest carbon sinks.

4.4. Limitations of the Study

This study did not account for the impact of stand density, which is also one of the important factors affecting forest carbon sequestration [44]; it can be included in the model to obtain a more accurate estimate of forest carbon sequestration. Second, in the estimation of forest carbon stock, this study assumed that stand structural diversity remains unchanged, while in fact, it will change as the stand age increases. In addition, this study only estimated the carbon sequestration of existing forests and ignored the carbon stock of newly planted forests [45], which may result in the study's results being lower than the actual value. It is also worth noting that the forest age increases by one every year during the simulation of forest carbon sequestration. For example, the stand with an age of 80 years in 2014 will be 126 years in 2060, but there is an age lag in mixed forests; that is, the stand age increment largely deviated from 1 year per year [46,47]. Future studies can construct a model system by supplementing the stand carbon storage model with density, structural diversity, and carbon sequestration grade, estimating the dynamic prediction model of stand structural diversity, estimating the dynamic prediction model of stand density, and quantifying the structure of stand age so as to more accurately estimate forest carbon sequestration.

5. Conclusions

Based on two theoretical growth models, the Logistic and Gompertz equations, the stand carbon growth model considering carbon growth grade and stand structural diversity was developed to improve the prediction accuracy of forest carbon sinks for natural coniferous–broadleaved mixed forests in Jilin province of China. The results showed that the classification of carbon growth grade and the stand structural diversity can solve the problem of different carbon stocks at the same stand age. The 5-grade classification of carbon growth grade based on the Gompertz equation and the stand structural diversity variable taking *N_Shannon_Size4_Div* into account have the best effect on improving the accuracy of the stand carbon growth model, both of which were better than the basic model considering only the stand age. Both R^2 and RMSE were improved, significantly improving the prediction accuracy of the model, which indicated that these two factors have significant effects on stand carbon storage. The prediction of carbon sinks in the future showed that the model prediction considering structural diversity produces a more positive carbon sink trend, which shows that by increasing the complexity of stand DBH structure, forest carbon storage can be maintained more effectively, and carbon sinks can be increased. In the future, it will be necessary to continue to develop models that take into account density in the structural diversity of the forest stand. The results of this study highlight the importance of considering carbon growth grade and structural diversity to improve the accuracy of forest carbon sink predictions and provide a scientific basis for the development of effective forest management and carbon sink strategies.

Author Contributions: Data curation, X.H., Y.L., H.G. and X.L.; formal analysis, X.H., H.G. and W.G.; conceptualization, H.G. and X.L.; supervision, H.G. and X.L.; methodology, X.H., H.G. and W.G.; visualization, X.H. and W.G.; software, X.H. and H.G.; writing—original draft, X.H., Y.L. and H.G.; writing—review and editing, X.H., H.G., X.L. and W.G. All authors have read and agreed to the published version of the manuscript.

Funding: This research was funded by the National Natural Science Foundation of China (Grant number No. 32301588) and the National Key R&D Program of China (Grant No. 2022YFD2200501).

Data Availability Statement: The processed data required to reproduce these findings cannot be shared at this time as the data form part of an ongoing study.

Conflicts of Interest: The authors declare no conflicts of interest.

References

1. Anjum, J.; Sheikh, M.A.; Tiwari, A.; Sharma, S.; Kumari, B. Carbon Sequestration: An Approach to Sustainable Environment. In *Microbial and Biotechnological Interventions in Bioremediation and Phytoremediation*; Springer: Cham, Switzerland, 2022; pp. 425–444.
2. Gao, Y.; Lai, G.; Wang, Y.; Wang, H.; Zhang, Y.; Shi, S.; Zhang, L. Growth Model of Carbon Storage in *Pinus tabuliformis* Plantation Based on National Forest Inventory Data. *For. Grassl. Resour. Res.* **2024**, 79–87.

3. Liu, C.L.C.; Kuchma, O.; Krutovsky, K.V. Mixed-species versus monocultures in plantation forestry: Development, benefits, ecosystem services and perspectives for the future. *Glob. Ecol. Conserv.* **2018**, *15*, e00419. [CrossRef]
4. Bauhus, J.; Forrester, D.I.; Gardiner, B.; Jactel, H.; Vallejo, R.; Pretzsch, H. Ecological stability of mixed-species forests. In *Mixed-Species Forests: Ecology and Management*; Springer: Berlin/Heidelberg, Germany, 2017; pp. 337–382.
5. Osuri, A.M.; Gopal, A.; Raman, T.S.; DeFries, R.; Cook-Patton, S.C.; Naeem, S. Greater stability of carbon capture in species-rich natural forests compared to species-poor plantations. *Environ. Res. Lett.* **2020**, *15*, 034011. [CrossRef]
6. Cook-Patton, S.C.; Leavitt, S.M.; Gibbs, D.; Harris, N.L.; Lister, K.; Anderson-Teixeira, K.J.; Briggs, R.D.; Chazdon, R.L.; Crowther, T.W.; Ellis, P.W. Mapping carbon accumulation potential from global natural forest regrowth. *Nature* **2020**, *585*, 545–550. [CrossRef] [PubMed]
7. Piao, S.; He, Y.; Wang, X.; Cheng, F. Estimation of China's terrestrial ecosystem carbon sink: Methods, progress and prospects. *Sci. China Earth Sci.* **2022**, *64*, 641–651. [CrossRef]
8. Zhu, K.; Zhang, J.; Niu, S.; Chu, C.; Luo, Y. Limits to growth of forest biomass carbon sink under climate change. *Nat. Commun.* **2018**, *9*, 2709. [CrossRef]
9. Xu, B.; Guo, Z.; Piao, S.; Fang, J. Biomass carbon stocks in China's forests between 2000 and 2050: A prediction based on forest biomass-age relationships. *Sci. China Life Sci.* **2010**, *53*, 776–783. [CrossRef] [PubMed]
10. He, N.; Wen, D.; Zhu, J.; Tang, X.; Xu, L.; Zhang, L.; Hu, H.; Huang, M.; Yu, G. Vegetation carbon sequestration in Chinese forests from 2010 to 2050. *Glob. Change Biol.* **2017**, *23*, 1575–1584. [CrossRef]
11. Peng, B.; Zhou, Z.; Cai, W.; Li, M.; Xu, L.; He, N. Maximum potential of vegetation carbon sink in Chinese forests. *Sci. Total Environ.* **2023**, *905*, 167325. [CrossRef]
12. Qin, J.; Liu, P.; Martin, A.R.; Wang, W.; Lei, Y.; Li, H. Forest carbon storage and sink estimates under different management scenarios in China from 2020 to 2100. *Sci. Total Environ.* **2024**, *927*, 172076. [CrossRef] [PubMed]
13. Bukoski, J.J.; Cook-Patton, S.C.; Melikov, C.; Ban, H.; Chen, J.L.; Goldman, E.D.; Harris, N.L.; Potts, M.D. Rates and drivers of aboveground carbon accumulation in global monoculture plantation forests. *Nat. Commun.* **2022**, *13*, 4206. [CrossRef]
14. He, X.; Li, H.; Zhang, Y.; Huang, J. Growth model of carbon storage and driving force of carbon sequestration capacity of natural secondary forests. *J. Beijing For. Univ.* **2023**, *45*, 1–10.
15. Bennett, A.C.; Penman, T.D.; Arndt, S.K.; Roxburgh, S.H.; Bennett, L.T. Climate more important than soils for predicting forest biomass at the continental scale. *Ecography* **2020**, *43*, 1692–1705. [CrossRef]
16. Zeng, W.; Chen, X.; Yang, X. Developing national and regional individual tree biomass models and analyzing impact of climatic factors on biomass estimation for poplar plantations in China. *Trees* **2021**, *35*, 93–102. [CrossRef]
17. Lei, X.; Wang, W.; Peng, C. Relationships between stand growth and structural diversity in spruce-dominated forests in New Brunswick, Canada. *Can. J. For. Res.* **2009**, *39*, 1835–1847. [CrossRef]
18. Ali, A.; Wang, L.-Q. Big-sized trees and forest functioning: Current knowledge and future perspectives. *Ecol. Indic.* **2021**, *127*, 107760. [CrossRef]
19. Ali, A.; Yan, E.-R.; Chen, H.Y.; Chang, S.X.; Zhao, Y.-T.; Yang, X.-D.; Xu, M.-S. Stand structural diversity rather than species diversity enhances aboveground carbon storage in secondary subtropical forests in Eastern China. *Biogeosciences* **2016**, *13*, 4627–4635. [CrossRef]
20. Del Río, M.; Pretzsch, H.; Alberdi, I.; Bielak, K.; Bravo, F.; Brunner, A.; Condés, S.; Ducey, M.J.; Fonseca, T.; von Lüpke, N. Characterization of the structure, dynamics, and productivity of mixed-species stands: Review and perspectives. *Eur. J. For. Res.* **2016**, *135*, 23–49. [CrossRef]
21. Zhai, L.; Will, R.E.; Zhang, B. Structural diversity is better associated with forest productivity than species or functional diversity. *Ecology* **2024**, *105*, e4269. [CrossRef]
22. Mura, M.; McRoberts, R.E.; Chirici, G.; Marchetti, M. Estimating and mapping forest structural diversity using airborne laser scanning data. *Remote Sens. Environ.* **2015**, *170*, 133–142. [CrossRef]
23. Ercanli, İ.; Kahriman, A. The evaluation of different forest structural indices to predict the stand aboveground biomass of even-aged Scotch pine (*Pinus sylvestris* L.) forests in Kunduz, Northern Turkey. *Environ. Monit. Assess.* **2015**, *187*, 90. [CrossRef] [PubMed]
24. Wang, T.; Wang, G.; Innes, J.L.; Seely, B.; Chen, B. ClimateAP: An application for dynamic local downscaling of historical and future climate data in Asia Pacific. *Front. Agric. Sci. Eng.* **2017**, *4*, 448–458. [CrossRef]
25. Lei, X.; Tang, M.; Lu, Y.; Hong, L.; Tian, D. Forest inventory in China: Status and challenges. *Int. For. Rev.* **2009**, *11*, 52–63. [CrossRef]
26. Zeng, W.; Tomppo, E.; Healey, S.P.; Gadow, K.V. The national forest inventory in China: History-results-international context. *For. Ecosyst.* **2015**, *2*, 23. [CrossRef]
27. Wang, C. Biomass allometric equations for 10 co-occurring tree species in Chinese temperate forests. *For. Ecol. Manag.* **2006**, *222*, 9–16. [CrossRef]

28. He, X.; Lei, X.; Xu, Q.; Lan, J.; Wu, B.; Guo, H.; Gao, W.; Liu, D. Combining multiple feature selection methods and structural equation model for exploring factors affecting stand biomass of natural coniferous-broad leaved mixed forests. *Sci. Total Environ.* **2024**, *954*, 176669. [CrossRef]
29. Yu, Z.; Liu, S.; Li, H.; Liang, J.; Liu, W.; Piao, S.; Tian, H.; Zhou, G.; Lu, C.; You, W. Maximizing carbon sequestration potential in Chinese forests through optimal management. *Nat. Commun.* **2024**, *15*, 3154. [CrossRef] [PubMed]
30. Fang, J.; Guo, Z.; Hu, H.; Kato, T.; Muraoka, H.; Son, Y. Forest biomass carbon sinks in East Asia, with special reference to the relative contributions of forest expansion and forest growth. *Glob. Change Biol.* **2014**, *20*, 2019–2030. [CrossRef]
31. Jagodziński, A.M.; Dyderski, M.K.; Gęsikiewicz, K.; Horodecki, P.; Cysewska, A.; Wierczyńska, S.; Maciejczyk, K. How do tree stand parameters affect young Scots pine biomass?—Allometric equations and biomass conversion and expansion factors. *For. Ecol. Manag.* **2018**, *409*, 74–83. [CrossRef]
32. Dong, L.; Zhang, L.; Li, F. Evaluation of Stand Biomass Estimation Methods for Major Forest Types in the Eastern Da Xing'an Mountains, Northeast China. *Forests* **2019**, *10*, 715. [CrossRef]
33. Zhang, C.; Ju, W.; Chen, J.; Fang, M.; Wu, M.; Chang, X.; Wang, T.; Wang, X. Sustained biomass carbon sequestration by China's forests from 2010 to 2050. *Forests* **2018**, *9*, 689. [CrossRef]
34. Zhang, Y.; Cheng, K.; Yang, Z.; Chen, Y.; Yang, H.; Ren, Y.; Wan, J.; Guo, Q. Spatio-temporal dynamics of future aboveground carbon stocks in natural forests of China. *For. Ecosyst.* **2025**, 100293. [CrossRef]
35. Li, H.; Zhao, P. Improving the accuracy of tree-level aboveground biomass equations with height classification at a large regional scale. *For. Ecol. Manag.* **2013**, *289*, 153–163. [CrossRef]
36. Gonçalves, A.C. Stand structure impacts on forest modelling. *Appl. Sci.* **2022**, *12*, 6963. [CrossRef]
37. LaRue, E.A.; Knott, J.A.; Domke, G.M.; Chen, H.Y.; Guo, Q.; Hisano, M.; Oswalt, C.; Oswalt, S.; Kong, N.; Potter, K.M. Structural diversity as a reliable and novel predictor for ecosystem productivity. *Front. Ecol. Environ.* **2023**, *21*, 33–39. [CrossRef]
38. Fu, X.; Zhang, Y.-x.; Wang, X.-j. Prediction of forest biomass carbon pool and carbon sink potential in China before 2060. *Sci. Silvae Sin.* **2022**, *58*, 32–41.
39. He, X.; Lei, X.-D.; Dong, L.-H. How large is the difference in large-scale forest biomass estimations based on new climate-modified stand biomass models? *Ecol. Indic.* **2021**, *126*, 107569. [CrossRef]
40. Yuan, Z.; Ali, A.; Jucker, T.; Ruiz-Benito, P.; Wang, S.; Jiang, L.; Wang, X.; Lin, F.; Ye, J.; Hao, Z. Multiple abiotic and biotic pathways shape biomass demographic processes in temperate forests. *Ecology* **2019**, *100*, e02650. [CrossRef]
41. Forrester, D.I.; Ammer, C.; Annighöfer, P.J.; Barbeito, I.; Bielak, K.; Bravo-Oviedo, A.; Coll, L.; del Río, M.; Drössler, L.; Heym, M. Effects of crown architecture and stand structure on light absorption in mixed and monospecific Fagus sylvatica and *Pinus sylvestris* forests along a productivity and climate gradient through Europe. *J. Ecol.* **2018**, *106*, 746–760. [CrossRef]
42. Niinemets, Ü. A review of light interception in plant stands from leaf to canopy in different plant functional types and in species with varying shade tolerance. *Ecol. Res.* **2010**, *25*, 693–714. [CrossRef]
43. Tilman, D.; Lehman, C.L.; Thomson, K.T. Plant diversity and ecosystem productivity: Theoretical considerations. *Proc. Natl. Acad. Sci. USA* **1997**, *94*, 1857–1861. [CrossRef]
44. Brunner, A. Stand volume growth varies with age, site index, and stand density–Comments on Stokland (2021). *For. Ecol. Manag.* **2021**, *495*, 119329. [CrossRef]
45. Huang, L.; Liu, J.; Shao, Q.; Xu, X. Carbon sequestration by forestation across China: Past, present, and future. *Renew. Sustain. Energy Rev.* **2012**, *16*, 1291–1299. [CrossRef]
46. Shang, R.; Chen, J.M.; Xu, M.; Lin, X.; Li, P.; Yu, G.; He, N.; Xu, L.; Gong, P.; Liu, L. China's current forest age structure will lead to weakened carbon sinks in the near future. *Innovation* **2023**, *4*, 100515. [CrossRef]
47. Tian, H.; Zhu, J.; Lei, X.; Jian, Z.; Chen, X.; Zeng, L.; Huang, G.; Liu, C.; Xiao, W. Models considering the theoretical stand age will underestimate the future forest carbon sequestration potential. *For. Ecol. Manag.* **2024**, *562*, 121982. [CrossRef]

forests

MDPI

Article

Study on the Natural Regeneration Characteristics and Influencing Factors of Typical *Quercus* Forests in Northern China

Xuefan Hu [1], Guangshuang Duan [2,3,4], Yingshan Jin [1,*], Yuxin Cheng [2], Fang Liang [1], Zhenghua Lian [1], Fang Li [1], Yuerong Wang [1] and Hongfei Chen [1]

1 Beijing Key Laboratory of Greening Plants Breeding, Beijing Academy of Forestry and Landscape Architecture, Beijing 100102, China; hufanzi@163.com (X.H.)
2 College of Mathematics and Statistics, Xinyang Normal University, Xinyang 464000, China; duans@xynu.edu.cn (G.D.)
3 Henan Dabieshan National Field Observation & Research Station of Forest Ecosystem, Zhengzhou 450046, China
4 Xinyang Academy of Ecological Research, Xinyang 464000, China
* Correspondence: j_ysh@163.com

Abstract: This study aims to analyze the natural regeneration characteristics and the key factors of *Quercus* forests, providing a theoretical foundation for maintaining the ecological stability of *Quercus* forests in northern China. In June and July 2023, 17 square plots of five *Quercus* species in Beijing were surveyed, and seedling regeneration and environmental factors (site, stand and soil factors) were measured. Pearson correlation and random forest algorithms were used to identify the relevant and key environmental factors affecting seedling regeneration density (Seedling 1, Seedling 2, Seedling 3). The natural regeneration capabilities of the five *Quercus* species in the Beijing area vary, with *Quercus aliena* and *Quercus variabilis* being stronger, while *Quercus mongolica*, *Quercus acutissima* and *Quercus dentata* are relatively weaker. Correlation analysis showed that Seedling 1 has no significant correlation with environmental factors; Seedling 2 is significantly negatively correlated with Pielou's evenness (J) and exchangeable calcium (ECa) ($p < 0.05$); Seedling 3 is significantly positively correlated with species richness (S), Shannon–Wiener index (H), stand volume (M), and litter layer thickness (LT) ($p < 0.05$), and significantly negatively correlated with Pielou's evenness (J) ($p < 0.01$). The random forest algorithm indicated that the regeneration of Seedling 1 is mainly affected by stand factors, while the regeneration of Seedling 2 and Seedling 3 is more influenced by soil and site factors. The *Quercus* forests in the Beijing region exhibit a rich species composition and demonstrate a certain capacity for natural regeneration. However, seedling growth is more constrained by soil and site factors in the later stages. Therefore, in the management of *Quercus* forests, environmental factors can be regulated during the seedling growth stage to create more suitable conditions for regeneration.

Keywords: *Quercus* species; regeneration diversity; importance values; correlation analysis; random forest

1. Introduction

The natural regeneration of forests is crucial for maintaining ecological balance, stabilizing community structures, and conserving biodiversity [1–3]. It plays a pivotal role in the natural restoration of forest ecosystems and in preserving the genetic diversity of species in the regeneration layer [4,5]. The regeneration of forests is influenced by human and habitat factors, the latter of which include site factors [6–8], stand factors [9–11], and soil factors [12–14]. However, the dominant factors affecting natural regeneration vary among different forest types. In fact, studies have shown that site factors (e.g., altitude, slope, and aspect [15–17]), stand factors (e.g., tree diversity, competition index, and understory vegetation [18–20]), and soil factors (e.g., soil nutrients, soil organic matter and total phosphorus [21–24]) are major factors influencing the density of regenerating seedlings, respectively. Furthermore, the forest regeneration stage (sapling stage, seedling stage, young tree stage) is the most sensitive period to environmental factors during tree growth and a crucial period determining the success of natural forest regeneration [25,26]. The factors influencing regeneration vary across different forest types and species, and even for the same forest type, differences at regeneration stages can result in varying factors influencing regeneration. Therefore, a comprehensive understanding of the regeneration characteristics of juvenile plants and the heterogeneity of environmental conditions is of practical significance for forest ecosystem protection and management.

The area of *Quercus* forests accounts for 11.95% of China's total forest area, and the volume of *Quercus* forests accounts for 9.98% of China's forests, making them a major component of China's natural forests [27]. *Quercus* species are also an important species in the natural zonal forests of northern China, providing vital ecological services such as water conservation, soil erosion control, and forest fire prevention, thereby contributing to the country's dual carbon goals. In Beijing, *Quercus* forests cover 122,700 hectares, accounting for 19.74% of the city's arboreal forests area. Natural and secondary *Quercus* forests are mainly distributed in the mountainous regions of Beijing, with *Q. aliena*, *Q. mongolica*, *Q. acutissima*, *Q. variabilis*, and *Q. dentata* being the predominant species. However, the connections between natural regeneration characteristics and environmental factors across different *Quercus* forest regeneration stages remain unexplored. Hence, exploring the natural regeneration characteristics and its influencing factors of the typical *Quercus* forests could improve the quality of forest resources and their management, directly affecting the overall quality of Beijing's forest resources.

Therefore, this study focuses on *Quercus* forests in the Beijing region. We addressed the following research contents. (1) The species composition and regeneration characteristics of different *Quercus* stands are analyzed to understand the current regeneration status of *Quercus* forests. (2) Through correlation analysis, the study explores the impact of environmental factors on seedling regeneration and (3) applies random forest algorithms to rank the importance of these factors in influencing regeneration density. The findings aim to provide a theoretical basis for the efficient utilization and management of *Quercus* forest resources in Beijing, contributing to the maintenance of biodiversity and ecological balance.

2. Materials and Methods

2.1. Study Site

Beijing (39°24′~41°36′ N, 115°42′~117°24′ E) is located in the northern part of the North China Plain. It covers an area of approximately 16,410.54 km^2, with a forested area of 1.1062 million hectares, a forest coverage rate of 44.4%, a total forest resource area of 848,300 hectares, and a forest volume of 25.2 million cubic meters [28]. The elevation of the Beijing Plain ranges from 20 to 60 m, while the mountains generally have elevations between 1000 and 1500 m, with an average elevation of 43.5 m. Beijing has a warm temperate semi-humid continental monsoon climate, with an annual average temperature of 10–13 °C and a frost-free period of 180–200 days, with most rainfall occurring in summer. The main soil types include mountain brown soil, cinnamon soil, and mountain meadow soil [29]. The predominant vegetation in the study area is composed of tree species such as *Quercus acutissima*, *Quercus variabilis*, *Quercus mongolica*, *Quercus aliena*, *Quercus dentata*,

Pinus tabuliformis, *Larix principis-rupprechtii*, *Acer truncatum*, *Ailanthus altissima*, *Ginkgo biloba*, *Populus tomentosa*, and *Fraxinus chinensis*; shrub species include *Leptopus chinensis*, *Spiraea cantoniensis*, *Syringa oblata*, *Lonicera maackii*, and *Jasminum nudiflorum*; herbaceous species include *Allium ochotense*, *Polygonatum odoratum*, *Sedum spectabile*, *Hemerocallis fulva*, *Chrysanthemum indicum*, and *Potentilla chinensis* [30].

2.2. Experiment Design

Field surveys were conducted in June and July of 2023 in typical *Quercus*-dominated communities in Beijing. Plots were established in areas where *Quercus* species were relatively concentrated, covering five types of *Quercus* stands: *Q. aliena* (Bailongtan Forest Farm), *Q. variabilis* (Xishan Experimental Forest Farm), *Q. mongolica* (Beijing Western Forest Farm), *Q. acutissima* (Shangfangshan National Forest Park), and *Q. dentata* (Xishan Ex-perimental Forest Farm). The plot locations are shown in Figure 1.

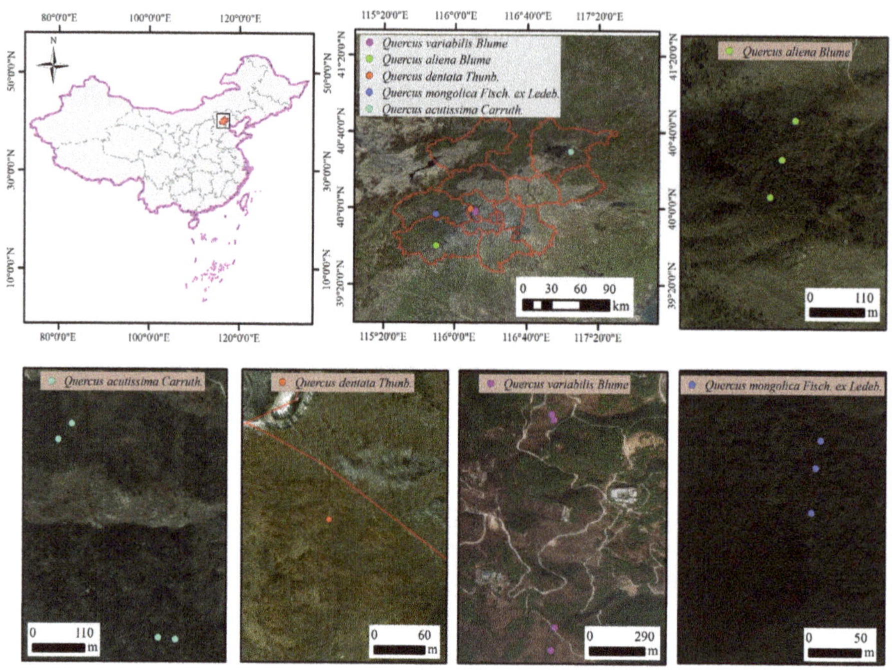

Figure 1. Location of survey plots for five *Quercus* species in Beijing: *Q. aliena*, *Q. acutissima*, *Q. dentata*, *Q. variabilis*, and *Q. mongolica*.

A total of 17 square plots were set up. Trees with a diameter at breast height (DBH) greater than 1 cm were measured for tree height, DBH, origin and dry biomass, and tree locations were also recorded with a real-time kinematic (RTK). For each plot, 5 m × 5 m subplots were set up at the four corners and the center for seedling and shrub surveys, and 1 m × 1 m subplot was set up for herbaceous vegetation surveys. These subplots recorded species names, quantities, height, and coverage. GPS was used to measure altitude, slope aspect, slope position, and other information. Basic plot information is summarized in Table 1.

Table 1. Basic information of the study plots.

Stand Type	Plot Number	Plot Area (m^2)	Altitude (m)	Slope Aspect	Slope Position	Stand Density (n/ha)
Q. aliena	1	900	483	N	U	1621
	2	400	470	N	U	899
	3	400	465	N	M	999
Q. mongolica	1	600	950	N	M	2581
	2	600	930	N	M	2914
	3	600	925	N	M	1832
Q. acutissima	1	600	284	S	M	566
	2	600	284	S	M	683
	3	600	284	S	L	549
	4	400	284	N	M	350
Q. variabilis	1	900	192	S	M	699
	2	900	195	S	M	533
	3	900	186	S	M	766
	4	2500	186	N	M	1139
Q. dentata	1	900	755	N	M	1365
	2	900	780	N	U	1365
	3	900	776	N	U	1543

Note: N is the north-facing slope, S is the south-facing slope; U is upper, M is middle, L is lower.

2.3. Index Measurement

Soil samples from the 0–30 cm soil layer were collected at the four corners and center of each fixed plot using aluminum boxes and soil augers for analysis of soil physical and chemical properties. Five additional 1 m × 1 m subplots near each sample point were established to measure the thickness of the litter layer and humus layer. Soil pH was measured using a pH meter, organic matter was determined by the potassium dichromate method, total nitrogen was measured by the Kjeldahl method, total phosphorus and available phosphorus were determined by the molybdenum-antimony colorimetric method, available potassium was measured using a flame photometer, available manganese, iron, copper, and zinc were measured by the DTPA extraction-atomic absorption spectrophotometry method, exchangeable calcium and magnesium were determined by ammonium acetate extraction-atomic absorption spectrophotometry, alkali-hydrolyzed nitrogen was measured by alkali diffusion, and electrical conductivity (EC) was measured using the conductivity method. The cation exchange capacity (CEC) was determined by the cobalt chloride extraction-spectrophotometric method [31–33].

In total, 36 environmental variables were selected for analysis, including 4 site factors (altitude, slope aspect, slope position, slope gradient), 14 stand factors (total tree count, species richness, Shannon–Wiener index, Pielou evenness index, crown density, stand volume, stand density, average DBH, shrub species richness, shrub density, shrub height, herbaceous species richness, herbaceous coverage, and herbaceous height), and 18 soil factors (soil thickness, litter layer thickness, humus layer thickness, organic matter, total nitrogen, total phosphorus, alkali-hydrolyzed nitrogen, available phosphorus, available potassium, pH value, EC value, cation exchange capacity, exchangeable calcium, exchangeable magnesium, available iron, available manganese, available copper, and available zinc).

2.4. Index Calculation and Selection

2.4.1. Seedling Density and Importance Value

Seedlings were divided into three categories: saplings (height < 30 cm), small seedlings (31 cm < height ≤ 51 cm), and young trees (height > 51 cm). The densities of each category were recorded as Seedling1 (S1), Seedling2 (S2), and Seedling3 (S3). The evaluation of seedling regeneration was carried out according to the technical regulations of the National Forestry and Grassland Administration (GB/T 38590-2020) [34]. The importance value (IV) for each species was calculated as follows [35]:

$$\text{IV} = (RA + RF + RD)/3$$

where *RA* represents relative abundance (proportion of species in the total number of individuals); *RF* represents relative frequency (proportion of species in the total number of plots); and *RD* represents relative dominance (proportion of species' average height in the total average height of all species).

2.4.2. Plant Species Diversity Indices

Species diversity in *Quercus* stands was measured using species richness (S), Shan-non–Wiener index (H), and the Pielou evenness index (J). The formulas for these indices are as follows [36]:

$$\text{H} = -\sum_{i=1}^{\text{S}} P_i \ln P_i,$$

$$\text{J} = \text{H}/\ln \text{S}$$

where S is the total number of species in the tree, and P_i is the proportion of all individuals belonging to the species in the sample to the total number of individuals.

2.4.3. Index Selection

To better understand the factors influencing *Quercus* stand regeneration, 16 environmental factors were selected after analyzing the correlation and collinearity between the variables. These include 3 site factors, 6 stand factors, and 7 soil factors. The distribution of the regeneration density and influencing factors in the *Quercus* stands is shown in Table 2.

Table 2. Environmental characteristics of the study plots.

Factor Type	Variable	Abbreviation	Mean	Range	Standard Error
Regeneration indicators	Seeding 1 (n/ha)	S1	1584	320–3836	955
	Seeding 2 (n/ha)	S2	917	160–3037	769
	Seeding 3 (n/ha)	S3	2177	16–6474	1710
Site factors	Altitude (m)	AL	495.82	186.00–950.00	291.70
	Slope aspect	SA	-	-	-
	Slope position	SP	-	-	-
Stand factors	Species richness (n)	S	8	4–20	4
	Shannon–Wiener index	H	1.48	1.09–2.34	0.33
	Pielou evenness index	J	0.21	0.12–0.29	0.05
	Stand volume (m^3/ha)	M	94.62	42.71–162.54	32.35
	Shrub density (10,000/ha)	SD	1.14	0.58–2.28	2.27
	Herbaceous coverage (%)	HC	13.17	0.6–36.6	9.72
Soil factors	Litter layer thickness (cm)	LT	5.65	2.00–10.00	2.52
	Total nitrogen (g/kg)	TN	2.01	0.70–2.99	0.59
	Available phosphorus (mg/kg)	AP	3.49	1.22–6.95	1.65
	Available potassium (mg/kg)	AK	101.07	38.15–161.91	33.75
	pH value	pH	6.46	5.41–7.05	0.48
	Exchangeable calcium (mg/kg)	ECa	2774.54	1197.11–4191.62	707.26
	Available manganese (mg/kg)	AM	25.99	9.43–71.80	16.39

2.5. Statistical Analysis

Correlation analysis was used to examine the relationships between the seedling regeneration density (S1, S2, and S3), site factors (altitude, slope aspect, slope position), stand factors (species richness, Shannon–Wiener index, Pielou evenness index, stand volume, shrub density, and herbaceous coverage), and soil factors (litter layer thickness, total nitrogen, available phosphorus, available potassium, pH value, exchangeable calcium, and available manganese). The seedling densities (S1, S2, and S3) were treated as continuous variables, the seedling regeneration stages (stage 1, stage 2, and stage 3) were treated as categorical variables, and a random forest model (random forest function) was used for processing. For continuous variables, the random forest regression algorithm was employed to rank the relative importance of variables based on node purity improvement. For categorical variables, the random forest classification algorithm was used to rank the relative importance based on the decrease in Gini index. Correlation analysis and random forest analysis were performed using R v4.4.1 software (R Development Core Team, 2024) [37], with the corrplot [38] and GGally [39] packages for correlation analysis and the randomForest package [40] for random forest analysis. Graphs were generated using the ggplot2 package [41].

3. Results

3.1. Characteristics of Quercus Stand Regeneration

3.1.1. Regeneration Quantity and Grade Evaluation in Quercus Stands

The regeneration quantity and grade evaluation for the five types of *Quercus* stands are shown in Table 3. From the regeneration quantity perspective, the numbers of regenerated individuals in *Q. aliena* stands and *Q. variabilis* stands were higher than that of *Q. dentata* stands, *Q. mongolica* stands, and *Q. acutissima* stands. In terms of growth status, at the <30 cm regeneration level, the average height of *Q. aliena*, *Q. acutissima*, *Q. mongolica*, and *Q. variabilis* stands was higher than that of *Q. dentata* stands. At the 31–50 cm regeneration level, the average height differences across the five *Quercus* stand types were minimal. However, at the >51 cm regeneration level, the average height of *Q. aliena* stands was lower than that of the other *Quercus* stands. Regarding regeneration density, at the <30 cm and >51 cm regeneration levels, the regeneration density of *Q. aliena* stands was significantly higher than that of the other *Quercus* stands. At the 31–50 cm regeneration level, the regeneration density of *Q. aliena* and *Q. dentata* stands was greater than that of the other three *Quercus* stands. From a regeneration grade perspective, at the <30 cm level, natural regeneration was poor in all five *Quercus* stand types. At the 31–50 cm level, *Q. aliena*, *Q. dentata*, and *Q. variabilis* had moderate regeneration, while *Q. acutissima* and *Q. mongolica* had poor regeneration. At the >51 cm level, *Q. aliena* and *Q. variabilis* showed good regeneration, while the other three *Quercus* stand types exhibited moderate regeneration. In summary, *Q. aliena* and *Q. variabilis* exhibited stronger regeneration potential, while *Q. dentata*, *Q. acutissima*, and *Q. mongolica* had relatively weaker regeneration abilities.

Table 3. Regeneration quantity, average tree height, regeneration density, and grade evaluation at different levels for various *Quercus* stands.

Level	Stand Types	Average Height (cm)	Regeneration Density (n/ha)	Grade
<30 cm	*Q. aliena*	21	1971	Poor
	Q. dentata	15	1678	Poor
	Q. acutissima	20	1379	Poor
	Q. mongolica	22	1545	Poor
	Q. variabilis	20	1459	Poor

Table 3. *Cont.*

Level	Stand Types	Average Height (cm)	Regeneration Density (n/ha)	Grade
31–50 cm	*Q. aliena*	41	1332	Moderate
	Q. dentata	40	1332	Moderate
	Q. acutissima	41	519	Poor
	Q. mongolica	42	453	Poor
	Q. variabilis	41	1039	Moderate
>51 cm	*Q. aliena*	100	4236	Good
	Q. dentata	117	1945	Moderate
	Q. acutissima	125	1039	Moderate
	Q. mongolica	112	1225	Moderate
	Q. variabilis	113	2657	Good

3.1.2. Seedling Regeneration Characteristics in Quercus Stands

Based on the seedling survey data from 17 sample plots, the species composition and importance values of the regenerated seedlings in the five *Quercus* stand types in Beijing are shown in Table 4.

Table 4. Species composition and importance values of regenerated tree species in *Quercus* stands.

Stand Types	Tree Species	*RA*/%	*RF*/%	*RD*/%	IV/%
Q. aliena	*Celtis koraiensis*	23.3	15.1	6.3	14.9
	Celtis sinensi	13.4	15.1	4.9	11.1
	Prunus triloba	9.8	9.6	8.0	9.1
	Fraxinus chinensis	9.4	6.8	4.3	6.8
	Fraxinus pennsylvanica	6.7	6.8	7.0	6.8
	Ulmus laevis	8.3	6.8	3.2	6.1
	Morus mongolica	3.9	6.8	5.7	5.5
	Sophora tomentosa	3.5	6.8	5.1	5.1
	Prunus padus	0.8	1.4	11.1	4.4
	Ailanthus altissima	0.8	1.4	10.9	4.4
	Pistacia chinensis	1.2	2.7	7.9	3.9
	Cotinus coggygria	8.6	1.4	1.6	3.9
	Morus alba	0.8	2.7	7.7	3.7
	Acer truncatum	1.6	4.1	5.5	3.7
	Quercus aliena	2.0	4.1	2.2	2.8
	Koelreuteria paniculata	2.0	2.7	3.1	2.6
	Celtis tetrandra	1.6	2.7	1.6	2.0
	Broussonetia papyrifera	0.7	1.5	2.6	1.7
	Quercus dentata	1.6	1.5	1.3	1.5
Q. dentata	*Quercus dentata*	19.4	21.7	3.5	14.9
	Fraxinus chinensis	15.6	18.3	10.6	14.8
	Morus mongolica	5.9	15.0	16.2	12.4
	Morus alba	8.4	13.3	11.6	11.1
	Diospyros lotus	24.0	3.3	4.1	10.5
	Koelreuteria paniculata	14.7	8.3	4.5	9.2
	Acer pictum	1.3	1.7	21.7	8.2
	Acer truncatum	3.2	5.0	13.8	7.3
	Celtis bungeana	3.8	5.0	5.0	4.6
	Cotinus coggygria	2.5	5.0	5.9	4.4
	Quercus variabilis	1.2	3.4	3.1	2.6

Table 4. *Cont.*

Stand Types	Tree Species	RA/%	RF/%	RD/%	IV/%
Q. acutissima	*Quercus acutissima*	16.7	26.8	3.3	15.6
	Quercus dentata	20.2	19.6	3.7	14.5
	Ziziphus jujuba	21.0	3.6	3.8	9.5
	Robinia pseudoacacia	7.9	12.5	7.6	9.3
	Prunus sibirica	7.9	8.9	7.4	8.1
	Celtis bungeana	3.5	1.8	19.1	8.1
	Ulmus pumila	7.0	7.1	6.7	6.9
	Morus alba	1.8	1.8	16.6	6.7
	Koelreuteria paniculata	6.1	7.1	5.0	6.1
	Acer truncatum	3.5	3.6	10.1	5.7
	Morus mongolica	1.8	1.8	11.6	5.1
	Quercus variabilis	2.6	5.4	5.1	4.4
Q. mongolica	*Fraxinus chinensis*	78.7	56.5	9.0	48.1
	Cotinus coggygria	2.4	4.3	67.5	24.7
	Quercus mongolica	14.2	34.8	7.2	18.7
	Crataegus pinnatifida	4.7	4.4	16.3	8.5
Q. variabilis	*Koelreuteria paniculata*	23.0	17.9	7.0	16.0
	Quercus variabilis	24.2	17.9	4.3	15.5
	Quercus aliena	11.7	13.4	4.7	9.9
	Cotinus coggygria	11.4	14.9	3.2	9.8
	Morus mongolica	6.4	7.5	10.7	8.2
	Broussonetia papyrifera	7.0	9.0	6.2	7.4
	Styphnolobium japonicum	1.8	3.0	17.3	7.4
	Diospyros lotus	2.9	6.0	9.8	6.2
	Morus alba	2.3	1.5	13.4	5.7
	Fraxinus chinensis	2.3	3.0	7.6	4.3
	Acer truncatum	1.2	1.5	9.3	4.0
	Ailanthus altissima	3.5	1.5	3.7	2.9
	Platycladus orientalis	1.2	1.5	2.2	1.6
	Pistacia chinensis	1.1	1.5	0.6	1.1

Note: *RA* represents relative abundance; *RF* represents relative frequency; *RD* represents relative dominance; IV represents importance value.

As shown in Table 4, there were differences in species richness and importance values of regenerated tree species among the five *Quercus* stands in Beijing. In *Q. aliena* stands, a total of 19 species were recorded, belonging to 9 families and 14 genera. The dominant species was *Celtis koraiensis*, with an importance value of 14.9%, and the main accompanying species were *Celtis sinensi* and *Prunus triloba*, both with importance values greater than 9%. In *Q. dentata* stands, a total of 11 species were recorded, belonging to 8 families and 9 genera. The dominant species were *Q. dentata* and *Fraxinus chinensis*, with importance values of 14.9% and 14.8%, respectively, and the main accompanying species were *Morus mongolica*, *Morus alba*, and *Diospyros lotus*, with importance values all greater than 10%. In *Q. acutissima* stands, a total of 12 species were recorded, belonging to 8 families and 9 genera. The dominant species were *Q. acutissima* and *Q. dentata*, with importance values of 15.6% and 14.5%, respectively, and the main accompanying species were *Ziziphus jujuba* and *Robinia pseudoacacia*, both with importance values greater than 9%. In *Q. mongolica* stands, a total of four species were recorded, belonging to four families and four genera. The dominant species was *Fraxinus chinensis*, with an importance value of 48.1%, and the main accompanying species were *Cotinus coggygria* and *Q. mongolica*, with importance values greater than 18%. In *Q. variabilis* stands, a total of 14 species were recorded, belonging to 11 families and 13 genera. The dominant species were *Koelreuteria paniculata* and *Q. variabilis*, with importance values of 16% and 15.5%, respectively, and the main accompanying species were *Quercus. aliena* and *Cotinus coggygria*, with importance values greater than 9%.

3.2. Correlation Analysis of Factors Affecting Natural Regeneration in Quercus Stands

3.2.1. Correlation Between Natural Regeneration and Site Factors in Quercus Stands

The pairwise relationship between the three site factors (altitude, slope aspect, and slope position) and three natural regeneration densities (S1, S2, and S3) in *Quercus* stands is shown in Figure 2 (continuous variables as scatter plots, discrete variables as box plots). The results show that altitude was positively correlated with S1 density, while negatively correlated with S2 and S3 densities, though these relationships were not statistically significant. Slope aspect and slope position, being categorical variables, did not show any obvious correlation with S1, S2, and S3 densities in the box plots.

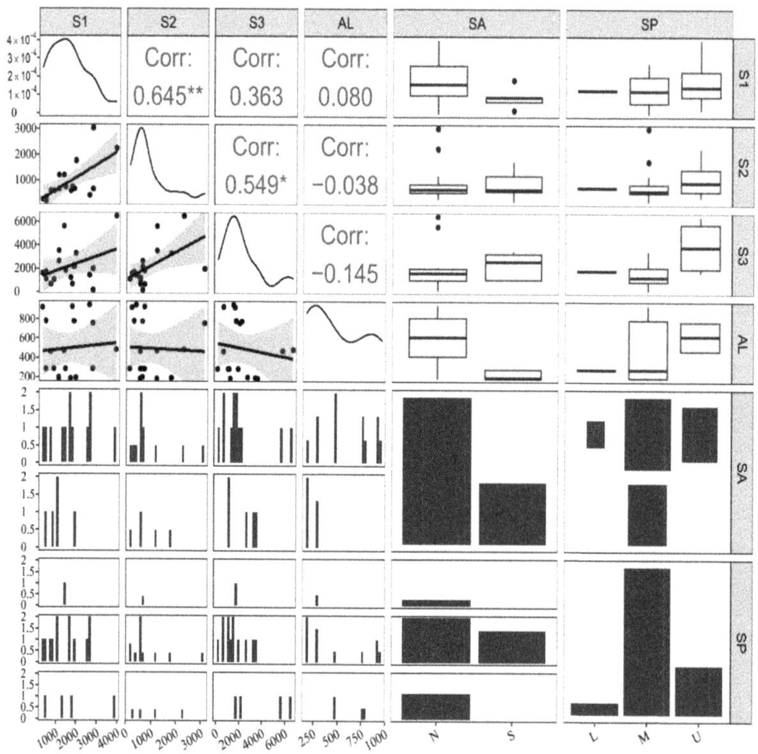

Figure 2. Pairwise relationship plots between natural regeneration density (Seedling 1 (S1), Seedling 2 (S2), and Seedling 3 (S3)) and three site factors. *: $p < 0.05$; **: $p < 0.01$; AL: altitude; SA: slope aspect; SP: slope position.

3.2.2. Correlation Between Natural Regeneration and Stand Factors in Quercus Stands

The correlation coefficients between six stand factors (S, H, J, M, SD, HC) and three natural regeneration densities (S1, S2, and S3) are shown in Figure 3. From the results, it is evident that S1 is negatively correlated with J, while the other five factors are positively correlated, but none of these correlations are significant. S2 is negatively correlated with J ($r = -0.5$, $p < 0.05$), while the other five factors were positively correlated but not significant. Seedling 3 showed a significant negative correlation with J ($r = -0.62$, $p < 0.01$), and a significant positive correlation with S ($r = 0.76$, $p < 0.001$), H ($r = 0.74$, $p < 0.001$), and M ($r = 0.5$, $p < 0.05$).

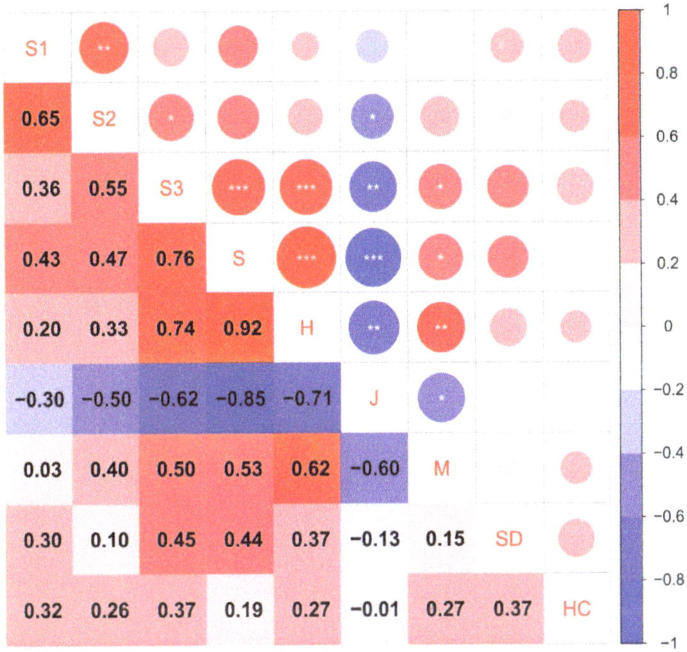

Figure 3. Heatmap of correlation coefficients between natural regeneration densities (Seedling 1 (S1), Seedling 2 (S2), and Seedling 3 (S3)) and stand factors. *: $p < 0.05$; **: $p < 0.01$; ***: $p < 0.001$; S: species richness; H: Shannon–Wiener index; J: Pielou evenness index; M: stand volume; SD: shrub density; HC: herbaceous coverage.

3.2.3. Spatial Distribution Pattern of Forest Stand

The correlation coefficients between seven soil factors (LT, TN, AP, AK, pH, ECa, and AM) and three natural regeneration densities (S1, S2, and S3) are shown in Figure 4. The results show that S1 density was negatively correlated with AK and AM, while positively correlated with the other variables, although none of these relationships were statistically significant. S2 density showed a positive correlation with litter LT but was not significant, while it was significantly negatively correlated with ECa ($r = -0.48$, $p < 0.05$), with other variables showing no significant correlation. S3 density showed positive correlations with LT, pH, and AM, with LT showing a significant positive correlation ($r = 0.49$, $p < 0.05$), while the other four variables showed negative correlations, none of which were significant.

Moreover, Figures 2–4 show that the relationships between the three natural regeneration densities (S1, S2, and S3) were different. S1 density is significantly positively correlated with S2 density ($r = 0.65$, $p < 0.01$), and S2 density is significantly positively correlated with S3 density ($r = 0.55$, $p < 0.05$), while the correlation between S1 and S3 density is positive but not significant.

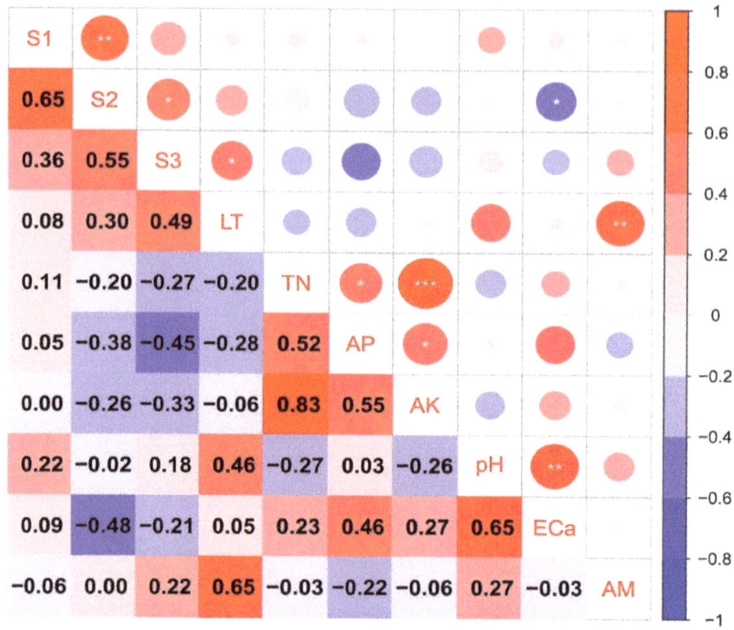

Figure 4. Heatmap of correlation coefficients between natural regeneration densities (Seedling 1 (S1), Seedling 2 (S2), and Seedling 3 (S3)) and soil factors. *: $p < 0.05$; **: $p < 0.01$; ***: $p < 0.001$; LT: litter layer thickness; TN: total nitrogen; AP: available phosphorus; AK: available potassium; pH: pH value; ECa: exchangeable calcium; AM: available manganese.

3.3. *Random Forest Algorithm for Ranking Important Factors*

3.3.1. Ranking of Important Factors Affecting Regeneration Grade in Quercus Stands

Using the random forest algorithm, the relative importance of the environmental factors affecting the natural regeneration grade (a categorical variable with three levels) in *Quercus* stands is ranked based on the Gini index reduction method, as shown in Figure 5. The top five important factors affecting S1 regeneration grade were H, SD, S, J, and AM, including four stand factors and one soil factor (Figure 5a). The top five important factors affecting S2 regeneration grade were AP, H, J, AL and S, including one site factor, three stand factors, and one soil factor (Figure 5b). The top five important factors affecting S3 regeneration grade were AK, H, TN, AP, and AL, including one site factor, one stand factor, and three soil factors (Figure 5c). The top five important factors affecting the regeneration grade of all seedlings (S1, S2, and S3) were AP, J, S, TN, and H, including three stand factors and two soil factors (Figure 5d). In summary, as the seedling growth stage progresses, the influence of site and soil factors on the regeneration grade increases.

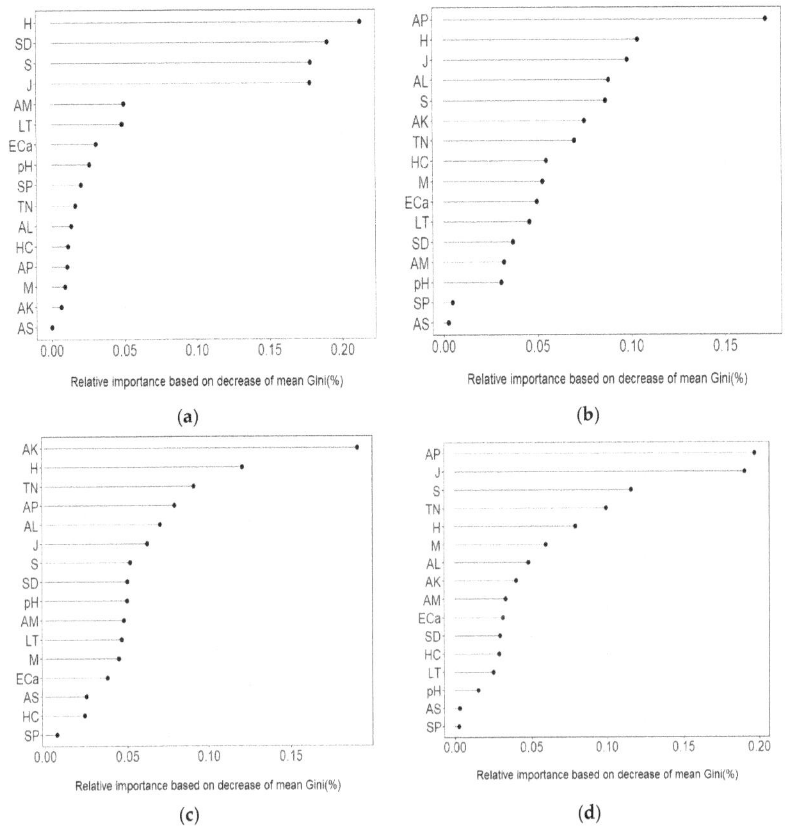

Figure 5. Relative importance ranking of environmental factors affecting regeneration grade of seedlings ((**a**): Seedling 1, (**b**): Seedling 2, (**c**): Seedling 3, (**d**): all seedlings) based on the Gini index reduction method. AL: altitude; SA: slope aspect; SP: slope position; S: species richness; H: Shannon–Wiener index; J: Pielou evenness index; M: stand volume; SD: shrub density; HC: herbaceous coverage; LT: litter layer thickness; TN: total nitrogen; AP: available phosphorus; AK: available potassium; pH: pH value; ECa: exchangeable calcium; AM: available manganese.

3.3.2. Ranking of Important Factors Affecting Regeneration Density in Quercus Stands

Using the random forest algorithm, the relative importance of the environmental factors affecting the natural regeneration density (a continuous variable) in *Quercus* stands is ranked based on the node purity improvement method, as shown in Figure 6. The top five important factors affecting S1 density were H, ECa, HC, J, and S, including four stand factors and one soil factor (Figure 6a). The top five important factors affecting S2 density were ECa, AP, M, AK, and J, including two stand factors and three soil factors (Figure 6b). The top five important factors affecting S3 density were H, S, J, AK, and AM, including three stand factors and two soil factors (Figure 6c). The top five important factors affecting the regeneration density of all seedlings (S1, S2, and S3) were J, H, AK, AP, and S, including three stand factors and two soil factors (Figure 6d). In summary, as the seedling growth stage progresses, the importance ranking of stand factors (reflecting species diversity H, S, and J) initially decreases and then increases, while the importance ranking of soil factors gradually increases.

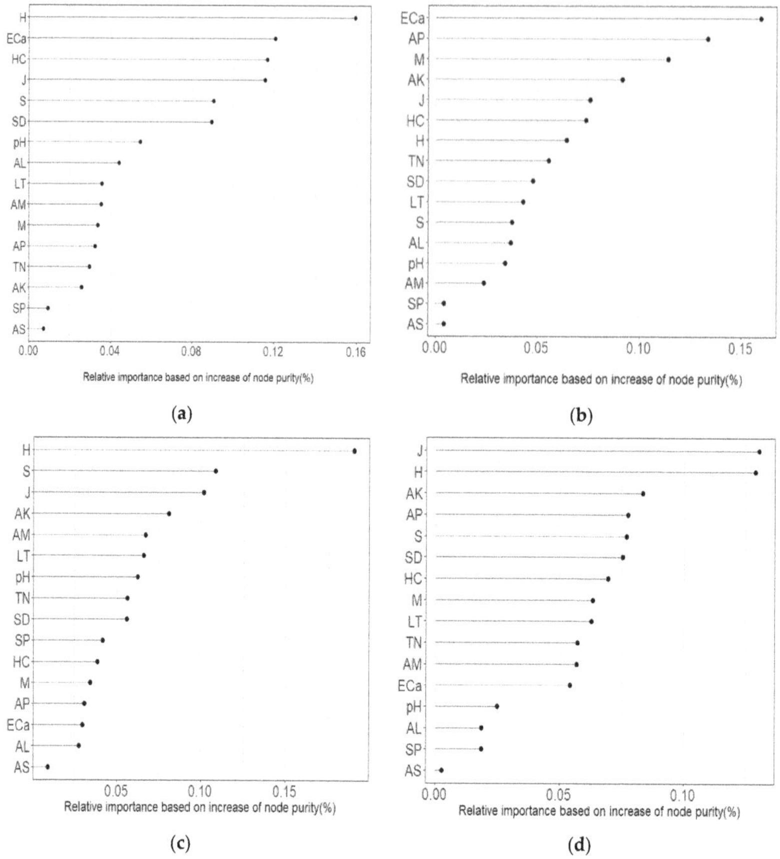

Figure 6. Relative importance ranking of environmental factors affecting regeneration density of seedlings ((**a**): Seedling 1, (**b**): Seedling 2, (**c**): Seedling 3, (**d**): all seedlings) based on the node purity improvement method. AL: altitude; SA: slope aspect; SP: slope position; S: species richness; H: Shannon–Wiener index; J: Pielou evenness index; M: stand volume; SD: shrub density; HC: herbaceous coverage; LT: litter layer thickness; TN: total nitrogen; AP: available phosphorus; AK: available potassium; pH: pH value; ECa: exchangeable calcium; AM: available manganese.

Additionally, the results of the random forest algorithm indicate that the factor importance rankings for continuous variables (S1, S2, and S3) and categorical variables (grade 1, grade 2, and grade 3) differ. Figures 5 and 6 show that the major factors affecting Seedling 1 (grade 1) regeneration density are mainly stand factors, while soil and site factors rank higher for Seedling 2 (grade 2) and Seedling 3 (grade 3) regeneration density.

4. Discussion

4.1. Analysis of Different Regeneration Characteristics in Quercus Stands

The survey of five types of *Quercus* stands in Beijing showed that *Quercus* stands with either excessively dense or sparse canopy layers had poor natural regeneration, while stands with moderate density showed good regeneration. This finding is consistent with the results of [42]. The possible reason for this could be the differences in understory vegetation competition intensity and soil physical–chemical properties under different stand density conditions. Furthermore, the species richness and composition of regenerated seedlings in the five *Quercus* stands in Beijing differed significantly, which is similar to [43]. Due to the different spatial distributions of the five *Quercus* stand types, there were variations in site factors, and differences in the dominant tree species and stand structure also led to significant differentiation in natural regeneration species. The species composition of the five *Quercus* stands was relatively complex, with a high species richness, and no single dominant species of regenerated seedlings was evident. Especially in *Q. mongolica* stands, the dominant regenerated seedling species was *Fraxinus chinensis*, with an importance value of 48.1%, and the number of regenerating species was limited, leading to a clear competitive relationship.

4.2. Effects of Environmental Factors on Natural Regeneration in Quercus Stands

The study indicated that site factors such as altitude, slope aspect, and slope position had no significant effect on the natural regeneration density in *Quercus* stands. This contradicts the findings of [44], which found slope aspect to significantly affect the regeneration density of *Q. aliena*. The lack of significant results in this study may be due to the relatively small number of plots sampled and the limited variation in site factors across those plots. As the seedlings' growth levels improve, the importance ranking of site factors (altitude, slope aspect, and slope position) on seedling regeneration density gradually increases. In the early stages of seedling growth, resources such as light, heat, and moisture are relatively abundant, allowing seedlings to access sufficient resources under varying site conditions. However, as seedling growth progresses, competition for resources intensifies, and site conditions influence the available resources within the stand, thereby affecting regeneration outcomes [45,46].

Plant species diversity index is one of the factors influencing natural regeneration in *Quercus* stands in Beijing. The correlation analysis revealed that seedling regeneration density was significantly positively correlated with Species richness and Shannon–Wiener index, but significantly negatively correlated with the Pielou evenness index. Higher species richness and Shannon–Wiener index suggest a greater number of ecological niches and more complex ecological relationships, providing more resources for seedling growth, which helps maintain ecosystem stability and enhance regeneration ability [47]. A higher evenness index indicates that, in areas with higher seedling density, the evenness of species distribution decreases, leading to dominance by a few species and suppression of regeneration of other species. This study also showed that stand volume directly and indirectly affects seedling regeneration density. Stands with larger volume may provide more nutrients, which benefit seedling growth and regeneration [48].

Soil factors are closely related to the regeneration density of *Quercus* stands. Some studies suggest that thicker litter layers may inhibit seed germination and seedling growth [20,49,50]. However, this study found that litter layer thickness was significantly positively correlated with the regeneration density of Seedling 2 and Seedling 3. Generally, a thicker litter layer provides a protective cover for seedlings, reducing soil surface temperature fluctuations and moisture evaporation, thus creating favorable conditions for seedling growth. Additionally, the study found that exchangeable calcium was significantly negatively correlated with Seedling 2 regeneration density and positively correlated with soil pH. In acidic soils, high concentrations of exchangeable calcium may neutralize soil acidity, which could affect seedling growth. Seedlings are better adapted to slightly acidic or neutral soils [51], and excessively high calcium levels may alter soil pH, negatively affecting seedling development.

Correlation analysis revealed that the primary environmental factors influencing seedling regeneration density were stand and soil factors. However, the random forest algorithm's ranking of important factors indicated that, as seedlings' growth levels increased, the importance of altitude as a site factor also increased. This may be due to the relatively few study plots and the small variation in site factors among the stands. Therefore, future research should expand the sample size and comprehensively consider both biotic and abiotic factors to deepen our understanding of the key environmental factors affecting natural regeneration in *Quercus* stands in northern China.

5. Conclusions

In Beijing, *Q. aliena* and *Q. variabilis* stands exhibited stronger regeneration abilities, while *Q. dentata*, *Q. acutissima*, and *Q. mongolica* stands had relatively weaker regeneration capacities. The main environmental factors influencing natural regeneration in *Quercus* stands in Beijing include altitude, species richness, Shannon–Wiener index, Pielou evenness index, stand volume, shrub density, litter layer thickness, total nitrogen, available phosphorus, exchangeable calcium, and available manganese. Among these, altitude was the most significant site factor affecting regeneration, species richness, Shannon–Wiener index, Pielou evenness index, stand volume, and shrub density were the most influential stand factors, and litter layer thickness, total nitrogen, available phosphorus, exchangeable calcium, and available manganese were the most significant soil factors. Therefore, we recommend augmenting the cultivated area of *Quercus* species in low-altitude regions by strategically supplementing *Quercus* seedlings within the forest understory. Additionally, enhancing the natural regeneration capacity of *Q. mongolica* forests can be prioritized by managing their stand density effectively. Overall, *Quercus* stand management should consider the impacts of site factors, stand factors, and soil factors comprehensively, creating conditions that are more favorable for natural regeneration in *Quercus* stands.

Author Contributions: Conceptualization, Y.J. and X.H.; methodology, G.D. and X.H.; software, G.D.; validation, Y.C.; formal analysis, G.D. and Y.C.; investigation, F.L. (Fang Li) and X.H.; resources, F.L. (Fang Liang); data curation, Y.W. and H.C.; writing—original draft preparation, X.H.; writing—review and editing, Y.J.; visualization, Z.L.; supervision, X.H.; project administration, X.H.; funding acquisition, X.H. All authors have read and agreed to the published version of the manuscript.

Funding: This research was funded by the Study on Natural Regeneration Characteristics and Influencing Factors of Oak Forests in Beijing, Phase II (YKYQN202502); Research on the multifunctionality and driving factors of oak forest ecosystems in Beijing (YZQN202405); the Postgraduate Education Reform and Quality Improvement Project of Henan Province (YJS2023SZ23); the Xinyang Academy of Ecological Research Open Foundation (2023XYMS10); and the key scientific research projects in universities of Henan (23A220003).

Data Availability Statement: The data are not publicly available due to proprietary rights.

Conflicts of Interest: The authors declare no conflicts of interest.

References

1. Li, Q.B.; Zhang, C.Y.; Zhao, X.H. Species diversity and influencing factors of coniferous and broad-leaved mixed forest communities in different successional stages of Changbai Mountain. *Acta Ecol. Sin.* **2022**, *42*, 7147–7155.
2. Chazdon, R.L.; Guariguata, M.R. Natural regeneration as a tool for large-scale forest restoration in the tropics: Prospects and challenges. *Biotropica* **2016**, *48*, 716–730. [CrossRef]
3. Rahman, M.H.; Khan, M.A.S.A.; Roy, B.; Fardusi, M.J. Assessment of natural regeneration status and diversity of tree species in the biodiversity conservation areas of Northeastern Bangladesh. *J. For. Res.* **2011**, *22*, 551–559. [CrossRef]
4. Xu, Z.B.; Dai, L.M.; Chen, J.Q.; Wang, Z.; Dai, H.; Cai, L.X. Natural regeneration condition in *Pinus koraiensis* broad-leaved mixed forest. *Acta Ecol. Sin.* **2001**, *9*, 1413–1420.
5. Dubravac, T.; Barčić, D.; Rosavec, R.; Španjol, Ž.; Vojniković, S. The Dynamics of Stand Structure Development and Natural Regeneration of Common Beech (*Fagus sylvatica* L.) in Plitvice Lakes National Park. *Forests* **2024**, *15*, 357. [CrossRef]
6. Redmond, M.D.; Kelsey, K.C. Topography and overstory mortality interact to control tree regeneration in spruce-fir forests of the southern Rocky Mountains. *For. Ecol. Manag.* **2018**, *427*, 106–113. [CrossRef]

7. Liu, M.G.; Yin, Y.; Kong, F.S.; Lu, G.Z.; Zhang, L.S. Effecting factors of natural regeneration of *Pinus tabulaeformis* plantation in semiarid region, Western Liaoning. *J. Shenyang Agric. Univ.* **2014**, *45*, 418–423.
8. Albrecht, M.A.; McCarthy, B.C. Seedling establishment shapes the distribution of shade-adapted forest herbs across a topographical moisture gradient. *J. Ecol.* **2009**, *97*, 1037–1049. [CrossRef]
9. Ali, A.; Dai, D.; Akhtar, K.; Teng, M.; Yan, Z.; Urbina-Cardona, N.; Mullerova, J.; Zhou, Z. Response of understory vegetation, tree regeneration, and soil quality to manipulated stand density in a *Pinus massoniana* plantation. *Glob. Ecol. Conserv.* **2019**, *20*, e00775. [CrossRef]
10. Zhang, Z.D.; Mao, P.L.; Liu, Y.H.; Li, Q.; Liu, S.; Xue, Q. Effects of forest structure on natural regeneration of *Pinus thunbergii* coastal shelter forest in Yantai region. *Acta Ecol. Sin.* **2010**, *30*, 2205–2211.
11. Bai, Z.Q.; Liu, H.; Zhang, X.P.; Liu, D.; Guo, Z.J. Analysis on influence factors of natural regeneration of young poplar seedlings in Irtysh River Basin, Xinjiang. *J. Northwest For. Univ.* **2011**, *26*, 98–102.
12. Sun, G.L.; Ji, P.P.; Ge, Z.X. Study on relationship of soil nutrients and stand growth of *Larix principis-rupprechtii* plantations with different ages. *Hebei J. For. Orchard. Res.* **2017**, *32*, 129–133.
13. Liu, S.C.; Chen, L.X.; Duan, W.B.; Zhang, C.; Li, S.B.; Li, Y.F.; Li, S.R.; Liang, W.W. Effects of soil characteristics on forest gap regeneration in different types of natural *Pinus koraiensis* mixed forest. *Acta Ecol. Sin.* **2017**, *37*, 4072–4083.
14. Bharathi, S.; Devi Prasad, A.G. Diversity, population structure and regeneration status of arboreal species in the four sacred groves of Kushalnagar, Karnataka. *J. For. Res.* **2017**, *28*, 357–370. [CrossRef]
15. Ye, Z.L.; Dou, X.W.; Tang, M.P. Impact of site factors on forest regeneration in Mount Tianmu mixed coniferous and broad-leaved forests. *J. Zhejiang AF Univ.* **2024**, *41*, 557–567.
16. Tiwari, O.P.; Sharma, C.M.; Rana, Y.S. Influence of altitude and slope-aspect on diversity, regeneration and structure of some moist temperate forests of Garhwal Himalaya. *Trop. Ecol.* **2020**, *61*, 278–289. [CrossRef]
17. Gou, Y.Y.; Zhang, W.H.; Zhou, J.Y.; He, J.F.; Li, Y.H. Characteristic of seedling regeneration of *Xanthoceras sorbifolia* in Hilly Areas of Loess Plateau. *Sci. Silvae Sin.* **2008**, *51*, 11–17.
18. Wang, Z.; Qin, K.; Fang, W.; Wang, H. Neighborhood Competition and Understory-Associated Vegetation Are Important Factors Influencing the Natural Regeneration of Subtropical Mountain Forests. *Forests* **2024**, *15*, 1017. [CrossRef]
19. Li, S.J.; Wang, X.J. Regeneration Characteristics and its contributing factors in *Castanopsis kawakamii* natural forests. *J. Northeast. For. Univ.* **2023**, *51*, 20–24+45.
20. Dong, L.L.; Liu, H.M.; Zhao, J.C.; Gao, Y.X.; Wang, C.C.; Fan, J.G. Effects of stand structure on natural regeneration of Mongolian oak forests in the Liaodong mountainous region of China. *For. Res.* **2021**, *34*, 104–110.
21. Lv, K.; Zhou, M.; Ding, Y.; Zang, R.; Yao, J.; Luo, Y.; Yan, D. Regeneration characteristics and influencing factors of woody plant on natural evergreen secondary broad-leaved forests in the subtropical, China. *Glob. Ecol. Conserv.* **2023**, *42*, e02394. [CrossRef]
22. Muscolo, A.; Sidari, M.; Mercurio, R. Influence of gap size on organic matter decomposition, microbial biomass and nutrient cycle in Calabrian pine (*Pinus laricio*, Poiret) stands. *For. Ecol. Manag.* **2007**, *242*, 412–418. [CrossRef]
23. Zeng, S.Q.; Zhang, M.; Xiao, H.S.; Huang, Y.J.; Gan, J.J.; Peng, Q.L. Study on regeneration and succession of mixed forest of *Schima superb* and *Cunninghamia lanceolata* in Qingshigang Forest Farm. *J. Cent. South Univ. For. Technol.* **2013**, *33*, 1–6.
24. Xu, E.; Niu, Y.; Zhao, W.; Jing, W.; Wu, X.; Zhao, J.; Ma, X.; Ren, X. Response of natural regeneration of *Picea crassifolia* forest to soil physical and chemical properties in the northern foot of Qilian mountains. *J. Cent. South Univ. For. Technol.* **2024**, *44*, 90–100.
25. Xiong, Z.; Sun, J.; Zhong, P.; Liang, L.; Dang, H.; Wang, G. The Impact of Natural Regeneration of *Phoebe bournei* in Anfu County, Jiangxi Province, on Community Diversity and Soil Nutrient Characteristics. *Forests* **2023**, *14*, 1783. [CrossRef]
26. Han, G.; Wang, G.; Mao, P.; Zhang, Z.; Yu, J.; Xu, J. Regeneration dynamics of young *Pinus thunbergii* and its influencing factors in the coastal protection forests in Northern Shandong Peninsula. *Sci. Silvae Sin.* **2010**, *46*, 158–164.
27. National Forestry and Grassland Administration. *China Forest Resources Report (2014–2018)*; China Forestry Press: Beijing, China, 2019; pp. 164–166.
28. Zhang, Y.M. Update and application of "One Map" for forest resource management in Beijing. *For. Investig. Des.* **2023**, *52*, 83–87.
29. Shi, N.N.; Han, Y.; Wang, Q.; Xiao, N.W.; Quan, Z.J. Spatial and Temporal Characteristics of Fractional Vegetation Cover and Its Response to Urbanization in Beijing. *Environ. Sci.* **2024**, *45*, 5318–5328.
30. Ren, Y.M.; Wen, Z.Y.; Wang, M.N.; Li, F.; Jia, Z.K. Evaluation of forest carbon sequestration capacity in Beijing. *J. Beijing For. Univ.* **2023**, *45*, 100–111.
31. Bao, S.D. *Soil Agricultural Chemical Analysis*, 3rd ed.; China Agricultural Press: Beijing, China, 2000; pp. 30–107.
32. Qin, W.X.; Si, G.C.; Lei, T.Z.; Zhang, S.Y.; Ma, J.Z. The impact of nitrogen fertilizer addition on soil microbial biomass and enzyme activity. *Jiangsu Agric. Sci.* **2021**, *49*, 170–175.
33. Wu, H.; Wang, L.D.; Song, D.C.; Guo, C.X.; Wang, F.L.; He, F.L.; Zhao, H.R. Soil properties and enzyme activities of abandoned farmland in different years in Minqin. *Agric. Res. Arid. Areas* **2021**, *39*, 191–199.
34. *GB/T 38590-2020*; Technical Regulations for Continuous Forest Resource Inventory. China Standards Press: Beijing, China, 2020.

35. Fang, J.; Wang, X.; Shen, Z.; Tang, Z.; He, J.; Yu, D.; Jiang, Y.; Wang, Z.; Zheng, C.; Zhu, J.; et al. Methods and protocols for plant community inventory. *Biodivers. Sci.* **2009**, *17*, 533–548.
36. Ma, K.P.; Huang, J.H.; Yu, S.L.; Chen, L.Z. Plant community diversity in Dongling Mountain, Beijing, China II: Species richness, evenness and species diversities. *Acta Ecol. Sin.* **1995**, *15*, 268–277.
37. R Core Team. *R: A Language and Environment for Statistical Computing*; R Foundation for Statistical Computing: Vienna, Austria, 2023; Available online: https://www.R-project.org/ (accessed on 27 January 2024).
38. Wei, T.Y.; Viliam, S. R Package 'corrplot': Visualization of a Correlation Matrix (Version 0.92). 2021. Available online: https://github.com/taiyun/corrplot (accessed on 27 January 2024).
39. Schloerke, B.; Cook, D.; Larmarange, J.; Briatte, F.; Marbach, M.; Thoen, E.; Elberg, A.; Toomet, O.; Crowley, J.; Hofmann, H.; et al. GGally: Extension to 'ggplot2'. R Package Version 2.1.2. 2021. Available online: https://CRAN.R-project.org/package=GGally (accessed on 31 January 2024).
40. Martin, L.; Breiman, L.; Friedman, T. Random forests. *Mach. Learn.* **2005**, *48*, 275–298.
41. Wickham, H. *ggplot2: Elegant Graphics for Data Analysis*; Springer-Verlag: New York, NY, USA, 2016.
42. Wei, S.P.; Liang, W.J.; Wei, X.; Bu, R.Y.; Yao, J.F. Natural regeneration of *Larix principis-rupprechtii* plantations with different densities and its influencing factors. *Chin. J. Appl. Ecol.* **2022**, *33*, 2687–2694.
43. Huang, L.; Zhu, G.Y.; Kang, L. Regeneration characteristics and related factors affecting saplings in *Quercus* spp. natural secondary forests in Hunan Province, China. *Acta Ecol. Sin.* **2019**, *39*, 4900–4909.
44. Ren, X.; Zhu, Y.; Chen, Z.; Ding, C.; Li, Y.; Yang, G. Regeneration of Arbor Trees and Its Contributing Factors in an *Oak* Forest in Taibai Mountain, China. *Sci. Silvae Sin.* **2019**, *55*, 11–21.
45. Yang, L.; Kang, Y.X.; Li, X.J.; Zhou, L.Y.; Wang, F. Natural regeneration of the ancient *Platycladus orientalis* in the mausoleum of the Yellow Emperor. *J. Northwest For. Univ.* **2015**, *30*, 82–86.
46. Li, H.P.; Wang, S.S. Regeneration capacity of *Spruce* forest in Xinjiashan forest area of Shaanxi. *Shaanxi For. Sci. Technol.* **2009**, *1*, 9–12.
47. Zhang, M.X.; Wang, D.X.; Peng, S.L.; Huang, Y.K.; Zhang, G.G. Community stability analysis for the oak-pine mixed forest in Qinling Mountains. *Acta Ecol. Sin.* **2015**, *35*, 2564–2573.
48. Wang, S.L.; Zhou, J.P. Coupling relationship between stand growth and impacting factors based on structural equation model. *J. Beijing For. Univ.* **2014**, *36*, 7–12.
49. Navarro-Cano, J.A. Effect of grass litter on seedling recruitment of the critically endangered *Cistus heterophyllus* in Spain. *Flora-Morphol. Distrib. Funct. Ecol. Plants* **2008**, *203*, 663–668. [CrossRef]
50. Zhang, S.Z.; Li, M.; Zhang, S.B.; Zhang, Z.D.; Huang, X.R. Factors affecting natural regeneration of *Larix principis-rupprechtii* plantations in Saihanba of Hebei, China. *Acta Ecol. Sin.* **2015**, *35*, 5403–5411.
51. Catovsky, S.; Bazzaz, F. Feedbacks between canopy composition and seedling regeneration in mixed conifer broad-leaved forests. *Oikos* **2002**, *98*, 403–420. [CrossRef]

forests

MDPI

Article

Construction and Comparison of Single-Tree Biomass Model for *Dendrocalamus brandisii*

Zuming Wang [1], Weisheng Zeng [2], Lei Guo [3], Zhihong Xu [3], Shaohui Fan [4], Chunjun Cai [4], Chaomao Hui [1,*] and Weiyi Liu [1,3,*]

1 Research Institute of Bamboo and Rattan, Cluster Bamboo Engineering Technology Research Center, College of Forestry, Southwest Forestry University, Kunming 650224, China; lighearting@foxmail.com
2 Academy of Forest Inventory and Planning, State Forestry Administration, Beijing 100714, China; zengweisheng0928@126.com
3 Centre for Planetary Health and Food Security, School of Environment and Science, Griffith University, Nathan, Brisbane, QSD 4111, Australia; lei.guo@griffith.edu.au (L.G.); zhihong.xu@griffith.edu.au (Z.X.)
4 International Center for Bamboo and Rattan, Key Laboratory of National Forestry and Grassland Administration, Beijing 100102, China; fansh@icbr.ac.cn (S.F.); caicj@icbr.ac.cn (C.C.)
* Correspondence: ynbamboo@163.com (C.H.); weiyiliu651@126.com (W.L.)

Abstract: *Dendrocalamus brandisii* (Munro) Kurz is a high-quality bamboo species for shoots, known for its sweet, tender, and crisp bamboo shoots, making it highly valuable for development. The biomass of bamboo forests is closely related to bamboo shoot yield, and studying biomass accumulation helps maintain the stability of artificial forest ecosystems. Biomass estimation facilitates the monitoring of stand dynamics and promotes the scientific management and sustainable development of *D. brandisii* plantations. This study collected biological data from 181 *D. brandisii* individuals in Changning County, Yunnan Province, to construct mathematical models for estimating single-plant biomass using the least squares method. The models were iteratively optimized using the quasi-Newton method. Based on performance indicators and residual analysis, six models were identified as suitable for estimating the biomass of *D. brandisii*, including multiple linear regression (MLR), linear, allometric growth, and cubic models. These models provide valuable references for biomass estimation and the management of *D. brandisii* plantations.

Keywords: *Dendrocalamus brandisii*; biomass modeling; model fitting optimization; bamboo forest management; plantation management; production estimation

1. Introduction

Forest biomass is a key indicator for assessing forest ecosystem functions and carbon budgets. It also serves as fundamental data for exploring ecological issues in forestry. Biomass estimation is a complex process that requires comprehensive consideration of factors such as tree species, stand density, and mixing ratios [1]. Studies have shown that mixed-species growth significantly enhances forest productivity [2], as the growth rates and proportions of different species directly influence overall forest output [3]. Currently, two primary methods are used for biomass estimation: direct felling measurements and indirect estimation [4,5]. Indirect methods include remote sensing combined with machine learning algorithms, suitable for large-scale rapid assessment of forest resources [6,7]; three-dimensional LiDAR modeling, which excels in volume estimation for complex terrains [8]; and statistical modeling based on biological characteristics, employing DBH and tree height to construct regression equations for precise small-scale estimations [9]. Many biomass equations for trees use diameter at breast height (DBH) and tree height as main variables, achieving high fitting accuracy [10].

While 3S technologies, encompassing remote sensing (RS), geographic information systems (GISs), and global positioning systems (GPSs), are widely used in forestry, the high costs of equipment and technical expertise pose challenges. Free remote sensing data also face limitations in resolution and timeliness [11]. As a result, empirical formula-based biomass models remain practical and cost-effective tools, particularly for plantation management. Accurate biomass estimation enables optimization of production strategies, stabilization of plantation ecosystems, and long-term development [12–14].

Bamboo, a globally significant non-timber forest product, plays an essential role as a perennial agricultural plant [12]. The accumulation of bamboo biomass directly affects shoot yields and economic returns. For instance, the large individual biomass of *Pleioblastus amarus* leads to higher fresh shoot yields under optimal density [13]. Allometric growth models based on DBH are considered effective for predicting bamboo biomass, allowing stand-level estimation by aggregating individual models. These models are also useful for carbon stock predictions [14–17]. In recent years, regression models incorporating DBH and bamboo height have demonstrated excellent performance in studies on *Bambusa chungii*, *D. tsiangii*, *B. blumeana*, *D. latiflorus*, and *P. amarus* [18–22]. Allometric growth models are widely applied to evaluate biomass across bamboo species [23]. Furthermore, studies on species such as *B. emeiensis*, *B. rigida*, *Phyllostachys edulis*, *P. bambusoides*, and *P. nigra* have enriched the theoretical and practical foundation of bamboo biomass models [24–26].

$$W = 0.008DH + 0.000403\,(DH)^2 \quad (1)$$

$$W = 0.372D^2 - 0.753D + 1.316 \quad (2)$$

Two biomass models are presented as examples: (1) The biomass model for *B. chungii*. (2) The biomass model for *D. tsiangii*. In these models, D denotes the diameter at DBH, and H denotes the height of the bamboo culm.

D. brandisii, a large clumping bamboo species in the family Gramineae and genus *Dendrocalamus*, holds significant development potential. It is widely distributed in Southeast Asia and southern China and is characterized by high yields, rapid growth, and dual-use for shoots and culms [27–29]. Its culms are suitable for pulp and engineered wood products, while its fresh shoots are rich in nutrients, can be eaten raw without the bitterness found in other bamboo species, and hold significant economic value [30,31]. The bamboo samples collected in this experiment exhibited stable phenotypic traits and the highest levels of various amino acids among multiple provenances in Yunnan, making them the most superior variety in terms of morphological characteristics and nutritional value [32]. Accurate biomass estimation of *D. brandisii* is essential for developing planting and harvesting plans, reducing costs, and maximizing returns. Among bamboo species, *Phyllostachys edulis* has been extensively studied. Although *Phyllostachys edulis* is a monopodial species, differing from the sympodial *D. brandisii*, both species are large bamboos with similar morphological traits, making *Phyllostachys edulis* biomass models valuable references. Models using DBH and bamboo height to estimate biomass for culms, branches, leaves, and roots in Moso bamboo have shown excellent performance [26]. However, precise measurement of culm height in mature Moso bamboo, which can reach 28 m, remains challenging without felling. Factors such as stand density, average DBH, competition indices, soil depth, and elevation significantly influence *Phyllostachys edulis* biomass [33]. Most of these studies focus on natural forests and offer limited reference value for plantations, which are key contributors to economic value. Currently, no biomass models specifically tailored for *D. brandisii* plantations exist. This study, based on data from *D. brandisii* plantations in Mengtong Town,

Changning County, Yunnan Province, constructs production-oriented biomass models, providing a scientific reference for bamboo forest management.

2. Materials and Methods

2.1. *Study Area*

The study area is located in Mengtong Township, Changning County, Yunnan Province, at longitudes 99°30′–99°41′ and latitudes 24°24′–24°38′. It covers an area of 298 km^2, with a width of 12.5 km from east to west and a length of 26 km from north to south. The elevation ranges from 1010 m to 2175 m, and the average annual rainfall is 1361 mm. The climate is diverse, with abundant rainfall and fertile soil. The area includes both subtropical lowland river valleys and temperate mountainous regions, making it suitable for various agricultural activities.

2.2. *Data Source*

2.2.1. Selection and Sample Collection of Standard Bamboo

The selection of sample bamboo and data collection took place from November to December 2023 in Mengtong Township, Changning County, Yunnan Province, with the specific location shown in Figure 1. Based on field surveys, 181 healthy, pest-free *D. brandisii* bamboos with good growth were selected from 41 typical plots in the *D. brandisii* plantation, representing different age classes and diameter classes. These included the following: age class I (bamboo age ≤ 1 year) with 41 plants, age class II (1 year < bamboo age ≤ 3 years) with 78 plants, and age class III (bamboo age > 3 years) with 62 plants. The classification of ages followed Ji Xing et al.'s work, and the management experience was accumulated over many years by the forest management staff. The moisture content varies significantly among different age classes, and the classification of age classes is a crucial step in the measurement of biomass [34].

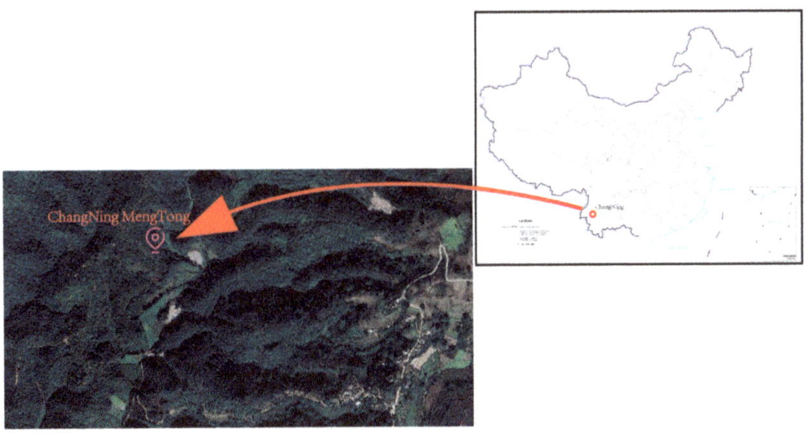

Figure 1. Map of the location of sampling sites.

2.2.2. Sample Indicator Measurement

Although most bamboos and hardwood trees have woody structures, their stem structures differ; bamboo plants are mostly hollow, while hardwood trees are generally solid [35]. Therefore, when data on wood density and moisture content are available, biomass accumulation can be predicted based on volume [36]. In bamboo plants, there are significant variations between individuals, such as differences in internode length, wall thickness, and other factors. These variations make it difficult to accurately calculate the biomass of bamboo plants using the typical hollow structure model. Thus, biomass determination in bamboo plants is typically based on the moisture content of each organ to calculate dry weight [34,37–39]. This method offers a more accurate reflection of the plant's actual growth condition.

The selected bamboo samples were felled at the base, and data such as DBH, internode length, and bamboo height were measured. Samples of bamboo stems, branches, and leaves were collected, with each organ sample weighing between 200 and 300 g. Bamboo stem samples were taken at four equal sections along the stem, with approximately 3 cm long bamboo rings collected to measure their fresh weight. Branch and leaf samples were taken from the upper, middle, and lower parts of the bamboo, selecting branches of varying thicknesses and leaves of different sizes. The collected samples were placed in an oven and dried at 80 °C for 72 h, then cooled to room temperature and weighed [40]. The moisture content of each organ was calculated, and the dry weight of each organ and the total dry weight of the bamboo were determined based on the moisture content. The formula is as follows:

$$MC = \frac{W_{fresh} - W_{dry}}{W_{fresh}}$$

$$W_{dry} = W_{dresh} \times (1 - MC)$$

MC represents the moisture content;
W_{dry} represents the dry weight of the organ (in kg);
W_{fresh} represents the fresh weight of the organ (in kg).

After data collection, all variables were initially compiled and organized for analysis. To ensure the accuracy and reliability of the dataset, the Mahalanobis distance method was applied to screen for potential outliers. By considering the correlations among variables and the covariance structure of the data, this method effectively identified anomalous data points [41]. Observations with a Mahalanobis distance exceeding the predefined threshold were excluded from further analysis. This screening process minimized the influence of outliers, ensuring that the dataset accurately represented the underlying patterns and relationships for subsequent modeling efforts.

2.3. Model Construction and Selection

2.3.1. Model Construction

The general form of the regression models commonly used for calculating individual tree biomass is as follows:

$$y = \beta_0 {x_1}^{\beta_1} {x_2}^{\beta_2} \cdots {x_n}^{\beta_n} + \varepsilon$$

In this formula, y typically represents biomass or volume, while x pertains to commonly employed morphological parameters such as DBH, tree height, and other related factors. Since the variables DBH and tree height are correlated, incorporating both DBH and tree height as independent predictors in an allometric biomass model inevitably introduces some degree of multicollinearity. To mitigate the adverse effects of multicollinearity, some models also adopt the ratio of DBH to tree height as an independent variable [42].

Existing biomass modeling research indicates a significant correlation between tree aboveground biomass and their DBH and tree height [43]. However, bamboo, as a member of the family Gramineae, differs greatly from trees in growth and development patterns. In *D. brandisii* plantations, bamboo height is strongly influenced by artificial management. For example, at the plantation base in Mengtong Township, Changning County, Yunnan Province, the tips of most mother bamboos are artificially pruned for easier management. As a result, we opted to use the length of the internode at breast height, denoted as L, as a calculation variable instead of bamboo height. This approach was adopted to more accurately estimate the biomass of *D. brandisii* and enhance the model's predictive performance.

In the process of modeling, it was observed that the contribution of internode length at breast height to total plant biomass was minimal. Therefore, in most bivariate models, this variable was excluded during regression analysis, resulting in biomass models that solely used DBH as the independent variable. However, a few models retained internode length at breast height as a variable. Ultimately, five models were identified that included both DBH and internode length at breast height as independent variables, along with five other models that used DBH alone. The specific formulas for these models are presented in Table 1. In the equations presented, W represents the biomass (in kilograms), D denotes the DBH (in centimeters), and L stands for the length of bamboo internodes at breast height (in centimeters). These variables are key predictors used to model and estimate the biomass of bamboo accurately.

Table 1. Mathematical model formula.

Model Name	Model Formula
Tree Biomass Model	$W = e^{\beta_0} D^{\beta_1} L^{\beta_2}$
Relative Growth Model	$W = a\left(D^2 L\right)^b$
MLR	$W = \beta_0 + \beta_1 D + \beta_2 L$
Height–Diameter	$W = aD^2 L + b$
Schumacher–Hall	$W = \beta_0 + \beta_1 \ln D + \beta_2 lnL$
Quadratic	$W = a + bD + cD^2$
Cubic	$W = a + bD + cD^2 + dD^3$
Compound	$W = ab^D$
Allometric	$W = aD^b$
Linear	$W = a + bD$

2.3.2. The Criteria for Selecting Models

Using SPSS 24 software, the correlation between various biomass indicators was determined to establish a biomass estimation model for *D. brandisii*. During the evaluation and comparison of biomass models, the following six metrics were used to assess the model: coefficient of determination (R^2), standard error of estimate (SEE), total relative error (TRE), mean system error (MSE), mean prediction error (MPE), and mean percentage standard error (MPSE) [44].

$$R^2 = 1 - \sum (y_i - \hat{y}_i)^2 / \sum (y_i - \overline{y})^2$$

$$SEE = \sqrt{\sum (y_i - \hat{y}_i)^2 / (n - p)}$$

$$TRE = 100 \times \sum (y_i - \hat{y}_i) / \sum \hat{y}_i$$

$$MSE = \sum \frac{(y_i - \hat{y}_i)}{\hat{y}_i} \times \frac{100}{n}$$

$$MPE = t_\alpha \cdot (SEE/\overline{y}) \cdot \sqrt{n} \times 100$$

$$MPSE = 100 \times \sum \left| \frac{y_i - \hat{y}_i}{\hat{y}_i} \right| / n$$

In these equations, y_i designates the actual observed values, $\hat{y}_i$ represents the model estimations, $\overline{y}$ constitutes the mean of the observed values, n indicates the sample size, and t_α represents the t-value at the confidence level of α. Six indices were computed to assess the developed model. Among them, R^2 and SEE are the most prevalent indicators for goodness of fit in regression models; conversely, TRE and MSE are crucial metrics that ought to be maintained within specific ranges; TRE should be within ±3% and MSE within ±5%, with values approaching zero being optimal. For a comprehensive assessment of applicability, TRE and MSE must also be independently verified using verification samples, with TRE ideally within ±3% or below MPE values and MSE within ±5%. If MSE exceeds ±5% but remains within ±10%, the model is regarded as having moderate systematic error; if MSE exceeds ±10%, the systematic error is considered high. MPE indicates the accuracy level of average biomass estimation, while MPSE indicates the accuracy of individual plant biomass estimation, with smaller values being preferred for both indices.

2.4. Analysis Tools

Preliminary data processing, including statistical analysis and filtering, was carried out using Excel spreadsheets, while model construction and validation were performed using SPSS 24 and Python 3.8.

First, the initial biomass model for *D. brandisii* was developed using the ordinary least squares (OLS) method based on measured data. To improve the model's accuracy, the minimize function from Python's SciPy library was employed for optimization. Specifically, the quasi-Newton method was applied [45].

The minimize optimization algorithm is a powerful tool for finding the minimum of a function under given constraints and is widely used in machine learning, signal processing, and system optimization. Various optimization algorithms, such as gradient descent, Newton's method, and genetic algorithms, are available within the minimize library. In this study, we adopted the BFGS (Broyden–Fletcher–Goldfarb–Shanno) method, a well-known quasi-Newton approach for unconstrained optimization. This method is particularly effective in approximating the Hessian matrix by using first-order derivatives, avoiding the computational cost of calculating second-order derivatives directly. The BFGS method iteratively searches for the function's minimum and is known for its computational efficiency, robustness, and ease of implementation, making it highly applicable in both theoretical and practical optimization tasks.

3. Results

3.1. Data Collection

3.1.1. Analysis of the Water Content in the Organs of *D. brandisii*

There are structural and functional differences among the various organs of *D. brandisii*, and their moisture content varies. As the bamboo's age increases, the moisture content in the different organs also changes, which is related to the accumulation of organic substances within the bamboo. The moisture content at different bamboo classes for each organ is shown in Figure 2.

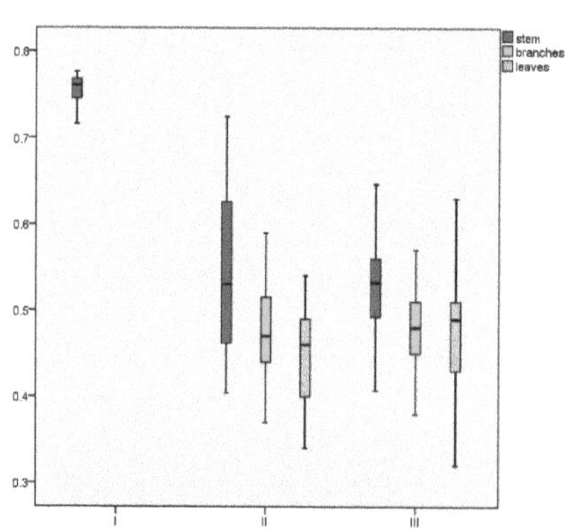

Figure 2. Moisture content of different organs of *D. brandisii* across age classes.

As shown in Figure 2, for *D. brandisii*, the average moisture content of the aboveground components (e.g., stems, branches, and leaves) decreases from 74.96% in age class I to 48.44% in age class II and 48.27% in age class III as biomass accumulates across the various organs. The changes in moisture content across organs show that in age class I, the bamboo has higher moisture content due to the shoot removal that year, with branches and leaves not yet developed and the bamboo culms being immature and tender, leading to less biomass accumulation. This results in a significant difference in moisture content compared to age classes II and III. In age classes II and III, the moisture content follows the pattern of stem > branch > leaf.

3.1.2. Aboveground Biomass Proportion of Each Organ in *D. brandisii*

The aboveground biomass of *D. brandisii* consists of stems, branches, and leaves. The proportion of each organ in the aboveground biomass at different age classes is shown in Figure 3.

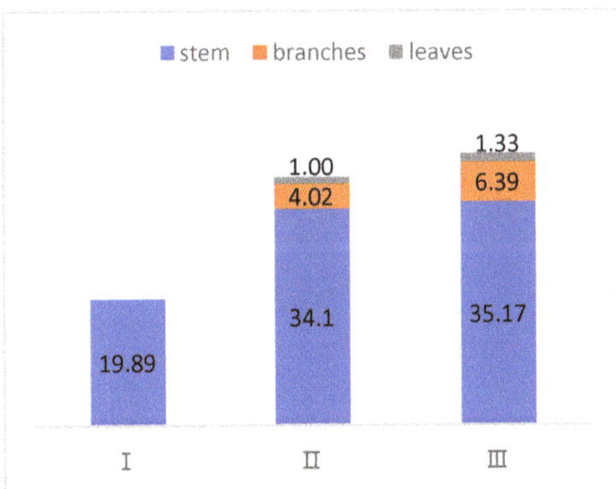

Figure 3. Distribution of each organ in the aboveground biomass of individual *D. brandisii* plants at age classes I, II, and III (kg).

According to the biomass allocation results of the aboveground parts of *D. brandisii* at different age classes, the average aboveground biomass of individual bamboo plants is 19.89 kg, 39.12 kg, and 42.89 kg per plant for age classes I, II, and III, respectively. In age class I, biomass accumulation is lower because the branches and leaves are not yet fully developed. As the bamboo grows from age class I to age class III, the biomass of each organ and the total biomass increase with age. Regarding the biomass proportions, as the bamboo matures, the branches and leaves become more abundant, and their biomass proportion increases. In contrast, the proportion of stem biomass decreases as the age class progresses.

3.2. Construction of Biomass Regression Model

The Mahalanobis distance method was applied to filter the dataset, and samples with a Mahalanobis distance greater than 10.596635 were removed, totaling 5 samples. The remaining 176 samples were used to randomly select 20% (35 samples) as the validation dataset, while the remaining 141 samples were used for model development. After data processing, regression analysis was conducted using the OLS method in SPSS. The preliminary results are shown in Table 2.

For the preliminary fitting results, R^2, SEE, TRE, MSE, MPE, and MPSE were used to compare the biomass model. Subsequently, the TRE and MSE of each model were calculated using the validation dataset. These eight metrics were then used to compare and select the best model. After comparison, it was found that only the MLR, cubic, and linear models met the standard range for all six metrics.

Table 2. Preliminary results of mathematical model fitting.

Model Name	Model Formula
Tree Biomass Model	$W = 0.1905D^{2.225}L^{0.059}$
Relative Growth Model	$W = 0.0079\left(D^2L\right)^{1.047}$
MLR	$W = -32.225 + 7.166D + 0.099L$
Height–Diameter	$W = 0.012D^2L + 1.471$
Schumacher–Hall	$W = -95.430 + 51.655lnD + 5.515lnL$
Quadratic	$W = -8.65 + 1.746D - 0.33D^2$
Cubic	$W = 32.572 - 15.62D + 2.573D^2 - 0.09D^3$
Compound	$W = 2.052 \times 1.339^D$
Allometric	$W = 0.232D^{2.228}$
Linear	$W = -29.072 + 7.186D$

Once the preliminary fitting results were obtained, the BFGS method from the minimize optimization library in Python was used for iterative optimization. The results after iteration are shown in Table 3. As shown in Table 3, the iterative method made adjustments to the MLR, height–diameter, Schumacher–Hall, quadratic, cubic, and linear models, while resulting in larger changes in the tree biomass model, relative growth model, compound, and allometric models.

Table 3. Optimized results of mathematical models.

Model Name	Model Formula
Tree Biomass Model	$W = 0.3290D^{2.0680}L^{0.0199}$
Relative Growth Model	$W = 0.0272\left(D^2L\right)^{0.9001}$
MLR	$W = -32.2251 + 7.1659D + 0.0993L$
Height–Diameter	$W = 0.0115D^2L + 1.4715$
Schumacher–Hall	$W = -95.4303 + 51.6547lnD + 5.5155lnL$
Quadratic	$W = -8.6498 + 1.7457D + 0.3298D^2$
Cubic	$W = 32.5724 - 15.6202D + 2.5733D^2 - 0.0904D^3$
Compound	$W = 4.4516 \times 1.2429^D$
Allometric	$W = 0.35509D^{2.0703}$
Linear	$W = -29.0719 + 7.1856D$

3.3. *Selection of Biomass Models*

3.3.1. Performance Metrics of the Model

R^2, SEE, TRE, MSE, MPE, and MPSE of the biomass model results obtained after iteration were compared. Then, the TRE and MSE of each model were calculated using the validation dataset. These eight metrics were used to compare and select the best models.

After iterative optimization, the indicators of several models reached the acceptable range and performed well. The models that met the standards include the following: tree biomass model, MLR, height–diameter, cubic, allometric model, and linear models for *D. brandisii* biomass. The results of the six models are shown in Table 4.

From the table, it can be seen that as the core indicator, R^2 shows decent performance in several models with little difference. Among the six models that meet the standards, the height–diameter model has a slightly lower R^2 and a slightly higher SEE compared to the other models. This suggests that while using the ratio of DBH to the internode length at breast height as model variables to calculate biomass has some merits, it is not the optimal solution. The TRE of the MLR, height–diameter, and cubic models is close to 0, indicating excellent performance, while the absolute value of the TRE for the tree biomass and allometric models is around 0.7%. The absolute value of the TRE for the linear model is 2.93%, significantly higher than the TRE of the other models. This indicates that the linear model may not be as suitable for calculating the biomass of *D. brandisii* as the other models. The MSE of the linear and cubic models performs significantly better than the other models. The MPE values of the six models exhibit minimal variation, whereas the MPSE values show considerable disparity. Notably, the MPSE for both the multiple linear regression and the linear model exceeds 50%, indicating poor performance. This suggests that linear models are not appropriate for estimating the biomass of *D. brandisii*.

Table 4. Performance indicators of the model.

Model Name	R^2	SEE	TRE	MSE	MPE	MPSE	TRE *	MSE *
Tree Biomass Model	0.745	10.6	−0.71	−3.16	6.29	3.92	5.35	0.34
MLR	0.737	10.7	1.76×10^{-6}	2.79	6.32	54.93	4.01	−1.30
Height–Diameter	0.692	11.5	5.21×10^{-6}	−3.11	6.80	26.17	8.71	−3.11
Cubic	0.753	10.4	-5.19×10^{-5}	0.24	6.09	24.06	3.87	−0.44
Allometric	0.745	10.5	−0.71	−3.13	6.19	23.97	5.32	0.30
Linear	0.736	10.7	−2.93	0.72	6.30	50.15	3.84	−1.21

R^2, SEE, TRE, MSE, MPE, and MPSE are the model's performance on this indicator in the modeling dataset. TRE * and MSE * represent the model's performance on this indicator in the validation dataset.

The performance of the six models in the validation dataset is relatively consistent. The TRE for all six models exceeds 3%, but all are lower than the MPE values for each model in the validation dataset. Among them, the height–diameter model has higher TRE and MSE compared to the other models, while the cubic model excels with both TRE and MSE indicators.

3.3.2. Fitting Results of Models

The six models selected after iterative optimization were plotted using the matplotlib.pyplot library and the mpl_toolkits.mplot3d library in Python, as shown in Figures 4 and 5.

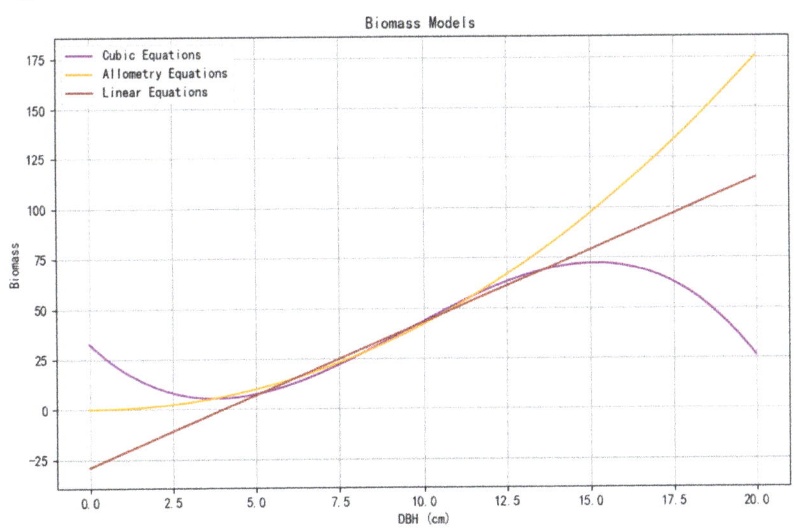

Figure 4. Curve fitting for cubic, allometry, and linear models.

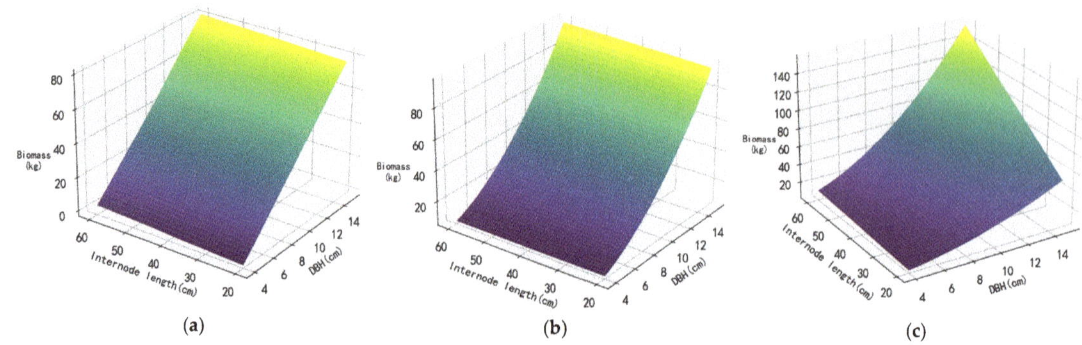

Figure 5. MLR model (**a**); tree biomass model (**b**); height–diameter model (**c**).

Among the models in Figure 4, the cubic model, allometric model, and linear model are univariate models, where biomass is only related to DBH. From the graph, it can be observed that the linear model fails to work effectively for small DBH values, as it predicts negative biomass when DBH is less than 4 cm, and the cubic model shows a local minimum around 4 cm and a local maximum at 15 cm. Therefore, it can be concluded that for the linear model, DBH should be greater than 4 cm, and for the cubic model, the applicable range of DBH is approximately between 4 cm and 15 cm. A histogram of the DBH values of all samples is shown in Figure 6. From the analysis, it can be seen that the majority of the sample bamboo has a DBH between 5 cm and 12.5 cm, which aligns with the optimal calculation range of the two models.

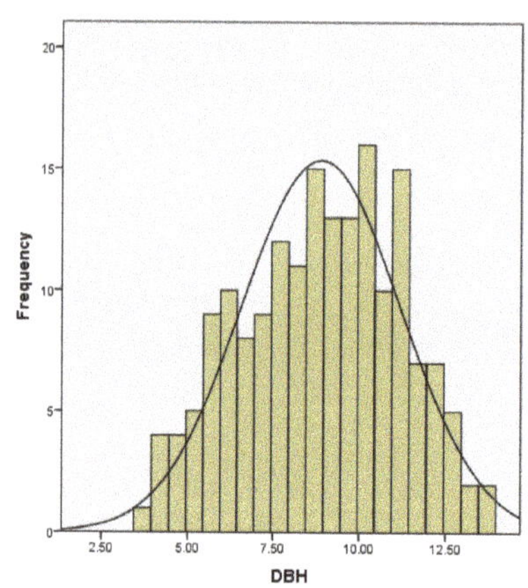

Figure 6. Histogram of the breast diameter distribution of *D. brandisii* samples.

Among the models in Figure 5, the MLR model, tree biomass model, and height–diameter model use the following two independent variables: DBH and the length of the bamboo segment at breast height. In the MLR model, DBH and bamboo segment length at breast height are treated as separate terms, with the coefficient for DBH much higher than that for the bamboo segment length. In the tree biomass model, DBH and bamboo segment length at breast height are combined in the same term. The 3D regression plot shows that in both the MLR model and the tree biomass model, the effect of bamboo segment length on biomass is much smaller than that of DBH. In contrast, the height–diameter model uses the D^2L as an independent variable, leading to a greater impact of bamboo segment length on biomass.

3.3.3. Residual Analysis of Models

Residual plots for six models were computed and plotted using a test dataset, as shown in Figure 7. The points on the residual plots represent the differences between the observed values and the model predictions, with the residuals on the vertical axis and DBH on the horizontal axis. The color bar in the figure indicates the variation in tree height. Residual plots for all six models show that the residuals are relatively dispersed within the large DBH range and do not exhibit significant systematic bias. This suggests that the models' fits are similar, with no significant trend error across different DBH ranges. As the DBH increases, the dispersion of biomass values also grows, leading to larger residuals. The residuals for all models are generally distributed between −40 kg and 30 kg, indicating comparable error magnitudes across the models. The residual distributions are similar with no significant systematic bias, indicating that the models' fits are close. The residual distribution for the MLR and allometric growth models is relatively uniform, especially in the middle DBH range (approximately 7–10 cm), where the residuals are small, suggesting that these models may perform better in predicting the middle DBH range. The tree biomass and height–diameter models show greater stability, but their residuals increase with higher DBH values, indicating that the models may overestimate actual values for larger DBH samples. The cubic model shows relatively consistent prediction errors across different DBH ranges but performs less well in the middle DBH range compared to other models. The linear model performs better in the small DBH range but underperforms in the large DBH range compared to the other models. Overall, MLR exhibits the broadest applicability. In contrast, the linear model, allometric growth model, and cubic model each demonstrate superior performance within specific intervals.

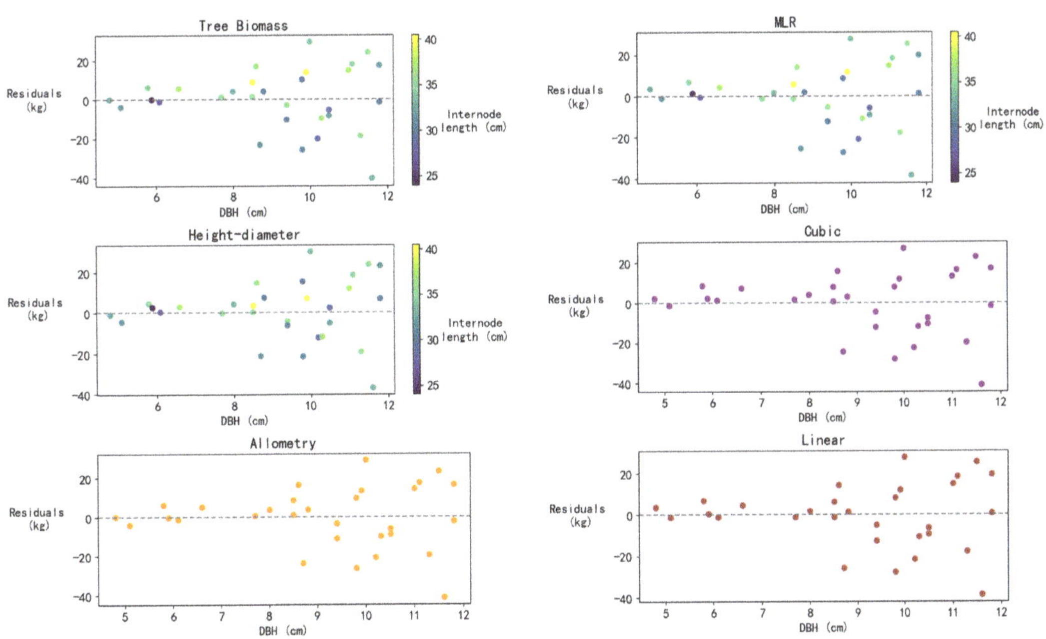

Figure 7. Residuals of tree biomass model, MLR model, height–diameter model, allometry model, and linear model (kg).

4. Discussion

4.1. Data

The experimental dataset comprised 181 samples, all collected from the same *D. brandisii* production site. During the process of mathematical modeling, substantial individual variability among samples was observed. With increasing DBH, the data showed greater dispersion. This phenomenon may be attributed to human-induced variations during the management and measurement of *D. brandisii*, as well as individual differences stemming from microhabitat variability among the bamboo plants [46]. All models exhibited low MPE but high MPSE. This typically indicates that the models perform well on average but show significant variability in error across different data points or larger errors for outliers. The following factors contribute to this phenomenon: Some samples have small actual values, leading to relatively larger proportional prediction errors, which in turn inflate MPSE. Despite low overall prediction error, specific numerical ranges may experience significant prediction deviations, elevating MPSE. Outliers in the dataset may disproportionately affect MPSE, causing higher variability in relative errors. To enhance model performance, collecting more individual biomass data for *D. brandisii* is necessary to achieve better fitting outcomes.

Among the ten commonly used mathematical models, six demonstrated strong fitting performance. These models relied solely on DBH and internode length at breast height, two easily accessible growth parameters. Future research should explore alternative model types and incorporate additional plant growth parameters to develop models more suitable for biomass estimation in *D. brandisii*. Comparing different models across various datasets would enable a more comprehensive evaluation of their accuracy and applicability, thereby providing more robust methods and references for biomass estimation in *D. brandisii*.

4.2. Model Building Method

When constructing biomass models, the method employed was the OLS, a classical regression approach characterized by its simplicity, efficiency, and wide applicability [47]. OLS minimizes the sum of squared differences between predicted and observed values to determine the optimal regression parameters. It is particularly effective for linear or approximately linear models and situations with large datasets and simple models. However, OLS assumes linearity of the model and independence of explanatory variable effects, which can lead to poor fitting performance for nonlinear relationships or data with noise and outliers, even potentially trapping the solution in a local optimum [48].

To address this issue, the quasi-Newton method was adopted as the optimization algorithm. The quasi-Newton method is effective for handling complex nonlinear optimization problems [44,45]. Compared to OLS, it offers significant advantages under conditions of nonlinear models and complex data, particularly in improving model accuracy and stability. Through quasi-Newton optimization, the model's accuracy is ensured while achieving biomass predictions more closely aligned with actual values [49].

In addition, future research could explore other iterative optimization methods, such as gradient descent, Newton's method, genetic algorithm, simulated annealing, and Lagrange multiplier method [50,51], to identify optimization approaches better suited to complex biomass models.

4.3. Comparison of Model Optimization Results

Comparing the changes in coefficients and constants across various models before and after iteration reveals that the quasi-Newton method induced minor adjustments in the MLR, height–diameter, Schumacher-Hall, quadratic, cubic, and linear models. In contrast, it caused significant changes in the tree biomass, relative growth model, compound, and allometric model. Notably, equations with substantial changes are primarily exponential or power functions.

An analysis of model evaluation metrics before and after iteration highlights the significant improvements achieved through optimization using the quasi-Newton method. Before optimization, only three models met the evaluation standards. After optimization, all models showed closer alignment with the evaluation criteria, and six models reached the standard range. This indicates that the quasi-Newton method has exceptional optimization capabilities for nonlinear mathematical models, such as the tree biomass model and allometric model, which are frequently used in plant biomass calculations [50].

Nevertheless, models such as the relative growth model, Schumacher–Hall, quadratic, and compound remained below the standards even after iterative optimization. This finding suggests that these models are unsuitable for predicting the biomass of *D. brandisii*.

5. Conclusions

Among the tested models, the allometric model consistently demonstrates superior performance metrics and is recommended as the default choice for biomass estimation under varying conditions. However, specific models are more appropriate for different DBH ranges, thereby enhancing accuracy and flexibility.

For smaller DBH ranges (5–7 cm), the linear model yields the highest accuracy. For medium DBH ranges (7–10 cm), the allometric growth model performs optimally. In larger DBH ranges (10–12 cm), the cubic model is most suitable, while for DBH values exceeding 12 cm, the allometric model remains the preferred choice. Additionally, when both DBH and internode length at breast height are available, the MLR model provides an effective means of biomass estimation.

This classification-based approach enables practitioners to select the most appropriate model according to specific DBH ranges, facilitating real-time adjustments and optimizations for biomass prediction. Although using multiple models may introduce complexity in traditional production processes, advanced applications can leverage computational tools to dynamically determine the most suitable equation based on real-time measurements, achieving optimal accuracy and flexibility. This methodology not only provides a scientific basis for production decision-making but also significantly improves the accuracy of biomass estimation for *D. brandisii*, thereby enhancing production efficiency and underscoring its strong practical applicability in bamboo forestry management.

Author Contributions: Conceptualization, Z.W. and W.L.; methodology, L.G.; software, Z.W.; validation, Z.W. and W.L.; formal analysis, C.C.; investigation, C.H.; resources, L.G.; data curation, Z.W.; writing—original draft preparation, Z.W.; writing—review and editing, W.Z., Z.X., and S.F.; visualization, L.G.; supervision, C.H. and W.L.; project administration, W.L.; funding acquisition, W.L. All authors have read and agreed to the published version of the manuscript.

Funding: This study was funded by the National Key R&D Program of China (2023YFD220120102), the Basic Research Special Project of Yunnan Province (202201AT070053), the Joint Special Project of Yunnan Province Agriculture (202301BD070001-123), the Monitoring Fund of "Dian Nan Bamboo Forest Ecosystem Positioning Observation Research Station" (2024-YN-15), and the Fund of the Changing Farmers Academician Workstation. It was also supported by the Bamboo and Rattan Research Institute of Southwest Forestry University.

Data Availability Statement: The data used and analyzed during the current study are not publicly available due to restrictions imposed by the funding institution. However, they can be made available from the corresponding author upon reasonable request.

Acknowledgments: I would like to express my deepest gratitude to all individuals and institutions who contributed to this research. Foremost, my sincere thanks go to my supervisor, for her invaluable guidance and academic insights that shaped this study. Special thanks to ChatGPT for its technical assistance in language polishing and translation, which significantly facilitated the cross-linguistic academic communication process.

Conflicts of Interest: The authors declare no conflict of interest.

References

1. Zianis, D.; Muukkonen, P.; Mäkipää, R.; Mencuccini, M. *Biomass and Stem Volume Equations for Tree Species in Europe*; Silva Fennica Monogr; Tammer-Paino Oy: Tampere, Finland, 2005.
2. Yang, H.; Guo, Z.; Chu, X.; Man, R.; Chen, J.; Liu, C.; Tao, J.; Jiang, Y. Comment on "Impacts of Species Richness on Productivity in a Large-Scale Subtropical Forest Experiment". *Science* **2019**, *363*, eaav9117. [CrossRef] [PubMed]
3. Návar, J. A Stand-Class Growth and Yield Model for Mexico's Northern Temperate, Mixed and Multiaged Forests. *Forests* **2014**, *5*, 3048–3069. [CrossRef]
4. Jyoti Nath, A.; Das, G.; Das, A.K. Above Ground Standing Biomass and Carbon Storage in Village Bamboos in North East India. *Biomass Bioenergy* **2009**, *33*, 1188–1196. [CrossRef]
5. Dong, L.; Zhang, L.; Li, F. Developing Two Additive Biomass Equations for Three Coniferous Plantation Species in Northeast China. *Forests* **2016**, *7*, 136. [CrossRef]
6. Esteban, J.; McRoberts, R.E.; Fernández-Landa, A.; Tomé, J.L.; Næsset, E. Estimating Forest Volume and Biomass and Their Changes Using Random Forests and Remotely Sensed Data. *Remote Sens.* **2019**, *11*, 1944. [CrossRef]
7. Millan, M.; Bonnet, A.; Dauzat, J.; Vezy, R. Advancing Fine Branch Biomass Estimation with Lidar and Structural Models. *Ann. Bot.* **2024**, *134*, 455–466. [CrossRef]
8. Lin, Y.; Jaakkola, A.; Hyyppä, J.; Kaartinen, H. From TLS to VLS: Biomass Estimation at Individual Tree Level. *Remote Sens.* **2010**, *2*, 1864–1879. [CrossRef]
9. Brown, S.; Gillespie, A.J.R.; Lugo, A.E. Biomass Estimation Methods for Tropical Forests with Applications to Forest Inventory Data. *For. Sci.* **1989**, *35*, 881–902. [CrossRef]
10. Jenkins, J.C. *Comprehensive Database of Diameter-Based Biomass Regressions for North American Tree Species*; United States Department of Agriculture, Forest Service, Northeastern Research Station: Amherst, MA, USA, 2004.
11. Seelan, S.K.; Laguette, S.; Casady, G.M.; Seielstad, G.A. Remote Sensing Applications for Precision Agriculture: A Learning Community Approach. *Remote Sens. Environ.* **2003**, *88*, 157–169. [CrossRef]
12. Wang, S.; Lin, S.; Li, W.; Ding, Y. Variations in Fiber Morphology and Chemical Components of *Dendrocalamus giganteus*. *For. Prod. J.* **2016**, *66*, 319–325. [CrossRef]
13. Wu, Y.; Cao, X.; Zhang, X.; Wei, J.; Xiong, Z.; You, J.; Gan, S.; Chen, D.; Zhang, H. The Influence Analysis of Bamboo Stand Structure on the Bamboo Shoot Yield of *Pleioblastus macutatus*. *J. Sichuan For. Sci. Technol.* **2019**, *40*, 55–59.
14. Yen, T.; Lee, J. Comparing Aboveground Carbon Sequestration between Moso Bamboo (*Phyllostachys heterocycla*) and China Fir (*Cunninghamia lanceolata*) Forests Based on the Allometric Model. *For. Ecol. Manag.* **2011**, *261*, 995–1002. [CrossRef]
15. Eggleston, H.S.; Buendia, L.; Miwa, K.; Ngara, T.; Tanabe, K. (Eds.) *2006 IPCC Guidelines for National Greenhouse Gas Inventories*; Institute for Global Environmental Strategies (IGES): Hayama, Japan, 2006.
16. Yen, T.; Sun, P.; Li, L. Predicting Aboveground Biomass and Carbon Storage for Ma Bamboo (*Dendrocalamus latiflorus* Munro) Plantations. *Forests* **2023**, *14*, 854. [CrossRef]
17. Yen, T.; Ji, Y.; Lee, J. Estimating Biomass Production and Carbon Storage for a Fast-Growing Makino Bamboo (*Phyllostachys makinoi*) Plant Based on the Diameter Distribution Model. *For. Ecol. Manag.* **2010**, *260*, 339–344. [CrossRef]
18. Huang, D.; Huang, D.; Wei, L.; Li, L.; Xu, Z. Culm Form Structure and Biomass of *Bambusa chungii*. *World Bamboo Ratt.* **2020**, *18*, 28–32.
19. Hou, Y.; Yao, W.; Yang, G.; Cui, Y.; Zhou, H.; Zhang, X. Study on the Culm Structure and Allocation of Aboveground Biomass of *Dendrocalamus tsiangii*. *J. Bamboo Res.* **2023**, *42*, 51–57.
20. Peng, Y.; Fan, S.; Su, W.; Xia, P. Distribution Pattern of Above-Ground Biomass and Culm Form Characteristics of *Bambusa blumeana*. *J. Sichuan Agric. Univ.* **2013**, *31*, 296–301.
21. Zheng, Y.; Xu, Z.; Jia, D.; Li, Z. Study on Allocation of Aboveground Biomass and Optimal Model of *Dendrocalamus latiflorus* Shoot-Producing. *J. Bamboo Res.* **2023**, *42*, 23–29.
22. Lin, D.; Zhao, H.; Li, Z.; Huang, C.; Xu, W. The Distribution Characteristics and Regression Model of Aboveground Biomass of *Pleioblastus amarus* in Nanling Mountain, Gangdong. *Chin. J. Ecol.* **2024**, *43*, 3205–3210.

23. Kaushal, R.; Islam, S.; Tewari, S.; Tomar, J.M.S.; Thapliyal, S.; Madhu, M.; Trinh, T.L.; Singh, T.; Singh, A.; Durai, J. An Allometric Model-Based Approach for Estimating Biomass in Seven Indian Bamboo Species in Western Himalayan Foothills, India. *Sci. Rep.* **2022**, *12*, 7527. [CrossRef] [PubMed]
24. Jiang, X.; Zhou, H.; Yan, Y.; Hou, Y.; Yang, Y.; Xie, T. Culm Structure and Above-Ground Biomass Allocation of *Dendrocalamus farinosus* and *Bambusa rigida*. *Guizhou For. Sci. Technol.* **2023**, *51*, 39–43.
25. Yuen, J.; Fung, T.; Ziegler, A.D. Carbon Stocks in Bamboo Ecosystems Worldwide: Estimates and Uncertainties. *For. Ecol. Manag.* **2017**, *393*, 113–138. [CrossRef]
26. Kong, F.; Zhong, Q.; Yu, L.; Kuang, X.; Huang, S.; Lu, Y.; Li, Y.; Gao, F. Regression Model of Biomass of *Phyllstachys edulis* at Different Age Classes in Ruijin, Jiangxi Province. *South China For. Sci.* **2022**, *50*, 39–42.
27. Shi, Z.; Hui, C.; Zhang, J.; Wu, C.; Qin, Y. Research on Chemical Composition and Fiber Morphology of *Dendrocalamus brandisii*. *Biomass Chem. Eng.* **2009**, *43*, 21–24.
28. Huang, D.; Li, L.; Huang, Y.; Lin, J.; Lu, Z.; Wei, C. Investigation on Germplasm Resources of Shoot-Producing Bamboo Species in Guangxi and Their Utilization. *World Bamboo Ratt.* **2024**, *22*, 48–53.
29. Ma, S.; Ouyang, W.; Liu, X.; Han, M.; Yang, Y.; Guo, A. Physicochemical Properties of Dietary Fiber from *Dendrocalamus brandisii* Leaves. *Food Res. Dev.* **2021**, *42*, 1–7.
30. Hui, C.; He, H.; Liu, W.; Shi, Z.; Sun, S. Promoting the Development of *Dendrocalamus brandisii* Industry for Rural Vitalization. *World Bamboo Ratt.* **2022**, *20*, 66–69.
31. Tan, H.; He, B.; Tan, R.; Zhou, T. Technical Approaches for Building High-Yield and High-Efficiency Bamboo Forest Models for Sweet Bamboo. *For. Sci. Technol.* **2000**, 32–33.
32. Jianwei, L.; Wanling, Q.; Chaomao, H.; Qian, C.; Yumeng, W.; Shuhong, Z.; Xiuting, Z.; Weiyi, L. Geographical Variation in Quality of *Dendrocalamus brandisii* Bamboo Shoots and Its Relationship with Site Conditions. *Sci. Silvae Sin.* **2024**, *60*, 75–83.
33. Yang, F.; Tang, M. Relationship between Site and Structure and Its Influence on Biomass in *Phyllostachys edulis* Forest. *J. Zhejiang A&F Univ.* **2020**, *37*, 823–832.
34. Ji, X.; Luo, Q.; Ding, Y.; Wang, Y.; Zhao, J.; Wang, S. A Study on the Aboveground Biomass Model of *Dendrocalamus brandisii*. *J. Bamboo Res.* **2015**, *34*, 49–53.
35. Zhang, X.; Zhang, F. A Study on the "Hollow Design" of Chinese Bamboo Product. In Proceedings of the 2008 9th International Conference on Computer-Aided Industrial Design and Conceptual Design, Kunming, China, 22–25 November 2008; pp. 729–733.
36. Casas, L.F.; Aldana, A.M.; Henao-Diaz, F.; Villanueva, B.; Stevenson, P.R. Specific Gravity of Woody Tissue from Lowland Neotropical Plants: Differences Among Forest Types. *Ecology* **2017**, *98*, 1474. [CrossRef] [PubMed]
37. Ghale, K.; Yadav, S.K.; Thapa, N.; Sah, J.N. Potentiality of Bamboos to Combat Climate Change Impacts and Biodiversity Conservation in the Annapurna Conservation Area, Nepal. *J. For. Nat. Resour. Manag.* **2020**, *2*, 79–91. [CrossRef]
38. Ordóñez-Prado, C.; Tamarit-Urias, J.C.; Nava-Nava, A.; Rodríguez-Acosta, M. Additive Equations System to Estimate Above-ground Biomass by Structural Component and Total of Three Giant Bamboo Species in Mexico. *CERNE* **2024**, *30*, e-103267. [CrossRef]
39. Li, Z.; Guan, F.; Zhou, X.; Liu, L.; Fu, D.; Zhang, X.; Li, M. Effect of Fertilization on Soil Fertility and Individual Stand Biomass in Strip Cut Moso Bamboo (*Phyllostachys edulis*) Forests. *Forests* **2024**, *15*, 252. [CrossRef]
40. Xayalath, S.; Hirota, I.; Tomita, S.; Nakagawa, M. Allometric Equations for Estimating the Aboveground Biomass of Bamboos in Northern Laos. *J. For. Res.* **2019**, *24*, 115–119. [CrossRef]
41. Etherington, T.R. Mahalanobis Distances for Ecological Niche Modelling and Outlier Detection: Implications of Sample Size, Error, and Bias for Selecting and Parameterising a Multivariate Location and Scatter Method. *PeerJ* **2021**, *9*, e11436. [CrossRef] [PubMed]
42. Osewe, E.O.; Dutcă, I. The Effects of Combining the Variables in Allometric Biomass Models on Biomass Estimates over Large Forest Areas: A European Beech Case Study. *Forests* **2021**, *12*, 1428. [CrossRef]
43. Zeng, W.; Tang, S. Goodness Evaluation and Precision Analysis of Tree Biomass Equations. *Sci. Silvae Sin.* **2011**, *47*, 106–113.
44. ArauÂjo, T.M.; Higuchi, N.; Carvalho, J.A., Jr. Comparison of Formulae for Biomass Content Determination in a Tropical Rain Forest Site in the State of ParaÂ, Brazil. *For. Ecol. Manag.* **1999**, *117*, 52.
45. Nocedal, J.; Wright, S.J. Quasi-Newton Methods. In *Numerical Optimization*; Springer: New York, NY, USA, 2006; pp. 135–163.
46. Chunquan, X.; Qihu, X.; Liping, L.; Xiao, H.; Yong, L.; Han, Z.; Lei, C.; Yuancai, L. Biomass Models with Breast Height Diameter and Age for Main Nativetree Species in Guangdong Province. *Sci. Silvae Sin.* **2019**, *55*, 97–108.
47. De Viliers, P. The Theory and Application of Regression Analysis and the Least-Sqaures Principle. *Sci. Mil. South Afr. J. Mil. Stud.* **2012**, *7*. [CrossRef]
48. Hongwei, J.; Jielai, X. Partial Least Square and Its Application. *J. Fourth Mil. Med. Univ.* **2003**, *24*, 280–283.
49. Berahas, A.S.; Jahani, M.; Richtárik, P.; Takáč, M. Quasi-Newton Methods for Machine Learning: Forget the Past, Just Sample. *Optim. Methods Softw.* **2022**, *37*, 1668–1704. [CrossRef]

50. Wei, H.; Wei, X. Research on Surveying and Mapping Data Processing Based on Nonlinear Mathematical Models and Deep Learning Optimization. *Scalable Comput. Pract. Exp.* **2024**, *25*, 2260–2265. [CrossRef]
51. Adediran, E.; Ameen, S. Machine Learning and Optimization Techniques for Solving Inverse Kinematics in a 7-DOF Robotic Arm. *arXiv* **2024**, arXiv:2406.13064.

MDPI AG
Grosspeteranlage 5
4052 Basel
Switzerland
Tel.: +41 61 683 77 34

Forests Editorial Office
E-mail: forests@mdpi.com
www.mdpi.com/journal/forests

www.ingramcontent.com/pod-product-compliance
Lightning Source LLC
Chambersburg PA
CBHW061447210926
51850CB00008B/1644
* 9 7 8 3 7 2 5 8 5 1 2 5 6 *